Computational Techniques for Analytical Chemistry and Bioanalysis

Theoretical and Computational Chemistry Series

Titles in the series:

1: Knowledge-based Expert Systems in Chemistry: Not Counting on Computers
2: Non-Covalent Interactions: Theory and Experiment
3: Single-Ion Solvation: Experimental and Theoretical Approaches to Elusive Thermodynamic Quantities
4: Computational Nanoscience
5: Computational Quantum Chemistry: Molecular Structure and Properties *in Silico*
6: Reaction Rate Constant Computations: Theories and Applications
7: Theory of Molecular Collisions
8: *In Silico* Medicinal Chemistry: Computational Methods to Support Drug Design
9: Simulating Enzyme Reactivity: Computational Methods in Enzyme Catalysis
10: Computational Biophysics of Membrane Proteins
11: Cold Chemistry: Molecular Scattering and Reactivity Near Absolute Zero
12: Theoretical Chemistry for Electronic Excited States
13: Attosecond Molecular Dynamics
14: Self-organized Motion: Physicochemical Design based on Nonlinear Dynamics
15: Knowledge-based Expert Systems in Chemistry: Artificial Intelligence in Decision Making

16: London Dispersion Forces in Molecules, Solids and Nano-structures: An Introduction to Physical Models and Computational Methods
17: Machine Learning in Chemistry: The Impact of Artificial Intelligence
18: Tunnelling in Molecules: Nuclear Quantum Effects from Bio to Physical Chemistry
19: Understanding Hydrogen Bonds: Theoretical and Experimental Views
20: Computational Techniques for Analytical Chemistry and Bioanalysis

How to obtain future titles on publication:
A standing order plan is available for this series. A standing order will bring delivery of each new volume immediately on publication.

For further information please contact:
Book Sales Department, Royal Society of Chemistry, Thomas Graham House, Science Park, Milton Road, Cambridge, CB4 0WF, UK
Telephone: +44 (0)1223 420066, Fax: +44 (0)1223 420247,
Email: booksales@rsc.org
Visit our website at www.rsc.org/books

Computational Techniques for Analytical Chemistry and Bioanalysis

Edited by

Philippe B. Wilson
Nottingham Trent University, UK
Email: philippe.wilson@ntu.ac.uk

and

Martin Grootveld
De Montfort University, UK
Email: mgrootveld@dmu.ac.uk

ROYAL SOCIETY OF **CHEMISTRY**

Theoretical and Computational Chemistry Series No. 20

Print ISBN: 978-1-78801-461-8
PDF ISBN: 978-1-78801-588-2
EPUB ISBN: 978-1-78801-985-9
Print ISSN: 2041-3181
Electronic ISSN: 2041-319X

A catalogue record for this book is available from the British Library

The Royal Society of Chemistry is a charity, registered in England and Wales, Number 207890, and a company incorporated in England by Royal Charter (Registered No. RC000524), registered office: Burlington House, Piccadilly, London W1J 0BA, UK, Telephone: +44 (0) 20 7437 8656.

For further information see our web site at www.rsc.org

Printed in the United Kingdom by CPI Group (UK) Ltd, Croydon, CR0 4YY, UK

Dedication

PBW wishes to dedicate this monograph to his supportive family, in particular Florence and Penelope, the two new additions who bring light to each day.

Theoretical and Computational Chemistry Series No. 20
Computational Techniques for Analytical Chemistry and Bioanalysis
Edited by Philippe B. Wilson and Martin Grootveld

Published by the Royal Society of Chemistry, www.rsc.org

Preface

With the advances in technology and the vast datastreams now produced by modern analytical instrumentation, computational models, systems and routines have been developed to better treat the information obtained. Indeed, we have seen such advances in applications of AI techniques to chemical data science that certainly the relevant chapter required updating multiple times throughout the publication process.

We have recruited a comprehensive group of colleagues internationally to develop this monograph. With chapters ranging from fundamental simulations of nuclear quantum effects, through to overviews of the various spectroscopic methods available and the computational techniques applied therein, this volume should appeal to colleagues both in the chemical and biosciences as a broad and comprehensive overview of *in silico* applications in analytical methodologies.

We begin in Chapter 1 with an overview of computational techniques as applied to metabolomic data analysis. Therein, Percival, Grootveld, and colleagues describe the significant advances over the past decade in data treatment, curation and availability, whilst describing well-known tools such as Wishart's *Metaboanalyst* packages.

Chapter 2, co-authored by Castro and Swart, consists of a broad and timely review of computational NMR spectroscopy for predictive purposes. The reproduction of NMR experiments in simulations is considered, as well as the numerous parameters contributing to spectral similarity with experiment.

In Chapter 3, Dario Estrin and colleagues from Universidad de Buenos Aires provide a comprehensive and contemporary perspective on computational vibrational spectroscopy. Indeed, they describe the state of the art within the field, covering the most recent developments in the context of

Theoretical and Computational Chemistry Series No. 20
Computational Techniques for Analytical Chemistry and Bioanalysis
Edited by Philippe B. Wilson and Martin Grootveld

Published by the Royal Society of Chemistry, www.rsc.org

the levels of theory available and applied. Discussions of (an)harmonicity complement the narrative which covers classical to fully QM approaches.

Paneth and Dybala-Defratyka consider the use of isotope effects, nuclear quantum effects and analytical probes in Chapter 4. They provide an overview of predicting isotope effects (both kinetic and equilibrium), and the information obtained which can be employed to consider the behaviour of a variety of chemical and biological systems. Applications of theoretical estimations of isotope effects are described, such as through reactions in solution, and the employment of a range of QM/MM procedures are considered in the context of enzyme-catalysed reactions.

Applications of artificial intelligence techniques in chemical and biochemical analysis are discussed in Chapter 5, where Gibson, Kamerlin, Wilson and colleagues first provide a holistic overview of theory and history in the field, prior to covering a snapshot of advances, particularly over the last 5 years. The developments by groups such as those of Aspuru-Guzik, Cronin and others in bringing AI techniques to the forefront of contemporary chemical analysis are described, whilst a perspective for future applications is offered.

Monari, Francés-Monerris, and co-workers explain the development of computational spectroscopic approaches for complex biological systems in Chapter 6. First offering a short review of hybrid QM/MM techniques, they continue to provide vital guidance in the simulation of photobiological phenomena, including the adequate sampling of chromophore conformations. The narrative develops the theory and application of computational techniques in circular dichroism spectra, and a discussion on modern methods for understanding the evolution of excited states.

In Chapter 7, Lucie Delemotte from KTH Stockholm considers bridging the gap between atomistic MD simulations and laboratory experiments for investigations of membrane proteins. The link between experimental characterisation and simulation is established, as alongside a consideration of the theoretical models developed for MD and recent advances. Moreover, a useful discussion is developed, considering the degree of automation of such studies, as well as the necessary comparability to interpretable experimental data.

Chapter 8, by Molinari and colleagues, provides a perspective of computational chemical analysis as applied to solid state chemistry, in particular, materials science. The methodologies and protocols applied to materials analysis are surveyed, considering surfaces and their interactions with the environment. Comparative assessments of computational methods and experimental techniques are offered, suggesting the substantial advantages combinatorial approaches have on contemporary studies.

We conclude by considering the computational approaches inherent in the analysis of ESR spectral data in Chapter 9, where Martín, Mather, Wilson and co-workers approach the topic pedagogically through a review of fundamental ESR theory, prior to describing recent developments in spectral analysis codes and algorithms.

Whilst we have collated a heterogenous collection of contributions, we hope readers will agree that these span the scope of computational approaches in analytical chemical and biosciences, and we are grateful to the Royal Society of Chemistry for co-creating this initiative.

Philippe B. Wilson and Martin Grootveld

Contents

Chapter 1 Univariate and Multivariate Statistical Approaches to the Analysis and Interpretation of NMR-based Metabolomics Datasets of Increasing Complexity 1
Benita Percival, Miles Gibson, Justine Leenders, Philippe B. Wilson and Martin Grootveld

1.1 Introduction 1
1.2 Brief Introduction to Metabolomics 5
1.3 Experimental Design for NMR Investigations 6
1.4 Preprocessing of Metabolomics Data 8
1.4.1 Spectral Prediction in Positive Signal Identification 13
1.4.2 Statistical Methodologies Applied to NMR Spectra 16
1.5 Univariate Data Analysis 16
1.5.1 ANOVA, ANCOVA and Their Relationships to ASCA 18
1.6 Multivariate Analysis 19
1.6.1 Principal Component Analysis 20
1.6.2 Partial-least Squares Discriminatory Analysis 22
1.7 Metabolic Pathway Analysis 28
1.8 Further Important Considerations: Accuracy and Precision 28
1.9 Multivariate Power Calculations 30
1.10 Future Scope of Metabolomics Research 33

Theoretical and Computational Chemistry Series No. 20
Computational Techniques for Analytical Chemistry and Bioanalysis
Edited by Philippe B. Wilson and Martin Grootveld

Published by the Royal Society of Chemistry, www.rsc.org

1.11 Conclusions 34
List of Abbreviations 34
Acknowledgements 35
References 35

Chapter 2 Recent Advances in Computational NMR Spectrum Prediction **41**
Abril C. Castro and Marcel Swart

2.1 Introduction and Scope 41
2.2 Calculation of NMR Chemical Shifts 43
2.3 Influence of Conformational Effects on NMR Shielding Constants 46
2.4 Influence of Environmental Effects on NMR Shielding Constants 47
2.5 Relativistic Effects on NMR Chemical Shifts 54
2.6 Molecular Orbital Analysis for Understanding Chemical Shieldings 56
2.7 Electron Paramagnetic Resonance Spectroscopy 58
2.8 Conclusions 59
Acknowledgements 60
References 60

Chapter 3 Computational Vibrational Spectroscopy: A Contemporary Perspective **69**
Diego J. Alonso de Armiño, Mariano C. González Lebrero, Damián A. Scherlis and Darío A. Estrin

3.1 Introduction 69
3.2 Fundamentals of Computational Vibrational Spectroscopy 70
3.2.1 The Harmonic Approximation 70
3.2.2 Intensities Calculations 72
3.3 Beyond the Harmonic Approximation: Advanced Techniques in Vibrational Spectroscopy 76
3.3.1 Motivation 76
3.3.2 The Watson Hamiltonian 78
3.3.3 Vibrational Self-consistent Field 80
3.3.4 Vibrational Configuration Interaction 83
3.3.5 Vibrational Perturbation Theory 85
3.3.6 Vibrational Coupled Clusters 87
3.3.7 Latest Developments and Applications 89

3.4 Computational Vibrational Spectroscopy in Complex Environments 97
3.4.1 Multiscale Methods for General Vibrational Spectroscopy 97
3.4.2 Hybrid Models for Surface-enhanced Raman Spectroscopy 100
3.5 Concluding Remarks 103
References 103

Chapter 4 Isotope Effects as Analytical Probes: Applications of Computational Theory 125
Piotr Paneth and Agnieszka Dybala-Defratyka

4.1 Introduction 125
4.1.1 Isotope Effects: What Are They? 125
4.1.2 Theory of Isotope Effects: How Do We Compute Them? 130
4.1.3 Beyond Transition State Theory 132
4.2 Examples 136
4.2.1 Kinetic Isotope Effects in Multipath VTST: Application to a Hydrogen Abstraction Reaction 137
4.2.2 Comparison of Different QM/MM Approaches: Nitrobenzene Dioxygenase 139
4.2.3 Hydrolysis of β-Hexachlorocyclohexane by LinB 142
4.2.4 Binding Isotope Effects 143
4.2.5 Vapor Pressure Isotope Effects (VPIEs) Predicted Using the Path Integral Formalism 146
4.2.6 Isotope Effects Associated with Adsorption on Graphene 148
Acknowledgements 150
References 150

Chapter 5 Applications of Computational Intelligence Techniques in Chemical and Biochemical Analysis 155
Miles Gibson, Benita Percival, Martin Grootveld, Katy Woodason, Justine Leenders, Kingsley Nwosu, Shina Caroline Lynn Kamerlin and Philippe B. Wilson

5.1 Historical Use of Artificial Intelligence 155
5.2 Early Adoption in the Chemistry and Bioanalysis Areas 159

5.3 Recent Applications of Machine Learning in Chemistry 161
5.3.1 Support Vector Machines 162
5.3.2 Neural Networks 166
5.3.3 *k* Nearest Neighbours 170
5.3.4 Decision Trees 171
5.3.5 Naïve Bayes Classifiers 173
5.3.6 Linear Regression 174
5.3.7 *k*-means Clustering 176
5.3.8 Self-organising Maps 178
5.3.9 Hierarchical Clustering 179
5.3.10 Independent Component Analysis 182
5.3.11 Deep Learning 183
5.3.12 Quantum Computing and Applications in Chemical Analysis 186
5.3.13 Particle Swarm Optimisation 187
5.4 Conclusions 191
List of Abbreviations 191
Acknowledgements 193
References 193

Chapter 6 Computational Spectroscopy and Photophysics in Complex Biological Systems: Towards an *In Silico* Photobiology 202
Antonio Francés-Monerris, Marco Marazzi, Vanessa Besancenot, Stéphanie Grandemange, Xavier Assfeld and Antonio Monari

6.1 Introduction 202
6.2 Computational Modelling and QM/MM Techniques 205
6.2.1 QM/MM Frontier 208
6.2.2 QM/MM Embedding 209
6.3 Linear Spectroscopy in Chemical and Biological Systems 209
6.3.1 Analysing the Excited States PES: A Computational Perspective 210
6.3.2 Practical Guidelines to Model Light Absorption and Emission in Complex Environments 213
6.3.3 Practical Case Study: Interaction Between Drugs and DNA 220
6.4 Modelling Circular Dichroism Spectroscopy 220

6.5 Photochemistry and Photobiology in Complex Environments 223
6.5.1 Computing Potential Energy Surfaces in Complex Environments 223
6.5.2 Photochemistry and Photobiology of Complex Systems Under the Dynamic Approach 225
6.6 Conclusion 238
Acknowledgements 239
References 239

Chapter 7 Bridging the Gap Between Atomistic Molecular Dynamics Simulations and Wet-lab Experimental Techniques: Applications to Membrane Proteins 247
Lucie Delemotte

7.1 Classical Molecular Dynamics Simulations 247
7.1.1 MD Simulation Algorithm 248
7.1.2 Force Fields and the Potential Energy Function 252
7.1.3 Enhanced Sampling Simulations Schemes 254
7.2 Interfacing MD Simulations and Experimental Results 258
7.2.1 Forward Modelling 258
7.2.2 Reweighting Schemes 259
7.3 Case Studies: Applications to Membrane Protein Function 262
7.3.1 Testing Predictions from MD Simulations 263
7.3.2 Using MD Simulations to Interpret Experimental Measurements 270
7.3.3 Guiding Processes Using Experimental Data-driven Enhanced Sampling MD Simulations 278
7.4 Conclusions 278
Acknowledgements 279
References 279

Chapter 8 Solid State Chemistry: Computational Chemical Analysis for Materials Science 287
Estelina Lora da Silva, Sandra Galmarini, Lionel Maurizi, Mario Jorge Cesar dos Santos, Tao Yang, David J. Cooke and Marco Molinari

8.1 Introduction 288

8.2 Computational Spectroscopy: Interaction Between Matter and Electromagnetic Radiation 288
8.2.1 Physical and Chemical Properties of Biomaterials 289
8.2.2 Absorption Spectroscopy: Opto-electronical Properties 291
8.2.3 Inelastic Scattering: Lattice Dynamics and Raman Spectroscopy 294
8.2.4 Resonance Spectroscopy 298
8.3 Computational Techniques for Biocompatible Materials 305
8.3.1 Overview of Surface Properties 305
8.3.2 Surface Adsorption and Surface Functionalization 312
8.3.3 Special Case of Nanoscale Materials 315
8.4 Summary and Conclusions 318
References 319

Chapter 9 Electron Spin Resonance for the Detection of Paramagnetic Species: From Fundamentals to Computational Methods for Simulation and Interpretation 335
Inocencio Martín, Leo Martin, Anwesha Das, Martin Grootveld, Valentin Radu, Melissa L. Mather and Philippe B. Wilson

9.1 Introduction 335
9.2 Theory 337
9.3 Fine Structure in ESR Spectroscopy 342
9.4 Quantum Mechanical Consideration of ESR Theory 346
9.5 Interpretation 348
9.6 Early Work in the Simulation of ESR Spectral Parameters 353
9.7 Progress in Density Functional Theory Approaches 354
9.8 Spectral Simulation and Fitting 356
9.9 Conclusions 359
Acknowledgements 359
References 359

Subject Index 362

CHAPTER 1

Univariate and Multivariate Statistical Approaches to the Analysis and Interpretation of NMR-based Metabolomics Datasets of Increasing Complexity

BENITA PERCIVAL,[a] MILES GIBSON,[a] JUSTINE LEENDERS,[a] PHILIPPE B. WILSON[b] AND MARTIN GROOTVELD*[a]

[a] Leicester School of Pharmacy, Faculty of Health and Life Sciences, De Montfort University, The Gateway, Leicester LE1 9BH, UK; [b] School of Animal, Rural and Environmental Sciences, Nottingham Trent University, Brackenhurst Campus, Southwell NG25 0QF, UK
*Email: mgrootveld@dmu.ac.uk

1.1 Introduction

Although some statistical approaches were developed much earlier, such as the pioneering Bayesian statistics conducted in the 18th century,[1] the interdisciplinary usage between science and statistics has still not been fully established. At present, there is a strong affinity between statistics and science, which dates back to the late 19th century and early 20th century. Works by Karl Pearson and Francis Galton, explored regression towards the mean, principal

Theoretical and Computational Chemistry Series No. 20
Computational Techniques for Analytical Chemistry and Bioanalysis
Edited by Philippe B. Wilson and Martin Grootveld

Published by the Royal Society of Chemistry, www.rsc.org

component analysis (PCA), and Chi-squared contingency table testing and correlation.[2] Later, Spearman also developed his theory on factor analysis, namely Spearman's rank correlation coefficient, and applied it to the social sciences research area.[3] William Gosset was responsible for the discovery of the *t*-test, which is embedded in most statistical testing applied in scientific fields to date, and which unfortunately remains the most widely abused, misapplied univariate form of statistical analysis test.[4] Ronald Fisher tied the aforementioned ideas together, observing Gaussian distribution accounting for both chi-squared and the *t*-test to formulate the infrequently used *F*-distribution test. Fisher later developed analysis-of-variance (ANOVA), and defined *p*-values for determining the level of statistical significance.[5] Fisher furthered these works by applying his knowledge to genetics, in particular the observation of alleles, specifically the frequency and estimation of genetic linkages by maximum likelihood methods within a population.[6] Basic statistical hypotheses, such as H_0 and H_1, which still stand to date, were then established[7] and are still fundamental to all experimental designs.

These statistical tools are now applied to just about every field possible; however, in science every research area has an element of statistical interpretation, from genomics in diseases diagnostics, forensic science and hereditary studies, the microbiome, and the discovery of biomarkers using biological immunoassays and 'state-of-the-art' bioanalytical techniques. In this chapter, modern advanced statistical methodologies will be explored through a major, now commonly employed multicomponent analytical/bioanalytical chemistry technique, namely nuclear magnetic resonance (NMR) spectroscopy. Statistical approaches and challenges which are associated with the collection of large or very large NMR datasets run in parallel with those of other multicomponent analytical techniques, such as liquid-chromatography–mass spectrometry (LC/MS), and hence the NMR examples provided serve as appropriate test models. Despite some problems occasionally being experienced with low sensitivity (especially at biofluid concentrations $<10\ \mu mol\ L^{-1}$) and untargeted analyses, which may result in xenobiotic resonances overlapping with those of endogenous metabolites, NMR provides many laboratory advantages over LC/MS in view of its high reproducibility, non-destructive methodology with minimal sample preparation, and the simultaneous detection and determination of 100 or more metabolites at operating frequencies greater than or equal to 600 MHz.[8] NMR is a suitable analytical technique across many different fields and sample types, including food chemistry, geological studies, drug discovery, forensics and an increasingly expanding number of metabolomics areas, for example lipidomics and fluxomics. Both solid and liquid samples can be analysed, which is obviously advantageous. Moreover, major recent advances in the computational capability have enhanced the applications of metabolomics-linked statistical approaches, and current software modules and their applications will also be discussed here.

Indeed, NMR as a technique advanced much later than the statistical methods, therefore combining statistical tools with the data acquired occurred subsequently. Nuclear spin was first described by Pauli in 1926,[9] and these

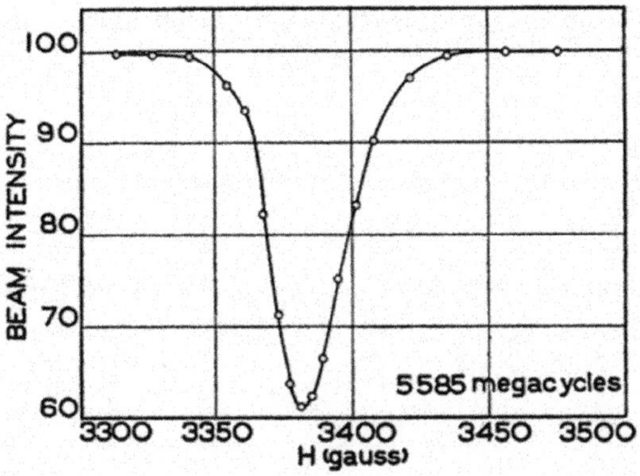

Figure 1.1 ^{7}Li nucleus NMR signal originally observed by Rabi *et al.*[10] Reproduced from ref. 10, https://doi.org/10.1103/PhysRev.55.526, with permission from American Physical Society, Copyright 1939.

ideas were then further developed by Rabi in 1939 who created the first radio frequencies for this purpose involving a beam of lithium chloride (Figure 1.1).[10] Overhauser in 1953 observed dynamic nuclear magnetisation.[11] Redfield then explored the theory of NMR relaxation processes,[12] and NMR was then developed from these principles by Bloch and Purcell from 1945–1970, both of whom then won several Nobel Prizes.[13,14] Continuous Wave (CW) methods were used to observe the nuclei spin, experiments which use a permanent magnet/electromagnet and a radio frequency oscillator to produce two fields, B_0 and B_1 respectively. To produce a resonance, CW methodologies were used to vary the B_1 field or the B_0 field to achieve resonance. In essence, the magnetic field is continuously varied and the peak signal (resonance) is recorded on an oscilloscope or an x–y recorder. However, these methodologies have substantially advanced, and at present radiofrequency pulse sequences are applied to nuclei within a magnetic field, B_1.

A range of NMR facilities are currently available, with the highest operating frequency reaching 1200 MHz, which requires a Gauss safety line and regular cryogen fills, in addition to more accessible permanent non-cryogen-requiring benchtop magnets operating at a frequency of up to 80 MHz. Progress in low-field NMR spectroscopy and their analytical applications, albeit very uncommon in metabolomic applications, has been recently reviewed.[15] A whole plethora of spectrometers exist between these two extreme frequencies, and these all have the capacity and capability to acquire a wide range of molecular analyte data, which can subsequently be employed for statistical evaluations and comparisons. Biological fluids have been examined using both low-[16] and high-frequency[17] NMR technologies for monitoring of a range of endogenous metabolites and xenobiotics, although the use of high-frequency spectrometers is often the preferred approach because of the much-enhanced level of resolution and deconvolution of NMR signals. An example of the differences in resolution observed between low- (60 MHz) and high-field (400 MHz) NMR analysis is shown in Figure 1.2.

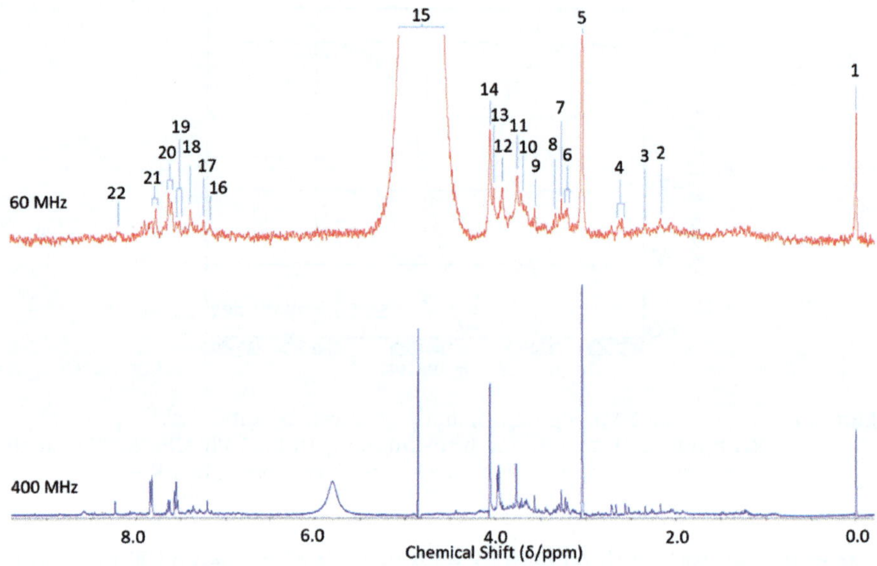

Figure 1.2 400 and 60 MHz NMR of the same urine sample from a control participant. Assignments include; [1] trimethylsilylpropanoic acid–CH_3; [2] acetone–CH_3; [3] pyruvate–CH_3; [4] citrate–$CH_{2A/B}$; [5] creatinine/creatine–CH_3; [6] *cis*-aconitate–CH_2; [7] taurine–CH_2; [8] trimethyl-*N*-oxide–CH_3; [9] glycine–CH_3; [10] taurine–CH_2; [11] unassigned-CH_2; [12] creatine–CH_2; [13] glycolate–CH_2; [14] creatinine–CH_2; [15] H_2O-OH; [16] histidine–CH; [17] indoxyl sulphate–CH; [18] indoxyl sulphate–CH; [19] hippurate–CH; [20] hippurate–CH; [21] hippurate–CH; and [22] formate–CH.

The majority of biological fluids are predominantly composed of water, such as urine and saliva, and therefore appropriate pulse sequences have been developed to suppress these resonances in order to focus on those of interest. Pulse sequences such as water excitation technique (WET), nuclear Overhauser effect spectroscopy (NOESY),[18] PRESAT and WATERGATE[19] are highly suitable for the analysis of spectra containing such broad signals arising from the ^1H nuclei in H_2O. The water signal can be irradiated at its characteristic frequency, and hence is unable to resonate, and this strategy serves to reveal metabolites at smaller concentrations that are located at similar frequencies (chemical shift values, δ).

Furthermore, all biological fluids such as blood serum or plasma contain low-molecular-mass metabolites in addition to large macromolecules, usually proteins and lipoproteins. These metabolite signals are then superimposed on the broad signals of the macromolecules, leading to signal loss and broadening. In this specific case, applying a CPMG pulse sequence makes it possible to overcome this problem by exploiting the differences in the relaxations of metabolites and macromolecules. This sequence deletes the fast-relaxing signals arising from large macromolecules, such as proteins, from the spectrum by applying a spin-echo pulse.[20]

Analysis of biofluids has been used for both pharmacokinetic and biomarker detection in many studies, and requires multidimensional data analysis using highly sophisticated, but nevertheless comprehensive, statistical techniques.[21]

1.2 Brief Introduction to Metabolomics

Although the metabolome was first explored centuries ago through organoleptic analysis, such as the smelling of urine as a means of diagnosing diseases, applications of bioanalytical techniques to analyse and determine molecules in urine were first performed by Pauling *et al.* (1971),[22] and the human metabolome was described by Kell and Oliver (2016).[23] Nicholson and Sadler's groups' pioneering work first detected drug metabolites using ^{1}H NMR analysis in the early 1980s, observing acetaminophen and its corresponding metabolites.[24] In addition, this group was the first to monitor metabolic conditions in urine samples.[24,25] Over the last 20 years, the metabolome itself, and the means for the application of multianalyte bioanalytical techniques to be applied to it, have been further defined by others such as Nicholson and Lindon.[26] Many developments have been explored in the literature with a progressive history for techniques employed for metabolomics studies, along with their establishment as key investigatory tools, as documented by Oliver and Kell.[23] The study of the metabolome allows for the monitoring of metabolic processes within a system by identifying and determining low-molecular-mass metabolites within biofluids, tissues or cells. Indeed, the metabolome can be affected by many biological processes. These can either be through external stimuli such as an intervention, such as a medication, diet or exercise regimen, or alternatively through internal stimuli. An internal stimulus can be introduced *via* the modification of gene expression using techniques such as cell transfection, both *in vivo* and *in vitro*. Moreover, metabolomics techniques assess these changes by providing a 'snapshot' of the status of biological processes occurring at a specific point in time. This responsive information can provide a high level of detail regarding cellular metabolic processes, and can facilitate phenotypic evaluations, and hence yield an overall 'picture' or 'fingerprint' of the chemopathological status of a disease. Even more valuably, metabolomics is able to probe the changing disease status, for example, the effects of a drug treatment, the removal of a tumour, regression of an inflammatory condition, and so forth. Hence, these strategies may be successfully employed to monitor the severity and progression of a disease; information which further increases our understanding of the aetiology, together with the manifestation and progression of particular conditions.[27]

Two approaches can be undertaken in metabolomics; they are chosen primarily by the objectives of the study and the hypotheses formulated:[28]

- A targeted approach focused on the quantitative analysis of a limited number or class of metabolites that are linked by one or more specific biochemical pathways. This makes it possible to compare variations in metabolites in a precise and specific way.

- The objective of an untargeted approach is to detect "all" metabolites present in biological samples. It involves simultaneous analysis of as many metabolites as possible.

Targeted analysis is usually performed by mass spectrometry (MS), or more specifically MS as a detection system for liquid- or gas-chromatographically-separated metabolites, the sensitivity of which designates it as an excellent method to quantify and identify a set of specific metabolites. By not requiring prior knowledge about compounds present in a sample, and focusing on the global metabolic profile, NMR is ideally suited for an untargeted approach, but thanks to its easy quantification capacity, it can also be used in a targeted manner for selected metabolites.

NMR is able to detect and quantify metabolites in a high-throughput, simultaneous, non-destructive manner and requires minimal sample preparation. LC–MS strategies can target specific metabolites more sensitively than NMR, whilst lacking the specificity of the latter for absolute metabolite determinations.[29] Therefore, a combination of both LC–MS and NMR methodologies is important for a global understanding of the metabolic effects involved in disease processes, and can be essential for thorough metabolite determinations; a description of the full strengths and weaknesses of these techniques can be found in Nicholson and Lindon.[26]

A number of other analytical techniques can be applied, such as Fourier transform infrared spectroscopy (FTIR), ultraviolet-visible spectroscopy (UV–Vis) and derivatives of NMR and LC–MS.[30] However, their applications remain limited and specific, each one having its own advantages and limitations.

When handling colossal and complex datasets from untargeted metabolomics approaches, which include a plethora of metabolites at a range of concentrations, it can be difficult to interpret and understand the significance of the datasets acquired. Multivariate (MV) statistics have been integrated into multicomponent NMR analysis in view of the number of datapoints produced from the output for analysing complex mixtures such as biofluids. MV statistics aids in the processing of large datasets into visual formats, which are more comprehensive for the analyst *via* the recognition of patterns or signatures in multianalyte datasets. The combination of using a scientific technique alongside MV and univariate statistical analysis strategies in this manner is termed 'chemometrics'.

1.3 Experimental Design for NMR Investigations

Any metabolomic approach is usually initiated by one or more biological/clinical questions to which the clinician or the biologist wishes to respond. Whether monitoring the effects of treatment or improving diagnostic tools, the experimental design must be carefully considered by assessing all sources of variation in order to reduce bias and avoid the introduction of irrelevant variability. Samples should be collected with appropriate research ethics approval and informed consent of all participants (both healthy and diseased, where relevant), and in metabolomics studies it is important to maintain a consistent cohort. Physiological factors, such as age, fasting or non-fasting protocols, gender, diet, physical activity, medical conditions, family medical history, genotype, and so forth, should all be taken into account prior to sample collection.[8] Moreover, analytical factors should also be considered; for example, samples should be collected in a uniform manner, for example using the same collection tubes, whilst remaining mindful that some collection vessels may have signals with NMR resonances and hence interfere with the spectra acquired. This is common with anti-coagulants when collecting blood samples, for example ethylenediaminetetraacetic acid (EDTA), will provide not only the signals from the EDTA chelator itself, but also those of its slowly-exchanging Ca^{2+}- and Mg^{2+}-complexes.[31] Likewise, the citrate anticoagulant also has intense ^{1}H NMR signals. Lithium heparin tubes are often recommended for plasma collection in order to avoid interfering signals. Contamination is possible in the early stages of sample collection to ensure sterility, and this also needs to be considered with sample transportation.

Sample stability is also an important experimental factor. Some samples are unstable when exposed to ambient temperature, which can occur if samples are on the autosampler. This can cause degradation, changes in the concentration or a complete loss of metabolites. Common pitfalls of biofluid storage include microbiological contamination of samples at ambient or even lower temperatures. Biological fluids should be stored at low temperatures, typically -80 °C prior to analysis.[32] Sodium azide can be added to samples to ensure that microbes do not infiltrate, grow and interfere with metabolite levels in the sample whilst samples are maintained at ambient temperature on an NMR sample belt.[33] Furthermore, freeze–thaw cycles should be minimised; indeed, it has been shown that no more than three freeze–thaw cycles are suitable for plasma sample analysis.[32]

A suitable internal or external standard will provide a reference signal outside the area of analyte interest without interference with the sample. For example, an internal standard can be added to samples of saliva and urine, predominantly sodium 3-(trimethylsilyl)[2,2,3,3-d_4] propionate (TSP). However, TSP can bind to proteins in plasma, serum, and synovial and cerebrospinal fluid samples,[34] and therefore is not added in these cases. A more suitable internal standard could be added such as 4,4-dimethyl-4-silapentane-1-ammonium trifluoroacetate which has been shown to have limited interactions with proteins,[35] or an external standard such as the use of a capillary. 4,4-Dimethyl-4-silapentane-1-ammonium trifluoroacetate has also been proposed as a suitable internal standard that does not interact with cationic peptides. Other useful internal standards which have been used for $C^{2}HCl_3$-based biofluid and tissue biopsy lipid extracts include 1,3,5-trichlorobenzene and tetramethylsilane, although the latter is not recommended as it readily evaporates. Electronic Reference To access In vivo Concentrations (ERETIC) can also be used as an electronic standard for biofluid or tissue extract NMR samples.[34]

Buffering of the sample is also important in NMR as changes in the pH can modify the chemical shift values.[34] Indeed, biofluid pH values vary significantly between sample classes *in vivo*. Biofluids are typically buffered to pH values of 7.0 or 7.4.[34] For example, in blood plasma the ^{1}H NMR profiles of histidine, tyrosine and phenylalanine are all affected by pH, and NMR-invisibility of both tyrosine and phenylalanine is possible at neutral pH in view of their binding to albumin.[36]

Appropriate extraction techniques can be performed for solid samples such as leaves, seeds, biological tissues, cells, foods, drugs and so forth. However, it is critically important that no solids are retained in liquid sample extracts as this will interfere with the homogeneity of the magnetic field. Freeze-drying and/or drying with liquid nitrogen, followed by vortexing and centrifuging, are often necessary to ensure there is no retention of any solid sample debris.[34]

Acquisition parameters should, of course, be maintained as uniform throughout the acquisition process. Full recommendations for such parameters are available in Emwas *et al.*[33] The temperature in the NMR room and, more importantly, within the NMR probe should be consistent. Pulse parameters from the number of scans, acquisition time and number of acquisition points should be kept constant. The NMR instrument should be shimmed, tuned and matched appropriately.[33] Occasionally, backward linear prediction (BLIP) and forward linear prediction (FLIP) can be appropriate in order to remove artefacts in the spectra, or if the acquisition time is too short giving rise to a truncated free induction decay (FID) respectively, processes that result in improved resolution. In metabolomics studies, these need to be used consistently throughout the post-processing stage in order to ensure that signals are not present as part of a "ringing pattern" or noise.

The experimental design ensures that the results acquired are indeed statistically significant and are not present owing to an error in the early sampling stages. Guidelines for urinary profiling have been established in the literature.[33] However, there is no harmonisation throughout the field in view of the ranges in the NMR field strength and experimental parameters, such as a range of pulse sequences.

1.4 Preprocessing of Metabolomics Data

Before performing statistical analyses on spectral data, it is important to apply several pre-treatment steps that will ensure the quality of the raw data and limit possible biases. Indeed, in view of possible imperfection of the acquisition (noise acquisition), the signal processing, as well as the intrinsic nature of the biological samples (such as a dilution effect between the samples), it is very often necessary to apply some correctional measures to the spectra acquired.[37] Differential methods of treatment can be used and each of them has its advantages and disadvantages. The choice of a method depends on the biological issue to be addressed, on the properties of the samples analysed, and on the methods of data analysis selected. Most of these post-processing steps are applicable to not only NMR datasets, but also LC–MS ones.

(A) Raw NMR spectral data acquisition

One of the crucial steps is to ensure appropriate dilution of the metabolites so that the internal standard, if one is used, can be referenced. Furthermore, internal standards which are much lower in concentration than that of the monitored metabolite may indeed give rise to inaccurate results. Water suppression may also dampen the intensities of signals present in close proximity to the water signal; for example, it has been shown that both the α- and β-glucose proton resonances can both be significantly, or even substantially, affected by these suppression techniques if the power is not adjusted accordingly using certain pulse sequences, for example NOESY PRESAT.[16,33,38] Further important quality-control assessments may include the recognition of drug signals, corresponding metabolites, alcohol, and so forth which are commonly found in biofluid matrices. These resonances, if not properly identified, may interfere and lead to false levels of statistical significance, misassignment of biomolecule signals, and drug-induced modifications to the biofluid profiles explored. Therefore, positive signal identification is important in ensuring valid statistical significance and will be discussed in greater depth below.

(B) Phase and baseline corrections

Phase correction is crucial in order to ensure that signals are uniform, and no negative signals or baseline roll are present. These can cause elevated or decreased bucket intensity regions which could inflate the degree of statistical significance. Baseline correction can also ensure accurate signal integration.[33]

(C) Alignment

A regular problem encountered during data processing is a signal shift between the different NMR profiles of different samples. Several parameters can influence these peak shifts: instrumental factors, pH modifications, temperature variations, different saline concentrations or variable concentrations of specific ions. This problem is frequently encountered in urinary samples with pH values which are particularly variable, and which are subject to important variations in dilution.[39] Several algorithms are available to realign peaks, each method has its own advantages and drawbacks. By shifting, stretching or compressing the spectra along their horizontal axis, this method maximizes the correlation between them (Figure 1.3).

(D) Bucketing

Classically-acquired NMR spectra correspond to a set of several thousand points. NMR spectral data acquired on biological tissue extracts or biofluids contains information corresponding to about 50 to 100 metabolites. This gap between the number of variables available and the number of useful variables must be reduced before statistical

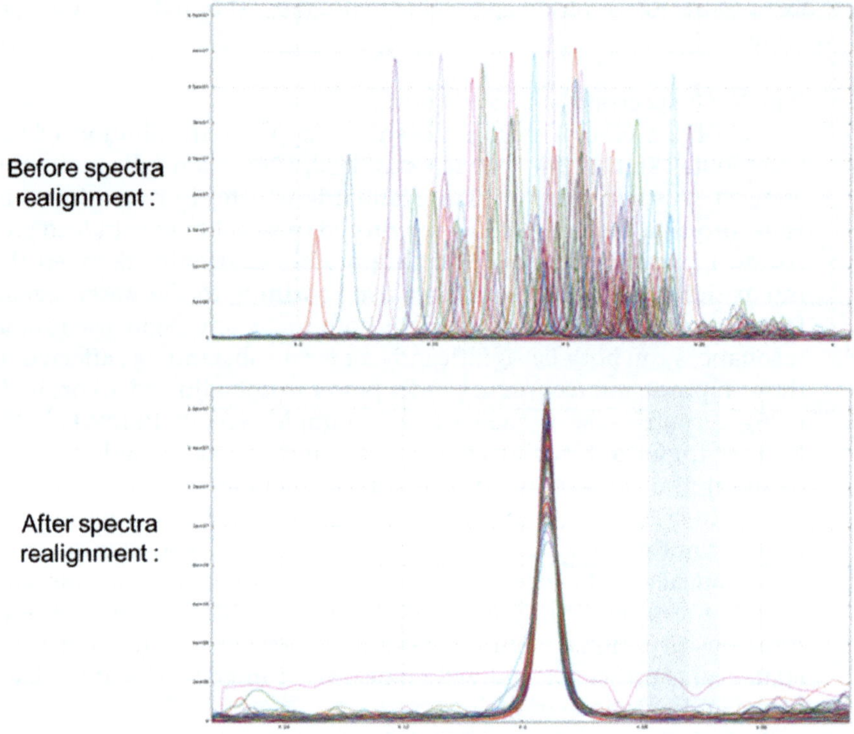

Figure 1.3 Effect of realignment on the creatinine signal: before (upper panel) and after (lower panel) realignment.

processing. This stage of segmentation, called bucketing (or binning) must firstly reduce the dimensionality of the dataset in order to extract N variables from each acquired metabolic profile (spectrum). This approach also diminishes the problem of spectral misalignment.

The most common segmentation technique comprises segmentation of the spectrum into N windows of the same width, otherwise known as bins or buckets. These buckets are usually of a size between 0.01 and 0.05 ppm. The total area within each bucket is measured instead of individual intensities, leading to a smaller number of variables. However, because of the lack of flexibility in the segmentation, some areas from the same resonance or peak could be split into two or more bins, dividing the chemical information between several bins, which could influence the data analysis subsequently conducted. To answer this problem, segmentation of variable intervals was developed. This technique, called *intelligent bucketing*, attempts to split the spectrum so that each bucket contains only one signal, peak or pattern (Figure 1.4). Of note, this method is highly sensitive to pH variations, and therefore the spectral realignment needs to be optimal before it is applied.[8]

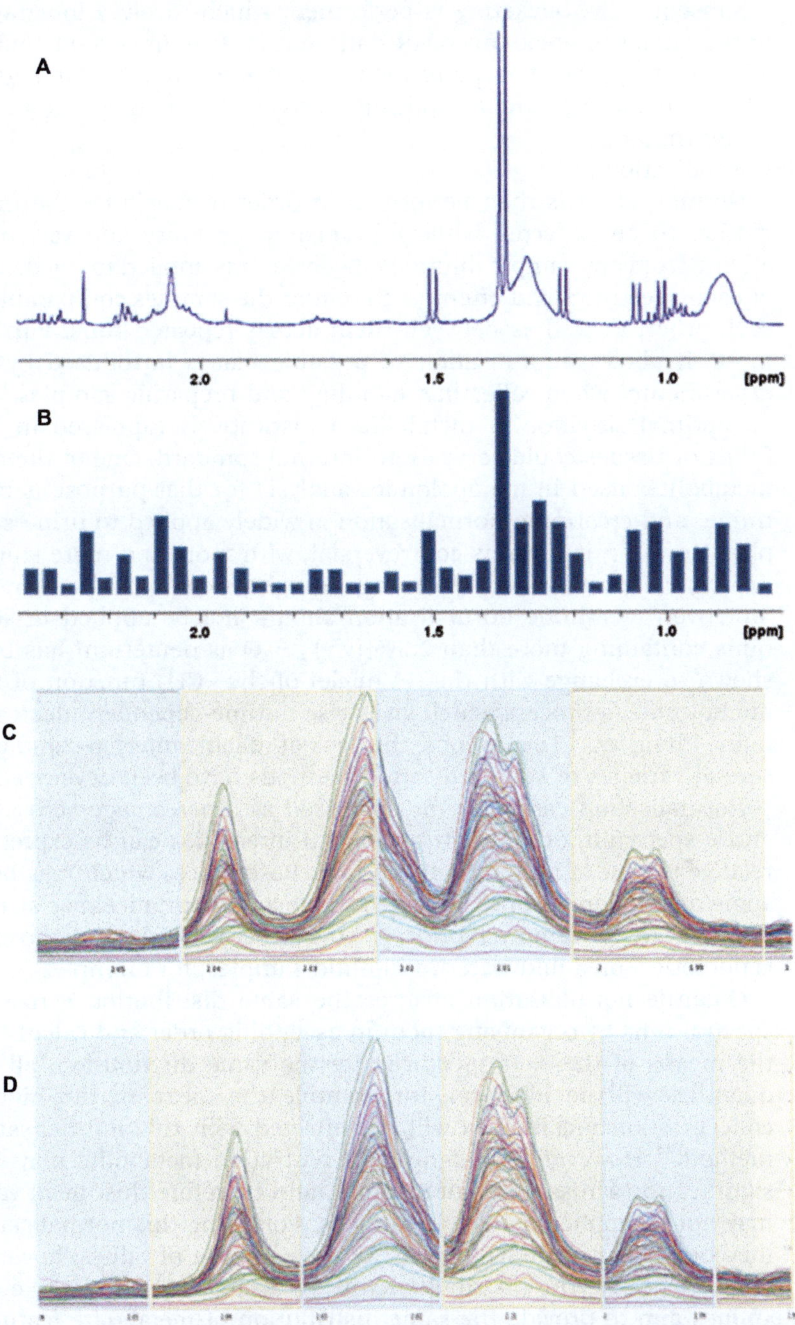

Figure 1.4 ^{1}H-NMR spectrum before (A) and after (B) equidistant bucketing. (C) Equidistant bucketing, in which the bucket size is constant. As shown, a single resonance could be erroneously divided into two buckets. (D) Intelligent bucketing, in which each signal is divided into only a single bucket.

Subsequently, bucketing is performed, which involves integration of the signal to create an NMR data matrix. It is important that an alignment approach is performed in order to ensure that regions which are bucketed are not splitting across two such integration areas rather than one.

(E) Normalisation

Normalisation is then performed in order to maximise the information to be extracted while minimising the noise and variability arising from any sample dilutions involved. It is applied to the dataset of each spectrum, and attempts to render the samples comparable to each other, as well as between them across repeated runs. Furthermore, it allows minimisation of possible biases introduced by the experimenter when collecting, handling and preparing samples.[40] In an optimal situation, a metabolite constitutively expressed in biofluids or tissues could serve as an internal standard. One of the only metabolites used in metabolomics analysis for that purpose is creatinine, and creatinine normalisation is widely applied to urine samples. However, it remains controversial, with more and more studies linking creatinine variations to age, weight, exercise or gender.[41,42] Moreover, creatinine normalisation should not be applied in solutions containing more than 2.5% (v/v) ^2H_2O as deuterium has been shown to exchange with the ^1H nuclei of the $-CH_2$ function of this biomolecule, a process which gives rise to time-dependent decreases in the intensity.[43] To overcome this lack of reliable internal standards, several varieties of standardisation methods have been developed.

Normalisation can either be expressed as a percentage across the entire spectrum, or alternatively, signal intensities can be expressed relative to that of an internal standard. Resonances which may be of some metabolomics or diagnostic/prognostic importance may also be required to be removed prior to this process, including those of xenobiotics, urea and water in biofluid samples, for example.

Quantile normalisation ensures the same distribution across all spectral bins by organising them in ascending order and calculating the means of these. If spectra share the same distribution, all the quantiles will be identical, for example the mean of the highest concentration metabolites will be reflected with this normalisation method.[44] However, the highest concentration metabolite may vary significantly across different samples, and therefore this mean value may not be applied across all samples. Following this normalisation method, each feature will consist of the same set of values; however, within features, the distribution will be different.[44] Similarly, cubic splines aim to provide the same distribution of metabolite features, however, both non-linear relationships are assumed between the baseline and individual spectra. In the cubic spline method, the geometric mean of all spectral features is calculated. A cubic spline is then fitted between the baseline and the spectral intensities several times in order to achieve normalisation. However, variance stabilisation normalisation (VSN) operates differently to the above described methods, and successfully maintains a constant variance for each predictor variable within the dataset.

Other methods of normalisation include probabilistic quotient normalisation, histogram matching, contrast normalisation and cyclic locally-weighted regression can be considered for use in metabolomics datasets, but are beyond the scope of this work.

(F) Scaling or bucket normalisation

Scaling can then be completed in order to standardise each bucket region. Indeed, the buckets associated with the most concentrated metabolites have a greater variance than others. Consequently, some buckets may have greater weights than those of others in variance-based multivariate data analyses. To avoid this bias, it is essential to rescale the weight of each variable. This can be performed using autoscaling, by subtracting the mean centre point from each observation and dividing by the variance, or the widely preferred Pareto scaling, which also subtracts the mean centre point, but is then divided by the square root of the standard deviation. Hence, Pareto-scaled variables do not have unit variance, but variances are all close to unity, albeit different. For example, urinary metabolites present in small concentrations such as formate will produce lower intensities, in view of the lower concentrations compared to that of creatinine, and therefore scaling these accordingly ensures normality for each variable and column (metabolite) variance homogeneity, despite their original concentrations.

Scaling methodologies have been reviewed by Gromski *et al.*,[45] and these suggest that VAST (variable stability) scaling is the best methodology for NMR data; this represents an extension of autoscaling. Small predictor variable metabolite variations are accounted for using this method as post autoscaling data are multiplied by a scaling factor and then divided by the standard deviation.[45] Other scaling methodologies include range-level-scaling which are not explored herein.

Transformation of data is also useful to ensure a bell-shaped data distribution, that is reducing distributional skewness. Indeed, logarithmic or cube root transformations are often recommended for metabolomics datasets.

Some authors recommend spectral or chromatographic smoothing to ensure noise reduction; however, clearly small signals need to be retained as much as possible by this process.[46] Overall, the quality of pre-processing spectral data prior to statistical analysis determines the quality and accuracy of the results.

1.4.1 Spectral Prediction in Positive Signal Identification

Positive signal identification can be performed without statistical approaches, and there is a plethora of metabolites and identification platforms such as the Human Metabolome Database (HMDB),[47] MetaboLights,[48] Biomagresbank (BMRB),[49] Spectral database for organic Compounds (SDBS), Madison-Qingdao Metabolomics Consortium Database (MMCD),[50] The Birmingham Metabolite Library (BML-NMR),[51] NMRshiftDB[52] and the metabolomics workbench,[53] which can markedly facilitate signal identification. It is important in NMR to account for the multiplicity, integral, *J* couplings, and chemical shift values prior to the assignment of signals. However, statistical approaches have also been used in conjunction with bioanalytical techniques in order to identify signals which are correlated to each other, a process also facilitating assignments. These methodologies have been demonstrated in model sample systems containing just a single molecule, and complex mixtures such as a biofluid sample, and hence provide a pseudo-two-dimensional NMR spectrum. Figure 1.5 shows confirmation of the identity of *n*-butyric acid using the most recently developed statistical total correlation spectroscopy (STOCSY) strategy as applied to faecal water, which shows the ability of this technique to tackle such complex mixtures.[54] Figure 1.6 shows the elucidation of a mixture of sucrose and glucose signals utilising the STOCSY approach.

STOCSY has also been used in conjunction with the statistical recoupling of variables giving rise to an R-STOCSY approach that shows correlations between distant clusters.[56] Previous methodologies of applying statistics to NMR spectra, predominantly from Nicholson's group, which include statistical heterospectroscopy (SHY) and accounts for covariance between signals, and which can be used to observe correlations between two applied analytical techniques such as MS and NMR, or STOCSY to MS, has been previously

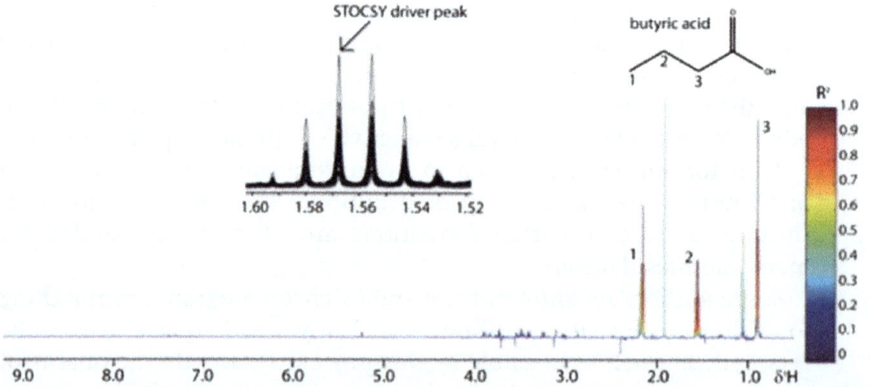

Figure 1.5 STOCSY analysis of faecal water, with a correlation matrix of $R^2 = 0.0–1.0$ (right-hand ordinate axis) which shows correlation with the β-CH_2 function $\delta = 1.562$ ppm driver signal, two other signals (those of the terminal-CH_3 (*t*) and α-CH_2 (*t*) functions) correlated with this resonance, which aids the positive identification of *n*-butyrate.[54]
Reproduced from ref. 54, https://doi.org/10.1016/j.csbj.2016.02.005, under a CC By 4.0 license, https://creativecommons.org/licenses/by/4.0/.

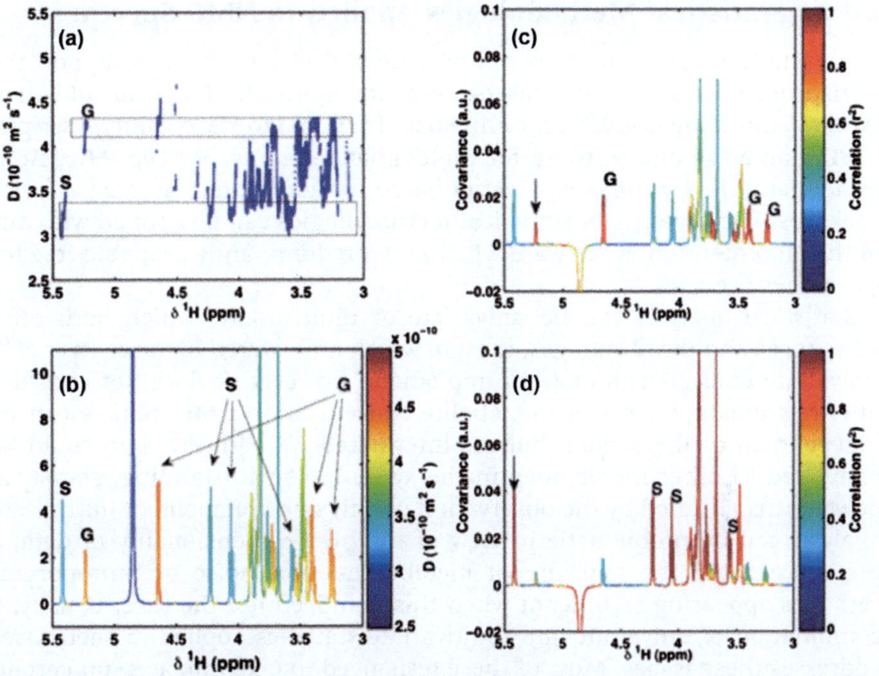

Figure 1.6 STOCSY analysis of a sucrose/glucose admixture (metabolites are represented by the letters S and G respectively).[55]
Reproduced from ref. 55 with permission from American Chemical Society, Copyright 2007.

demonstrated by Crockford *et al.*[57] and Nicholson *et al.*[58] However, sufficient computing power is required to perform these techniques.[57] Diffusion order (DO) and STOCSY have also been combined to yield an S-DOSY technique which can be employed for complex mixture analysis in order to facilitate assignments, the deconvolution of overlapping metabolite signals, and simple comparisons of the diffusional variances in signals.[55]

Other useful non-statistical techniques include 2D NMR such as ^{1}H–^{1}H COSY and ^{1}H–^{13}C HSQC to help with the assignment of NMR signals, without involving such statistical complexity. Metabolite prediction has also been trialled by observation of the chemical shift and the concentration of the biofluid itself, comparing relationships between these two elements in order to provide a chemical shift and concentration dataset matrix, From this a prediction model was constructed including an algorithm model and salient navigator signals, such as those of creatine, creatinine and citrate, to aid with prediction capability.[59] The idea of the chemical shift prediction remains in the early stages of development, and requires uniform sample preparation and operating frequencies in order to achieve successful assignments, as has been demonstrated for proteins[60] and multicomponent biofluids.[59]

1.4.2 Statistical Methodologies Applied to NMR Spectra

At present there are numerous computational packages that can support the statistical analysis of NMR datasets. These include *XLSTAT2019*, an add on to excel, *Metaboanalyst 4.0*,[61] an online user-friendly interface using R scripts, *SIMCA*, an all-in-one software for multivariate analysis, *MatLab*, *MVAPACK*, *Python* and *R Programming*, script-based programming languages with packages. The majority of statistical methodologies can be applied with any of the aforementioned software which, are predominantly available free for researcher use.

Statistical analysis can be univariate or multivariate, which both offer advantages and disadvantages, most of which are covered by Saccenti *et al.*[62] Univariate analysis is simple to implement; however, it does not consider inter-relationships between metabolite concentrations. Metabolites can be independent or dependent, but are interlinked *via* pathways and could be correlated to other metabolites in the system. Notwithstanding, statistical power is also limited by the observation of only one metabolite. Multivariate analysis can be problematic in view of the high dimensionality of data, a process causing the masking of metabolites, and noise or unimportant variables appearing significant when this is indeed not the case. Usually, a combination of univariate and multivariate statistics applied in such cases addresses these issues. Most of these tests need to take into account certain assumptions which can be found in any statistical textbook, for example that the data has been suitably preprocessed, normalised and scaled to unit or near-unit variance as discussed above.

Each statistical technique herein will be described, and a case study showing statistical applications in ^{1}H NMR spectral analysis will be considered. A range of applications will be explored to show the diversity of fields in metabolomics, but the predominant theme will be biofluids and liquid biopsies. A summary table showing the advantages and disadvantages of each technique respectively in metabolomics applications will be provided at the end of this chapter. Often, a combination of techniques will be used in order to classify and provide a statistical significance to the results acquired, that is 1D and 2D NMR spectra, LC–MS and so forth, which is required for validation. This chapter covers the most frequently applied statistical methods employed in metabolomics research investigations at present.

1.5 Univariate Data Analysis

Univariate data analysis is crucial in any metabolomics data analysis strategy. A variable may be insignificant in a multivariate model, but significant in a univariate context. This is because multivariate models can often miss/mask significant variables as all metabolites (and metabolite relationships) are simultaneously examined. Hence, it is important that univariate data analysis is integrated into metabolomics experimental designs. This is particularly salient for validation purposes for specific potential biomarkers.

Student's *t*-tests can be used in order to discover statistical significance in univariate datasets consisting of two sample comparisons, or more if suitable corrections are applied for a false discovery rate. There are several variations of this test which rely on similar concepts, including the unequal variance *t*-test derivative, and the unrecommended non-parametric Mann–Whitney *U* test. Typically, these tests can be paired and unpaired, and are used in conjunction with the variable type, whether this be dependent or independent respectively. An unpaired *t*-test will evaluate the statistical significance of any differences between mean values between two independent groups. Degrees of freedom are considered in order to establish statistical significance. As with all other parametric tests for evaluating differences between mean values, critical assumptions of normality, intra-sample variance homogeneity, and in cases of randomized blocks ANOVA without replications so that predictor variable interactions may not be considered, and additivity all apply.

$$t = \frac{\bar{x} - \mu_0}{s/\sqrt{n}} \quad \text{One} - \text{sample } t\text{–test} \tag{1.1}$$

$$t = \frac{\bar{x}_D - \mu_0}{s_D/\sqrt{n}} \quad \text{Dependent } t\text{–test for paired samples} \tag{1.2}$$

$$t = \frac{\bar{x}_1 - \bar{x}_2}{\sqrt{s^2\left(\frac{1}{n_1} - \frac{1}{n_2}\right)}} \quad \text{Two sample } t\text{–test} \tag{1.3}$$

$$s_2 = \frac{\sum_{i=1}^{n_1} (x_i - \bar{x}_1)^2 + \sum_{j=1}^{n_2} (x_j - \bar{x}_2)^2}{n_1 + n_2 - 2}$$

$$t = \frac{\bar{x}_1 - x_1}{s_\Delta} (4) s_\Delta = \sqrt{\frac{s_1^2}{n_1} + \frac{s_2^2}{n_2}} \quad \text{Welchs t} - \text{test} \tag{1.4}$$

$$W = \frac{\left(\sum_{i=1}^{n} a_i x_{(i)}\right)}{\sum_{i=1}^{n} (x_i - \bar{x})^2} \quad \text{Shapiro} - \text{Wilk test} \tag{1.5}$$

In which $\bar{x}$ represents the mean, μ_0 is the null hypotheses, s is the standard deviation, n is the sample size, ${s_1}^2$ and ${s_2}^2$ are the variance with the associated numerical value indicating the group number, n_1 and n_2 represent the sample size with the associated numerical value indicating the group number, and s^2 is the pooled sample variance.

In eqn (1.1) and (1.2) the degrees of freedom are $(n - 1)$. In eqn (1.2) when calculating the degrees of freedom, $(n - 1)$ is used in which n represents the number of paired samples. The Welch–Satterwaite equation is required for calculation of the degrees of freedom calculation in eqn (1.4).

Percival *et al.* applied a paired student *t*-test for a metabolomics investigation monitoring methanol and other metabolites in saliva using ^{1}H NMR analysis.[38] Two samples were taken from smoking participants prior and subsequent to smoking a single cigarette; thus, a paired test was appropriate. The paired student *t*-test showed highly significant differences between molecules such as methanol and propane-1,2-diol, which were significantly elevated post-smoking, with significance levels of $p = <10^{-6}$ and 2.0×10^{-4} respectively.

The Mann–Whitney-*U* test counts the number of times the null hypothesis is proven false and this process is completed for both sample groups. The U statistic is then calculated, and is equivalent to the area under the receiver operating characteristic (ROC) curve which will be described in more detail later.

Fold-change analysis can also be performed to assess the degree of change in variable levels, and can be used to describe an increase of "X-fold" per sample classification. It is simply a ratio of two mean values.

1.5.1 ANOVA, ANCOVA and Their Relationships to ASCA

Analysis of Variance has been successfully applied in metabolomics investigations such as the detection and determination of methanol in smokers' salivary profiles using ^{1}H NMR analysis;[38] one typical experimental design is shown in eqn (1.6). This analysis of covariance (ANCOVA) model included the between sampling time-points T_i, smoking/non-smoking groups S_j, between participants $P_{(j)}k$ and between gender sources of variation, G_l The mean value, μ and unexplained error, e_{ijkl} are also incorporated into this mathematical model, in addition to the first-order interaction effect between the smoking/non-smoking groups and sampling time points, that is TS_{ij} Participants were 'nested' within treatment groups.

$$Y_{ijklm} = \mu + T_i + S_j + P_{(j)k} + G_l + TS_{ij} + e_{ijklm} \quad \text{ANCOVA} \qquad (1.6)$$

This ANCOVA test complimented the results acquired in the aforementioned paired students *t*-test but is particularly advantageous as the ANCOVA model factored in all possible sources of variation, including interaction effects and unexplained errors. However, ANOVA or ANCOVA models can be applied in different manners which are also applicable to metabolomics applications. ANOVA-simultaneous component analysis (ASCA), for example, allows comparison of data which has been acquired on the same human participants at increasing time-points, or when considering alternative second variables. It can handle two experimental factors, but also observe the factors separately, along with their magnitudes by operating with the use of a combination of ANOVA factors with PCA, the latter of which is described below. It can also isolate ASCA contributions from statistical interaction effects, just as it can in univariate ANOVA and ANCOVA models.

Factorial ANOVA can handle more than one factor simultaneously at multiple levels. Repeated measurements of ANOVA can also be applied in longitudinal studies. ANOVA techniques are generally more applicable in targeted metabolomics using MS; however, there are a few examples seen in NMR-based metabolomics applications, as discussed below.

ANCOVA can account for qualitative and quantitative variables.

Ruiz-Rodado *et al.* successfully applied ANCOVA and ASCA to ^{1}H NMR metabolomics datasets from mice with Niemann–Pick Disease, Type C1.[63] The ANCOVA model accounted for three factors and six sources of primary variation as shown in eqn (1.7), to provide the univariate predictor variable, Y_{ijk}. Between disease classifications, D_i, between genders G_j and the experimental sampling time points, for example, for 3, 6, 9 and 11 week-old mice, T_k were incorporated into the experimental ANCOVA design. Interactions between each variable were also considered, for example DG_{ij}, DT_{ik} and GT_{jk}. Interaction effects are computed to assess the dependence of the effects or significance of one variable at different levels or classifications of another μ represents the mean value in the population in the absence of all sources of variation, and e_{ijk} is the residual (unexplained) error contribution.

$$Y_{ijk} = \mu + D_i + G_j + T_k + DG_{ij} + DT_{ik} + GT_{jk} + e_{ijk} \quad \text{ANCOVA} \qquad (1.7)$$

Once key features were identified using multivariate analysis, ANCOVA was applied in a univariate context in order to reveal information regarding significant metabolites that were time-dependent, such as 3-hydroxyphenylacetate, and gender-dependent such as tyrosine. Moreover, this tool was able to show significant metabolites for a combination of variables, for example the "time-point x disease" interaction effect revealed inosine as one of the significant biomarkers; in addition, the "gender x disease" interaction effect showed a combined lysine/ornithine resonance as one of the significant distinguishing spectral features. Thus, this technique can be used successfully in metabolomics across numerous markers, and provide distinct p values for each metabolite investigated. False discovery rates and power calculations can be applied, which will be discussed below.

An alternative to ASCA is multilevel simultaneous component analysis (MSCA), which can also allow for paired datasets and divides the data into two parts, for example age and sex, and then monitors the variance associated within and between each variable.[64] ASCA supersedes MSCA, as it is simply a MV extension with the benefit of ANOVA, and this explains why it is less commonly used in metabolomics studies, as other multilevel techniques are more frequently applied.

1.6 Multivariate Analysis

It is, of course, essential to incorporate multivariate data analysis into a metabolomics investigation. Univariate data analysis can consider some metabolites insignificant, but this is not the case in a multivariate context, generally because its effects only correlate, perhaps strongly, with one or more of a pattern of other metabolite variables. Moreover, the insignificance of a variable in univariate analysis could also be explicable owing to high levels of biological and/or measurement variation. Multivariate analysis may perhaps overcome this problem by further explaining classifications attributable to biological/measurement variations, and is able to combine variables together as components by their correlations and inter-relationships.

1.6.1 Principal Component Analysis

The most common unsupervised multivariate method is termed PCA, which is particularly useful for data mining and outlier detection. It summarises the variance contained in the dataset in a small number of principal components (PCs, latent variables). The principle consists of applying a rotation in the space of the N-dimensional variables, so that the new axis system, composed of the principal components, maximises the dispersion of the samples. The first principal component (PC1) represents the direction of the space containing the largest variance expressed in the analysed data. The second principal component PC2 represents the second direction of greater variance, in the orthogonal subspace at PC1, and so on. This procedure continues until the entire variance is explained and thus allows the essential information contained within the dataset to be synthesised by a limited number of PCs (usually $\ll$ N). Each PC corresponds to a linear combination of the N original metabolite variables, the weights represent the contribution of them to these components. The representation of these components makes it possible to visualise the associated metabolic signatures. One of the interests of the method is to identify, without *a priori* considerations, possible groupings of individuals and/or variables. However, it is possible that the primary sources of variance within a cohort of samples are not related to the effect studied, and therefore this unsupervised analysis attempts a sample/participant classification without any prior consideration of their classifications. Supervised analysis methods, however, may identify variations in metabolites which are or may be correlated with the parameters of interest of the study. Both approaches also allow the detection of samples with atypical behaviour ("outliers") when compared relative to the remainder of the population. Figure 1.7 shows 95% confidence intervals for a typical PCA scores plot. These may be established using a multivariate generalization of Student's *t*-test, known as Hotelling's T^2 test.[30] T^2 determines how far away an observation is from the centre of a PC. In Figure 1.7, the points highlighted with blue arrows are outliers.

These outliers could arise for a variety of reasons, such as xenobiotics and/or unusual or unexpected metabolites being detected in the urine or alternatively, the sample could display unexpected intensity alterations in a particular profile region. The PCA plot will not only indicate which samples are outliers, but also which principal component (PC) it is loading on (*via* a loadings plot), and also which other bucket regions are strongly loaded on

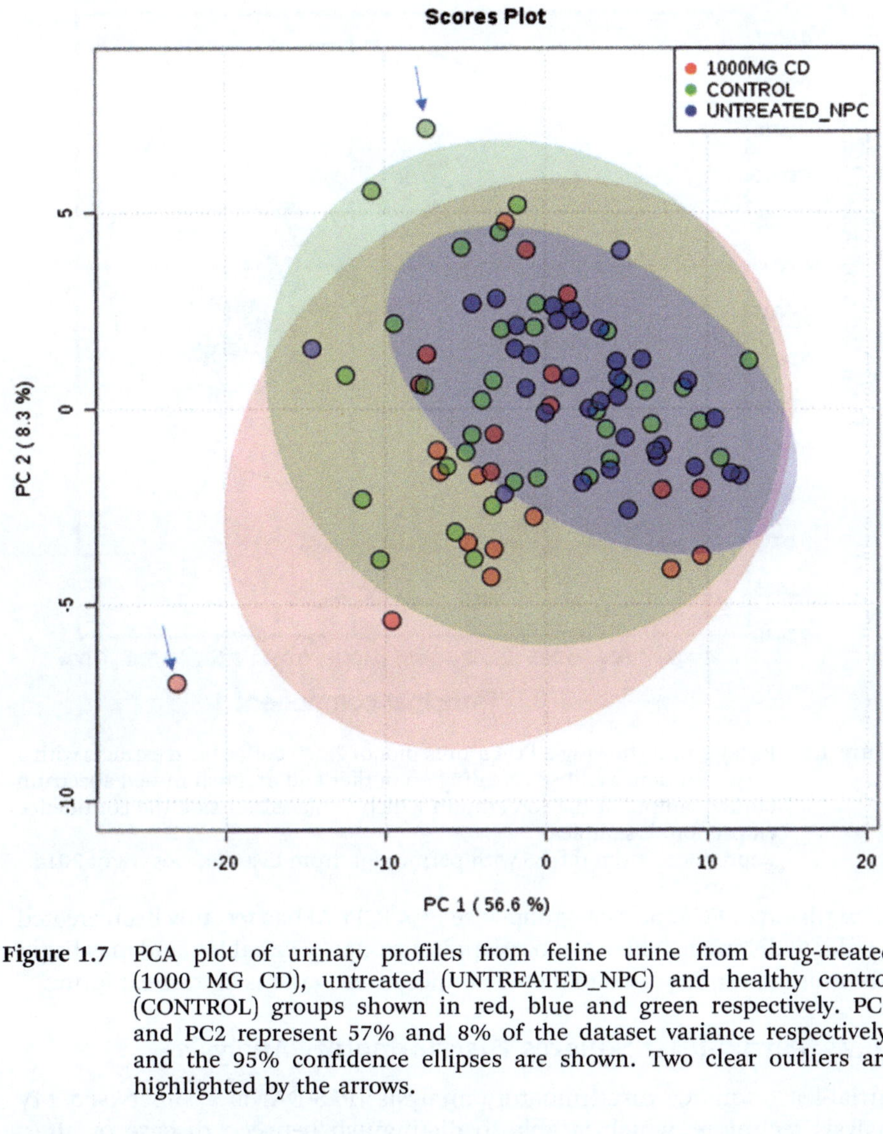

Figure 1.7 PCA plot of urinary profiles from feline urine from drug-treated (1000 MG CD), untreated (UNTREATED_NPC) and healthy control (CONTROL) groups shown in red, blue and green respectively. PC1 and PC2 represent 57% and 8% of the dataset variance respectively, and the 95% confidence ellipses are shown. Two clear outliers are highlighted by the arrows.

that component. This information aids in the identification of classifications for these samples/participant donors.

Figure 1.7 shows a typical PCA plot obtained from feline urine samples with two outliers also identified. An improved example is shown in Figure 1.8 (provided by Kwon *et al.*[65]), who assessed green coffee bean metabolites in which a sample was removed in view of poor spectral shimming, with the unshimmed sample is shown in the inset image.

An alternative to PCA analysis is simultaneous component analysis (SCA) which takes into account different sources of variation by separating datasets into sub-matrices;[64] however, PCA is more commonly employed in this field.

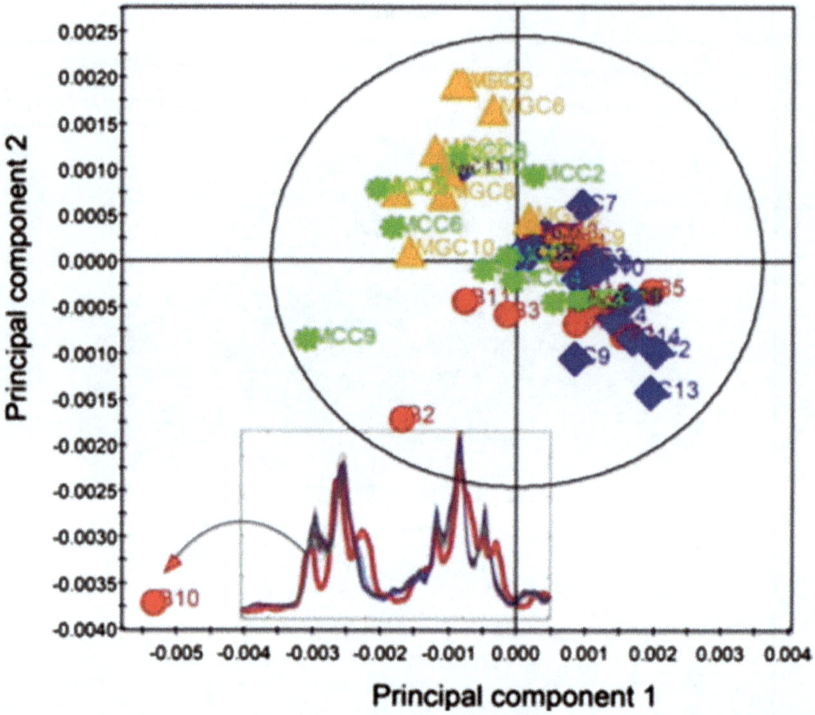

Figure 1.8 Background showing a PCA scores plot of green coffee bean extracts with a 95% confidence ellipse revealing an outlier and an unshimmed spectrum outlier sample in the foreground which is placed outside the confidence ellipse for all samples.[65]
Reproduced from ref. 65 with permission from Elsevier, Copyright 2014.

An extension of PCA, namely group-wise PCA (GPCA) has recently been created in order to distinguish between overlapping groups of variables, and may begin to be more commonly used in metabolomics investigations in the near future.[66]

1.6.2 Partial-least Squares Discriminatory Analysis

Partial-least squares discriminatory analysis (PLS-DA) is a supervised MV analysis technique which is able to distinguish between disease or alternative classifications, and focuses on 'between-class' maximisation. This method aims to predict a response variable Y (qualitative) from an explanatory data matrix X. The components of the PLS are composed to take into account the maximum variance of the data (X) which are the most correlated possible with Y. In this case, Y is a discrete variable that takes a value that depends only on the categorical class associated with the sample. PLS analysis allows the identification of the most important response-variables in the prediction of the variable Y, and thus makes it possible to highlight the most discriminant variables between the groups, and whether metabolites are upregulated or downregulated by creating latent structures and variable importance plots (VIPs). Similar to PCA, PCs can be plotted in order to observe clusterings, with each PC being orthogonal to each-other, and with PC1 again containing the highest sample variance, and PC2 containing the next highest sample variance, and so forth. PLS-DA has many variants, including orthogonal PLS-DA (O-PLS-DA), multilevel PLS-DA (M-PLS-DA), powered PLS-DA (PPLS-DA), and N-way-PLSDA. Each has its own advantages and disadvantages for metabolomics use. For example, O-PLS-DA can only handle two groups for comparative evaluations. In this case, the orthogonal signal correction filter applied enables separation between predictive variation and correlation.

1.6.2.1 Validation Methods

Supervised analysis, unlike PCA, can lead to biased data and overestimation of the predictive capabilities of the model. Indeed, the large amount of data generates a space with a large number of dimensions in which it is almost always possible to find a direction of separation between the samples. Therefore, it is essential to ensure the quality of the models established with validation methods such as permutation testing, cross-validation and ROC analysis.

- Cross-validation: Cross-validation is the most common validation method used in metabolomics. It is based on two parameters to evaluate the model's performance: R^2 and Q^2Y. R^2 (X and Y) represents the explained variance proportion of the matrix of the X and Y variables, and Q^2Y cumulative represents the predictive quality of the model. It can be interpreted as the estimation of R^2 for new data. The closer these values are to 1, the more the model is considered as predictive, and the results of the separation as significant.[67]
- Permutation test: The objective of this test is to confirm that the initial model is superior to other models obtained by permuting the class labels and randomly assigning them to different individuals. The initial model is statistically compared to all the other randomly-assigned models. Based on this, a *p*-value is then calculated. If the *p*-value is lower than 0.5, this indicates that the initial model performs better than 95% of the randomly assigned models.[68]
- Cross-validation and permutation tests are complementary, and both must be performed in order to validate a model. Indeed, cross-validation makes it possible to evaluate the capacity of the model to correctly predict in which class a new sample will be, while the test of permutation validates the model used.[68]
- ROC analysis: Area under the curve receiver operating characteristic (AUROC) value can then be used to monitor the sensitivity and specificity of singular metabolites, and the performance of the test system as a whole. Sensitivity and specificity are monitored, in which a correlation of 1.0 and 0.0 can be observed, with correlations of 1 representing a perfect distinction between classes, values greater than 0.5 being

considered discriminatory, and a value equivalent to 0.50 demonstrating that the model is as likely to correctly classify a sample as if one was tossing a coin.[69] PLS-DA is then validated using permutation testing, which is able to define the *p* value for the PLS-DA discriminatory ability. Further validation can be performed using leave-one-out cross validation (LOOCV) and 10-fold cross validation in order to obtain the Q^2 and R^2 values; Q^2 values greater than 0.5 are considered satisfactorily discriminatory. Advantageously, PLS-DA provides the VIPs, which are able to distinguish which metabolites are responsible for the distinction observed, and also whether these metabolites are up- or downregulated.

PLS-DA analysis was effectively applied to an ^{1}H NMR investigation of brain extracts obtained from the post-mortems of patients with Huntington's disease and control patients (Figure 1.9). Permutation testing was applied in order to validate the study using 2000 permutations for the frontal lobe analysis and striatum region, yielding *p* values of 0.003 and <0.001.[70] In addition, permutation testing was performed again using only 1000 permutations, the results showing that these values for the frontal lobe and striatum were 0.108 and 0.015 respectively, which indicated that the frontal lobe was less affected by the pathological implications of Huntington's Disease than the striatum.[70] VIPs were also useful for the identification of the metabolites causing these significant differences, and their up- or downregulation status (Figure 1.10). Values for AUROC, sensitivity and specificity values were 0.942, 0.869 and 0.865 respectively, using training/discovery, and 0.838, 0.818 and 0.857 respectively using 10-fold cross validation, the results demonstrate the success of the model.

OPLS-DA has been used to discriminate between two groups using orthogonal latent structures. The OPLS-DA plots revealed a loadings diagram with a S (sigmoidal)-shaped curve, and in which the validation and permutation tests can also be performed. This S-plot can show visualisation of the OPLS-DA loadings (Figure 1.10), which is useful in the identification of significant metabolites.

The successful use of O-PLS-DA has been demonstrated by Quansah *et al.* observing the effects of an anti-ADHD psychostimulant, methylphenidate (MPH), on brain metabolite levels.[71] Using this methodology, the researchers were able to establish significant and non-significant groupings using this approach. A significant difference was observed between the acute high 5.0 mg kg^{-1} dose MPH-treated and age-matched saline-treated control groups with an OPLS-DA model showing $R^2X = 0.60$, $R^2Y = 0.54$, and $Q^2 = 0.44$; with a permutation test p value = 0.0005. A lower acute dosage of 2.0 mg kg^{-1} MPH provided insignificant results when compared to the saline-treated control groups, showing $R^2X = 0.45$, $R^2Y = 0.05$, and $Q^2 < 0.1$; with a permutation p value = 0.93. The lower acute dose of 2.0 mg/kg given twice daily was not significantly different from that of the control group.

The significant results pertaining to the higher dosage were further analysed, and an S-plot (Figure 1.10) was obtained, and results acquired were complimentary to those obtained with ANOVA analysis and revealed significant

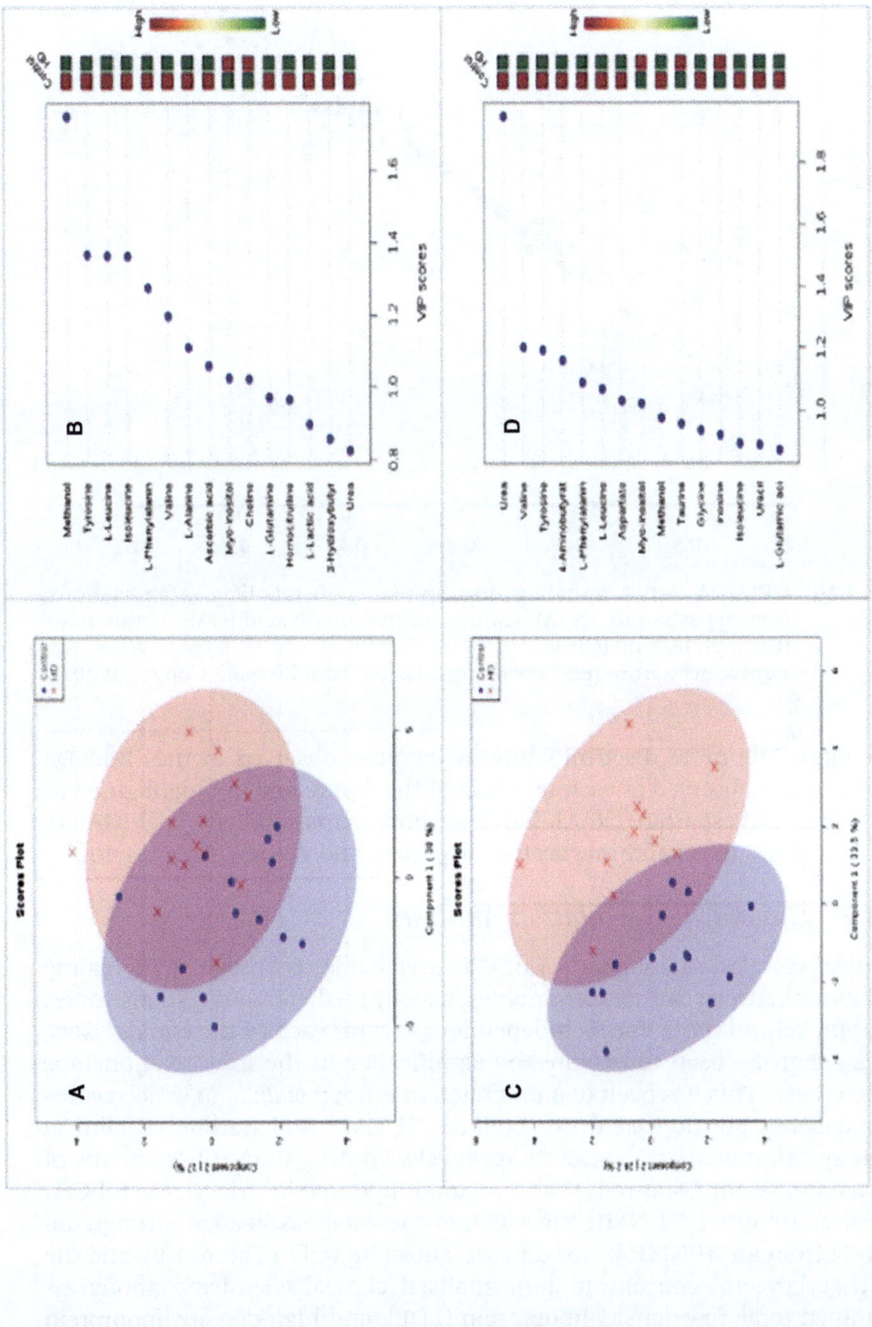

Figure 1.9 PLS-DA showing 95% confidence ellipses and corresponding VIPs of Huntington's disease (red) *versus* control (blue) frontal lobe extracts in (A) and (B), and the striatum region shown in (C) and (D).[70] Reproduced from ref. 70 with permission from Elsevier, Copyright 2016.

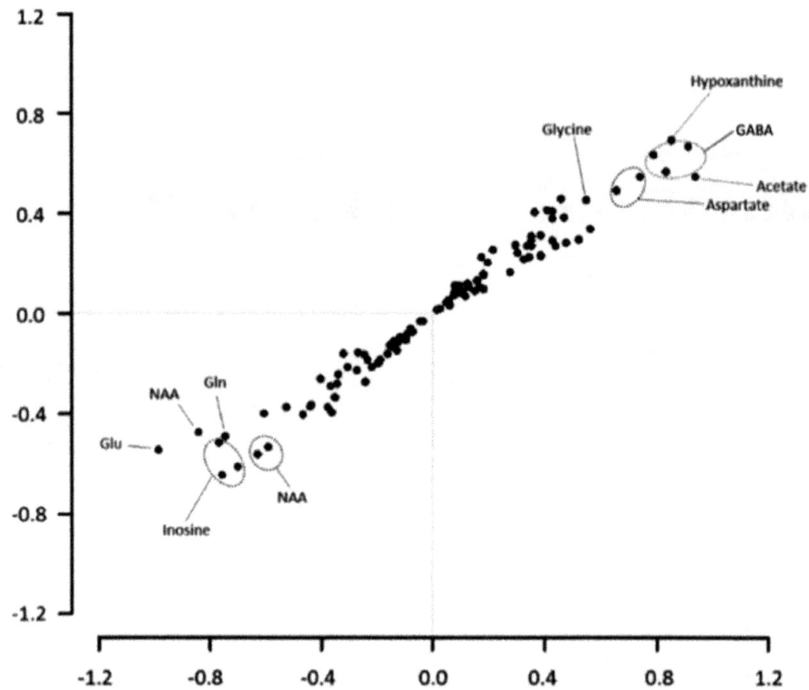

Figure 1.10 OPLS-DA S-plot revealing discriminatory metabolites. Abbreviations: *N*-acetyl-aspartate (NAA), gamma-aminobutyric acid (GABA), glutamate (Glu), glutamine (Gln).
Reproduced from ref. 71 with permission from Elsevier, Copyright 2018.

metabolites. The more discriminatory metabolites observed in the OPLS-DA analysis can be observed at each terminal of the S-plot, and are highlighted as glucose, *N*-acetyl-aspartate (NAA), inosine, gamma-aminobutyric acid (GABA), glutamine (Gln), hypoxanthine, acetate, aspartate and glycine (Figure 1.10).

1.6.2.2 Canonical Correlation Analysis

Canonical correlation analysis (CCorA) is a valuable technique for revealing correlations between two sets of variables, usually predictor and response ones. This approach primarily forms independent PCs for each of the two datasets, and can then be used to explore the significance of the inter-relationships between these. This has been demonstrated in Probert *et al.*,[31] in which scores vector datasets are derived from separate ^{1}H NMR and traditional clinical chemistry determination datasets respectively. For this study, observations of the loading vectors showed that the total lipoprotein triacylglycerol-CH_3 function-normalised ^{1}H NMR triacylglycerol resonances, loaded strongly on PC1–PC4 from an ^{1}H NMR-based dataset (shown in red in Figure 1.9), and the total triacylglycerol concentration-normalised clinical chemistry laboratory-determined total, low-density-lipoprotein (LDL)- and high-density-lipoprotein

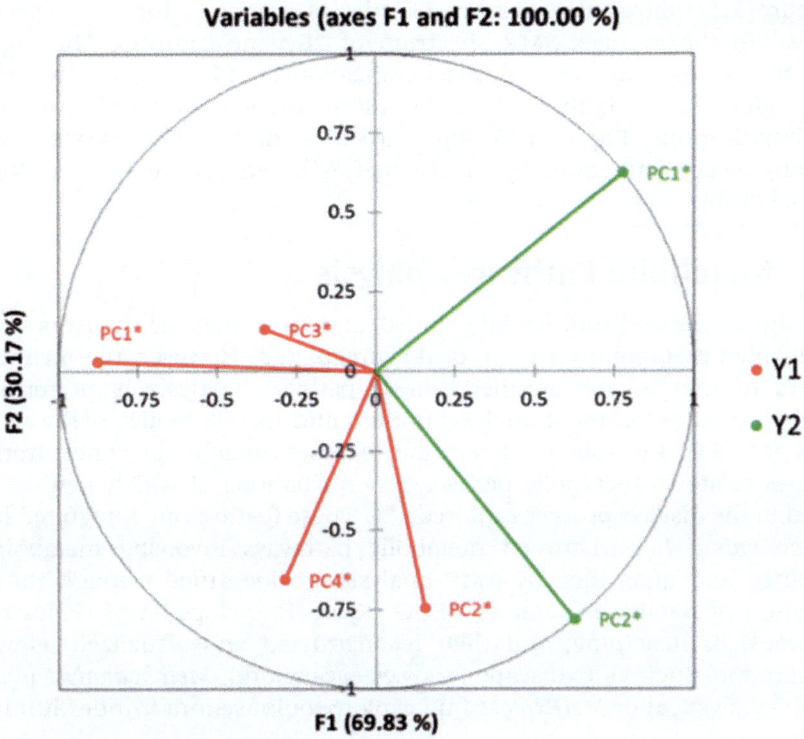

Figure 1.11 CCorA analysis from a PCA plot in which Y1 and Y2 represent scores vector datasets arising from the separate ^{1}H NMR and clinical chemistry datasets respectively.
Reproduced from ref. 31, https://doi.org/10.1038/s41598-017-06264-2, under the terms of a CC BY 4.0 license, https://creativecommons.org/licenses/by/4.0/.

(HDL)-associated cholesterol levels loaded on PC1* and PC2* (shown in green in Figure 1.11). This CCorA analysis demonstrated firstly that the PC2* scores vectors positively correlated with those of PC2, and this was consistent with their common HDL sources. Secondly, PC1* was negatively correlated with PC4, that is a linear combination of plasma triacylglycerol-normalised total cholesterol and LDL-cholesterol concentrations was anti-correlated with the ^{1}H NMR PC arising from the LDL-triacylglycerols.

1.6.2.3 Extended Canonical Variate Analysis

Extended canonical variate analysis (ECVA) uses a more complex supervised algorithm than PLS-DA, and is used in order to distinguish the maximum ratio of between-class variation to within-class variation[72]). ECVA observes individual metabolite regions, in addition to the dataset as a whole. The benefit of using ECVA is that it can discriminate between more than two groups without overfitting.

Figure 1.12 shows the number of misclassifications for each spectral interval in the average NMR spectrum of 26 wine samples. The region with the lowest numbers of misclassifications, and therefore the most discriminatory, is highlighted as the 100th interval, with only two such misclassifications. Figure 1.13 shows a scores plot of EVC3 *versus* EVC1 showing clear distinctions between wineries based on the selected 100th interval (Table 1.1).

1.7 Metabolic Pathway Analysis

Both univariate and multivariate statistical approaches enable users to explore which metabolites are up- or downregulated. However, the meaningfulness of this is not unveiled unless pathway analysis is performed. Metabolite set enrichment analysis (MSEA) and metabolomics pathway analysis (MetPA) are able to determine whether metabolite concentration changes relate to metabolic pathways, perturbations of which may be involved in the disease process explored.[74,75] These features are integrated into *MetaboAnalyst 4.0*. Disturbed metabolic pathways involving metabolites identified and quantified by NMR analysis are identified through the exploitation of databases such as KEGG (Kyoto Encyclopedia of Genes and Genomes) or Reactome, and then reconstructed and visualized using a software tool such as Cytoscape (www.cytoscape.org), *Metaboanalyst* (www.metaboanalyst.ca) or *MetExplore* (metexplore.toulouse.inra.fr/index.html/).

1.8 Further Important Considerations: Accuracy and Precision

It is important to appreciate the challenges of applying such statistical tools to metabolomics datasets. Data should be interpreted and described in an accurate manner. Bias can easily be introduced from both analytical and biological perspectives. Although analytical variances are addressed in previous sections of this chapter, it is important not to introduce bias from sample preparation techniques, whether this be by extraction, storage or the employed procedures and analytical conditions.[76] This can lead to metabolite assignment and/or concentration errors. Biological variance does need considering, and hence it is important to follow experimental design factors such as those noted above, for example potential metabolic differences between ages, genders, body mass indices, races, and so forth should also be incorporated into experimental designs, but unlike univariate analysis approaches, this is often very difficult to achieve in multivariate analysis models. Moreover, the statistical power of metabolomics experimental models can be poor in view of low sample numbers; power cannot yet be assessed *a priori* to a study in a multivariate sense, usually only *posteriori*. Validation of metabolomics datasets are required, for example ensuring that biomarkers are: (i) reproducible at another laboratory site; and (ii) decrease or elevate succinctly upon treatment, if they are indeed influenced by such

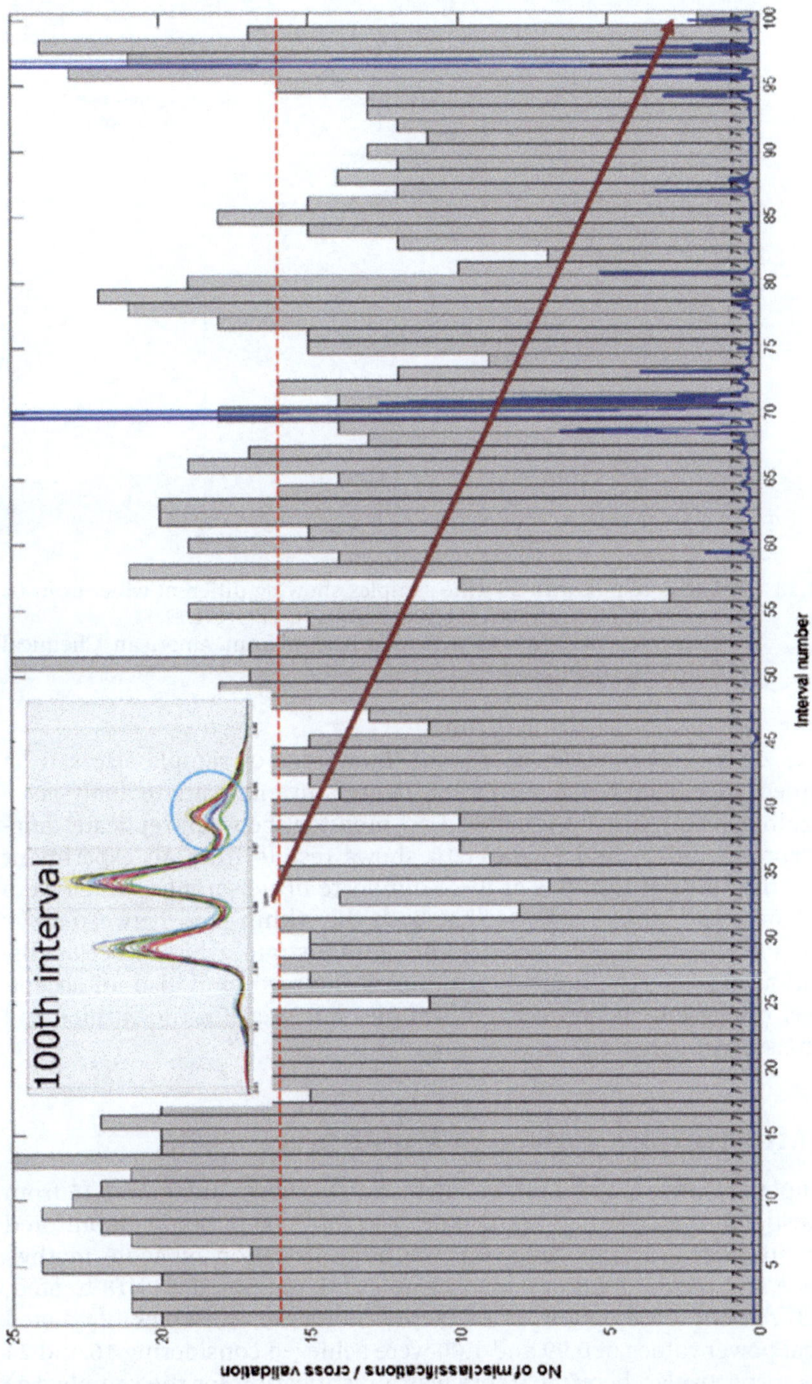

Figure 1.12 iECVA plot overlay with average NMR spectra from 26 wine samples from La Rioja.[73] Reproduced from ref. 73 with permission from American Chemical Society, Copyright 2012.

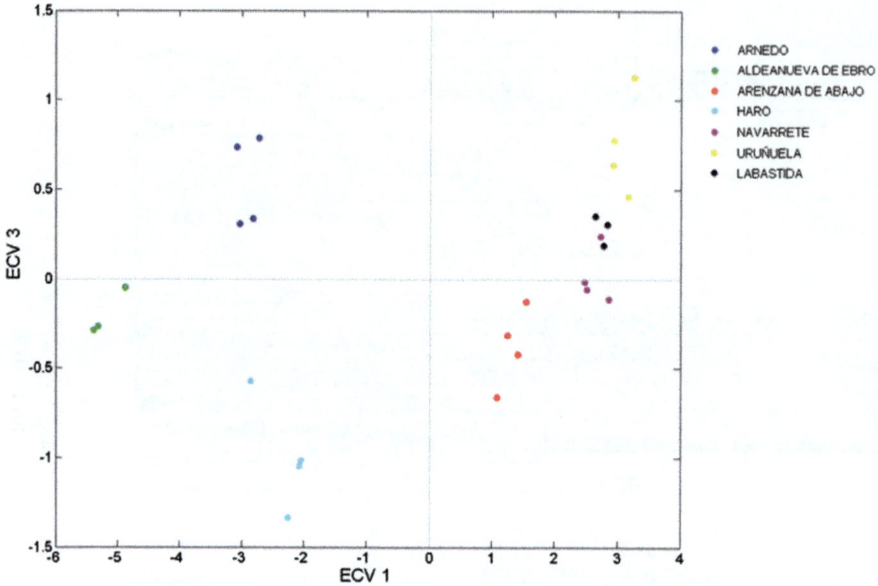

Figure 1.13 ECVA score plot with 26 wine samples showing different wineries in La Rioja based on the 100th interval shown in Figure 1.12.
Reproduced from ref. 73 with permission from American Chemical Society, Copyright 2012.

processes. In a univariate sense, a prior knowledge of sample size can be determined, as indeed it can with pilot datasets in multivariate analysis.

Intriguingly, PCA can be employed to monitor 'between-replicate' analytical reproducibility, and Figure 1.14 shows results from an experiment involving the ^{1}H NMR profiles of $n = 4$ duplicate urine samples which were analysed on two separate days so that both the within- and between-assay effects could be evaluated. These results demonstrate a high level of discrimination between the individual samples analysed, and also an acceptable level of agreement between the samples analysed for the within- and between-assays performed.

1.9 Multivariate Power Calculations

An example of a retrospective power calculation is shown in Figure 1.15 from the investigation performed by Quansah *et al.*[77] This study, monitored markers in murine brains following the administration of acute methylphenidate and used a total of $n = 36$ samples, 18 untreated and 18 treated, and featured the observation of 13 of the biomarker variables. Predicted statistical power values of 0.99 and 1.00 were achieved considering 16 and 24 samples respectively; therefore, there was a justification for the sample size of $n = 18$ being selected for this particular study.

Table 1.1 Statistical and Classifications Strategies Inclusive of Advantages and Disadvantages in NMR Metabolomics Applications.

Metabolomic Methodology	Classification or Statistical	Univariate or Multivariate	Supervision	Advantages (+) and Disadvantages (−)
ANOVA	Statistical	Univariate	Unsupervised	+ Hypothesis testing, with the ability to evaluate the statistical significance of a wide range of contributory variables, and their interactions, simultaneously. Partitions the total experimental variance into differential 'predictor' components, which may be fixed or random effects. Satisfaction of essential assumptions can be achieved with suitable transformations, for example logarithmic or square root ones.
Student's *t*-test	Statistical	Univariate	Unsupervised	+ Hypothesis testing, but without corrections for false discovery rate, is only appropriate for comparisons of the means of only two sample groups.
Mann–Whitney *U*- test	Statistical	Univariate	Unsupervised	+ Hypothesis testing – non-parametric equivalent of two sample *t* test. + Data does not require normalisation prior to use.
Fold-change Analysis	Statistical	Univariate	Unsupervised	+ Hypothesis testing; represents the ratio of two sample group mean values, and the significance of these indices may be tested.
ASCA	Statistical	Multivariate	Unsupervised	+ Can consider paired samples, for example from the same person at different time-points, or two or more possible predictor variables simultaneously.
PCA	Statistical	Multivariate	Unsupervised	+ Outlier detection + Unsupervised multivariate technique for dimensionality reduction and the preliminary exploration of 2D or 3D samples or participant clusterings.
PLS-DA	Statistical	Multivariate	Supervised	+ VIPs for significant metabolites − As it is subject to overfitting, permutation and validation testings are essential.
O-PLS-DA	Statistical	Multivariate	Supervised	+ S-Plot for significant metabolites − Can only consider two groups, validation and permutation testing required.

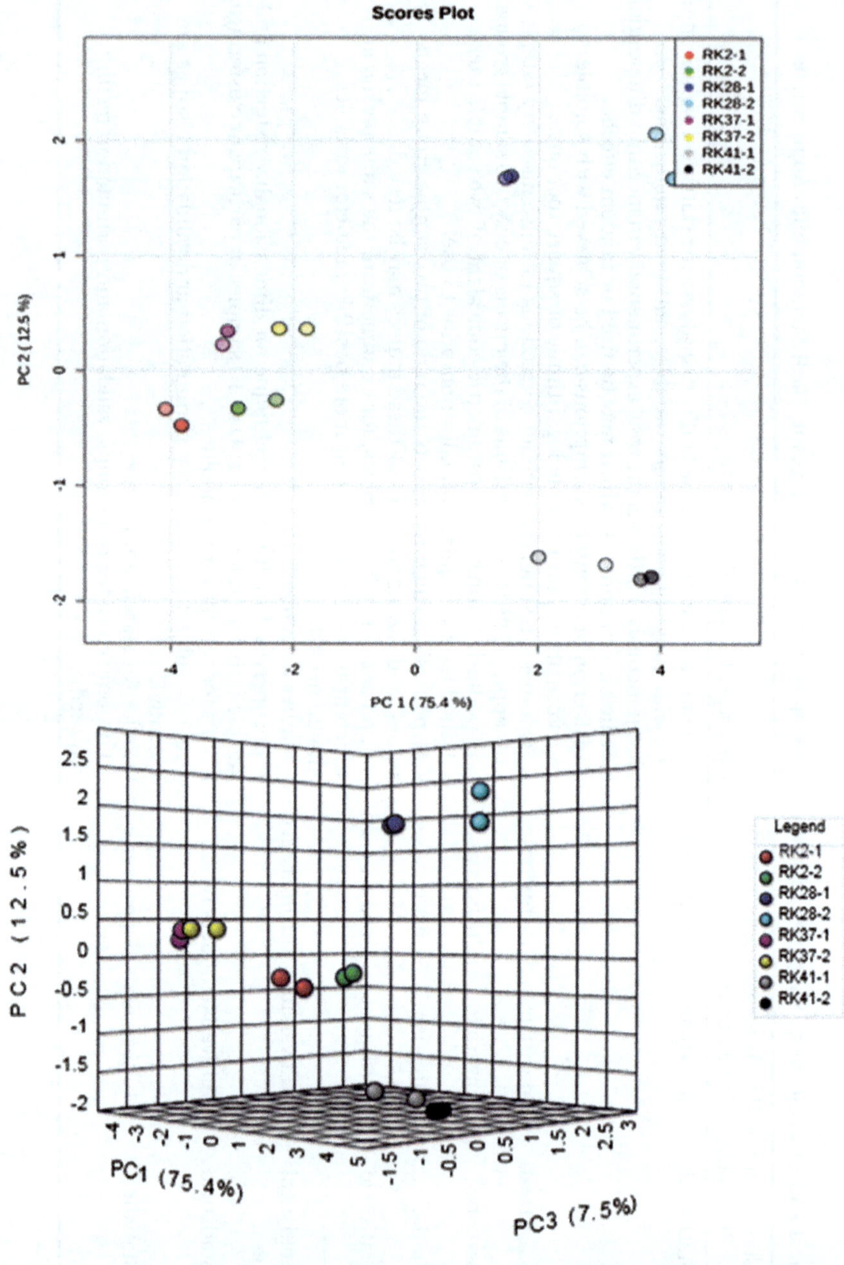

Figure 1.14 Two- and three-dimensional PCA scores plots showing clear distinctions between four sets of $n = 2$ replicate samples of four separate urine samples (coded RK2, RK28, RK37 and RK41) analysed using ^{1}H NMR spectroscopy at an operating frequency of 400 MHz. The colour codes provided on the right-hand side abscissa axes also provide information on the dates that the samples were analysed, for example the RK2-1 and RK2-2 samples correspond to the same sample analysed in duplicate on two separate occasions. Duplicate samples were prepared for analysis on each occasion.

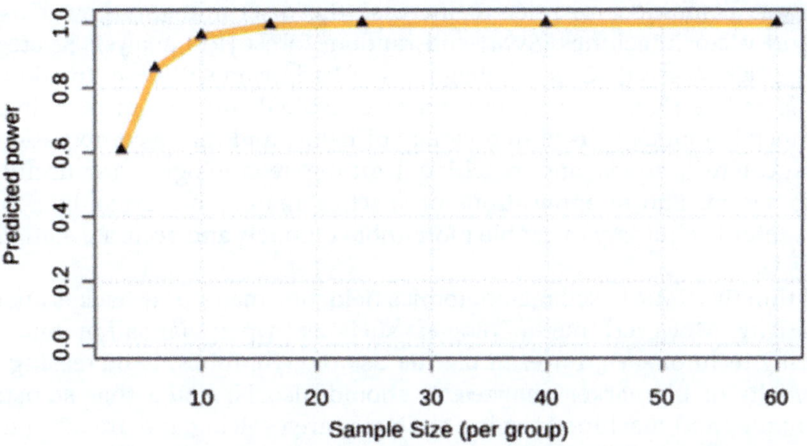

Figure 1.15 Power calculations performed by Quansah *et al.*,[77] in which the line reading 1.0 shows the optimum number of samples required for statistical significance.
Reproduced from ref. 77 with permission from Elsevier, Copyright 2017.

Power analysis ensures that ethical boundaries are implemented appropriately.

Some techniques work better with a larger number of variables, as is the case for PLS-DA.[78] Indeed, overfitting is also possible if the number of variables exceeds the number of samples, a now increasingly common experience in metabolomics research.[78]

Type I and Type II errors both need to be considered: the former is the improper rejection of the null hypothesis for example a false positive, and the latter assesses the rigour of the test which ensures that the null hypothesis is properly rejected, that is a false negative.[76]

Violation of statistical analysis approaches may sometimes import a false significance to a variable when it is not, and *vice versa*. Common misconceptions regarding p values and confidence interval values, and statistical power were explored by Greenland *et al.*[79] Bonferroni and Bonferroni–Holm and Sidak corrections can be applied for type I errors, and the Benjamin–Hochberg approach can be applied for false discovery rates in univariate analysis.

1.10 Future Scope of Metabolomics Research

Metabolomics is now becoming a more integral part of diagnostics to strengthen and predict disease conditions and their progression more rapidly. Several diseases present challenging diagnostic problems. Markers within disease metabolomics can be combined with data for enhanced discrimination. An example of this is shown by Glaab *et al.*,[80] in which metabolomics data and positron emission tomography brain neuroimaging data was combined in order to increase the discriminatory power using support vector machines (SVM) and random forest (RF) analysis strategies using LOOCV and ROC in the diagnosis of Parkinson's disease. In addition to this, it is evident in the publications consulted throughout this chapter that metabolomics is used in a variety of fields, and that a combination of statistical techniques and machine learning technologies are useful in combination. Future applications of spectral analysis are expanding to be more interdisciplinary to enable more robust models and accurate statistical analysis.

Within the NMR-based metabolomics field, one major drawback is analyte sensitivity. However, methodologies such as hyperpolarisation and enhancing technologies, for example the use of cryoprobes, is increasing the sensitivity of biomarker analytes. It should also be noted that statistical techniques and machine learning strategies are evolving and are often used in combination to effectively cope with the high dimensionality of datasets acquired in NMR-linked metabolomics. Indeed, enhancements in computer power are promoting faster turnaround times for data acquisition.

1.11 Conclusions

Statistical applications can successfully be applied to spectral and chromatographic datasets acquired in numerous fields, whether this be the diagnosis of diseases, the stratification of disease developmental stages and prognostics, or the creation of pseudo-two-dimensional spectra, as in STOCSY-type approaches. Sound relationships can be established using NMR datasets if correct standard operating procedures are followed, and which consider careful experimental design. Statistical methods can serve to distinguish between the metabolic patterns of different classifications of diseases and disease stages in both multivariate or univariate senses. Machine learning compliments statistical techniques, and aids further understanding of the clustering of the metabolites.

List of Abbreviations

ANCOVA	Analysis of Covariance
ANOVA	Analysis of variance
ASCA	Analysis of variance simultaneous component analysis
AUC	Area Under Curve
BLIP	Backward Linear Prediction
BML-NMR	Birmingham Metabolite Laboratory-Nuclear Magnetic Resonance
CCORA	Canonical correlation analysis
CV	Cross Validation
CW	Continuous Wave
DFA	Discriminant Function Analysis
ECVA	Extended Canonical Variate Analysis

EDNN	Ensemble Deep Neural Networks
EDTA	Ethylenediaminetetraacetic acid
FLIP	Forward Linear Predication
HCA	Hierarchical Clustering Analysis
HMDB	Human Metabolome Database
HPLC	High Performance Liquid Chromatography
KNN	K-Nearest Neighbour
LDA	Linear Discriminant Analysis
LOOCV	Leave one out cross validation
LS	Least Squares
MANOVA	Multivariate Analysis of Variance
MLR	Multiple Linear Regression
MMCD	Madison Metabolomics Consortium Database
MS	Mass Spectrometry
NMR	Nuclear Magnetic Resonance
O-PLS-DA	Orthogonal-Partial Least Square-Discriminant Analysis
PC	Principal Component
PCA	Principal Component Analysis
PLS-DA	Partial Least Squares Discriminant Analysis
RBF	Radial
RF	Random Forest
ROC	Receiver Operating Characteristic
SDBS	Spectral database for organic Compounds
SHY	Statistical Heterospectroscopy
STOCSY	Statistical Total Correlation Spectroscopy
SOM	Self-Organising Maps
SVM	Support Vector Machine
VIP	Variable Importance Plot

Acknowledgements

The authors are grateful to Katy Woodason for useful discussions. BCP would like to acknowledge De Montfort University for her fee waiver for her PhD studies.

References

1. T. Bayes and R. Price, An Essay towards solving a Problem in the Doctrine of Chances, *Philos. Trans. R. Soc. London*, 1763, **53**, 370–418.
2. K. Pearson, On lines and planes of closest fit to systems of points in space, *London, Edinburgh Dublin Philos. Mag. J. Sci.*, 1901, **2**(11), 559–572.
3. C. Spearman, General Intelligence Objectively Determined and Measured, *Am. J. Psychol.*, 1904, **15**(2), 201–292.
4. W. S. Gosset, The application of the law of error to the work of the brewery, *Guinness Internal Note*, 1904.

5. R. A. Fisher, *Statistical Methods for Research Workers Oliver and Boyd Edinburgh*, 1925.
6. R. A. Fisher and B. Balmukand, The estimation of linkage from the offspring of selfed heterozygotes, *J. Genet.*, 1928, **20**(1), 79–92.
7. J. Neyman, E. S. Pearson and K. Pearson, IX. On the problem of the most efficient tests of statistical hypotheses, *Philos. Trans. R. Soc., A.*, 1933, **231**, 694–704.
8. A.-H. M. Emwas, *et al.*, NMR-based metabolomics in human disease diagnosis: applications, limitations, and recommendations, *Metabolomics*, 2013, **9**(5), 1048–1072.
9. W. Pauli, On the hydrogen spectrum from the standpoint of the new quantum mechanics, *Z. Phys.*, 1926, **36**, 336–363.
10. I. I. Rabi, *et al.*, The Molecular Beam Resonance Method for Measuring Nuclear Magnetic Moments. The Magnetic Moments of 3Li6, 3Li7 and 9F19, *Phys. Rev. J. Arch.*, 1939, **55**, 526.
11. A. W. Overhauser, Polarization of Nuclei in Metals, *Phys. Rev.*, 1953, **92**, 411.
12. A. G. Redfield, On the Theory of Relaxation Processes, *IBM J. Res. Dev.*, 1957, **1**(1), 19–31.
13. E. M. Purcell, *et al.*, Resonance Absorption by Nuclear Magnetic Moments in a Solid, *Phys. Rev. J. Arch.*, 1945, **69**, 37.
14. F. Bloch, *et al.*, Nuclear Induction, *Phys. Rev.*, 1946, **69**, 127.
15. M. Grootveld, *et al.*, Progress in Low-Field Benchtop NMR Spectroscopy in Chemical and Biochemical Analysis, *Anal. Chim. Acta*, 2019, **1067**, 11–30.
16. B. Percival, *et al.*, Low-Field, Benchtop NMR Spectroscopy as a Potential Tool for Point-of-Care Diagnostics of Metabolic Conditions: Validation, Protocols and Computational Models, *High-Throughput*, 2019, **1**, 2.
17. J. K. Nicholson, *et al.*, 750 MHz ^{1}H and ^{1}H-^{13}C NMR Spectroscopy of Human Blood Plasma, *Anal. Chem.*, 1995, **67**(5), 783–811.
18. G. Lippens, C. Dhalluin and J. M. Wieruszeski, Use of a water flip-back pulse in the homonuclear NOESY experiment, *J. Biomol. NMR*, 1995, **5**(3), 327–331.
19. M. Piotto, V. Saudek and V. Sklenář, Gradient-tailored excitation for single-quantum NMR spectroscopy of aqueous solutions, *J. Biomol. NMR*, 1992, **2**(6), 661–665.
20. Le Guennec, *et al.*, Alternatives to Nuclear Overhauser Enhancement Spectroscopy Presat and Carr-Purcell-Meiboom-Gill Presat for NMR-Based Metabolomics, *Anal. Chem.*, 2017, **89**, 8582–8588.
21. J. Lindon, J. Nicholson and Holmes, *The Handbook of Metabonomics and Metabolomics*, Elsevier Science, 1st edn, 2008.
22. L. Pauling, *et al.*, Quantitative analysis of urine vapor and breath by gas-liquid partition chromatography, *Proc. Natl. Acad. Sci. U.S.A.*, 1971, **68**(10), 2374–2376.
23. D. B. Kell and S. G. Oliver, The Metabolome 18 Years on: A Concept Comes of Age, *Metabolomics*, 2016, **12**(9), 148.

24. J. R. Bales, *et al.*, Urinary-excretion of acetaminophen and its metabolites as studied by proton NMR-Spectroscopy, *Clin. Chem.*, 1984, **30**, 1631–1636.
25. J. K. Nicholson, *et al.*, Monitoring metabolic disease by proton NMR of urine, *Lancet*, 1984, **2**, 751–752.
26. J. Lindon and J. Nicholson, Spectroscopic and Statistical Techniques for Information Recovery in Metabonomics and Metabolomics, *Annu. Rev. Anal. Chem.*, 2008, **1**(1), 45–69.
27. J. B. German, *et al.*, Metabolomics: building on a century of biochemistry to guide human health, *Metabolomics*, 2005, **1**(1), 3–9.
28. S. Beisken, *et al.*, Getting the right answers: understanding metabolomics challenges, *Expert Rev. Mol. Diagn.*, 2015, **15**(1), 97–109.
29. Y. Sandlers, The future perspective: metabolomics in laboratory medicine for inborn errors of metabolism, *Transl. Res.*, 2017, **189**, 65–75.
30. J. C. Lindon, E. Holmes and J. Nicholson, Metabonomics techniques and applications to pharmaceutical research & development, *Pharm. Res.*, 2006, **23**(6), 1075–1088.
31. F. Probert, *et al.*, NMR analysis reveals significant differences in the plasma metabolic profiles of Niemann Pick C1 patients, heterozygous carriers, and healthy controls, *Nat. Sci. Rep.*, 2017, **7**, 6320.
32. J. Pinito, *et al.*, Human plasma stability during handling and storage: impact on NMR metabolomics, *Analyst*, 2014, **139**(5), 1168–1177.
33. A.-H. M. Emwas, *et al.*, Recommendations and Standardization of Biomarker Quantification Using NMR-Based Metabolomics with Particular Focus on Urinary Analysis, *J. Proteome Res.*, 2016, **15**, 360–373.
34. O. Beckonert, *et al.*, Metabolic profiling, metabolomic and metabonomic procedures for NMR spectroscopy of urine, plasma, serum and tissue extracts, *Nat. Protoc.*, 2007, **11**(2), 2692–2703.
35. M. F. Alum, *et al.*, 4,4-Dimethyl-4-silapentane-1-ammonium trifluoroacetate (DSA), a promising universal internal standard for NMR-based metabolic profiling studies of biofluids, including blood plasma and serum, *Metabolomics*, 2008, **4**, 122–127.
36. J. K. Nicholson and K. R. Gartland, ^{1}H NMR studies on protein binding of histidine, tyrosine and phenylalanine in blood plasma, *NMR Biomed.*, 1989, **2**, 2.
37. M. Martin, *et al.*, PepsNMR for ^{1}H NMR metabolomic data preprocessing, *Anal. Chim. Acta*, 2018, 1–13.
38. B. Percival, *et al.*, Detection and Determination of Methanol and Further Potential Toxins in Human Saliva Collected from Cigarette Smokers: A ^{1}H NMR Investigation, *JSM Biotechnol. Biomed. Eng.*, 2018, **5**(1), 1081.
39. A. Alonso, S. Marsal and A. Julià, Analytical methods in untargeted metabolomics: state of the art in 2015, *Front. Bioeng. Biotechnol.*, 2015, **3**(23), 1–20.
40. P. Giraudeau, I. Tea, G. S. Remaud and S. Akoka, Reference and normalization methods: Essential tools for the intercomparison of NMR spectra, *J. Pharm. Biomed. Anal.*, 2014, **93**, 3–16.

41. A. Scalabre, *et al.*, Evolution of Newborns' Urinary Metabolomic Profiles According to Age and Growth, *J. Proteome Res.*, 2017, **16**, 3732–3740.
42. C. M. Slupsky, *et al.*, Investigations of the Effects of Gender, Diurnal Variation, and Age in Human Urinary Metabolomic Profiles, *Anal. Chem.*, 2007, **78**(18), 6995–7004.
43. K. E. Haslauer, D. Hemmler, P. Schmitt-Kopplin and S. Sophie Heinzmann, Guidelines for the Use of Deuterium Oxide (D_2O) in ^{1}H NMR Metabolomics, *Anal. Chem.*, 2019, **91**(17), 11063–11069.
44. S. M. Kohl, *et al.*, State-of-the art data normalization methods improve NMR-based metabolomic analysis, *Metabolomics*, 2012, **8**(1), 146–160.
45. P. S. Gromski, *et al.*, The influence of scaling metabolomics data on model classification accuracy, *Metabolomics*, 2015, **11**, 684–695.
46. H. K. Liland, Multivariate methods in metabolomics – from pre-processing to dimension reduction and statistical analysis, *TrAC, Trends Anal. Chem.*, 2011, **30**(6), 827–841.
47. D. S. Wishart, *et al.*, HMDB 4.0 — The Human Metabolome Database for 2018, *Nucleic Acids Res.*, 2018, **46**, D608–D617.
48. K. Haug, *et al.*, MetaboLights– an open-access general-purpose repository for metabolomics studies and associated meta-data, *Nucleic Acids Res.*, 2013, 41 D1 D781–D786.
49. E. L. Ulrich, *et al.*, BioMagResBank, *Nucleic Acids Res.*, 2008, **36**, D402–D408.
50. Q. Cui, *et al.*, Metabolite identification via the Madison Metabolomics Consortium Database, *Nat. Biotechnol.*, 2008, **26**, 162–164.
51. C. Ludwig, *et al.*, Birmingham Metabolite Library: A publicly accessible database of 1D 1H and 2D 1H J-resolved NMR authentic metabolite standards (BML-NMR), *Metabolomics*, 2012, **8**(1), 8–12.
52. C. Steinbeck and S. Kuhn, NMRShiftDB – compound identification and structure elucidation support through a free community-built web database, *Phytochemistry*, 2004, **65**(19), 2711–2717.
53. M. Sud, *et al.*, Metabolomics Workbench: An international repository for metabolomics data and metadata, metabolite standards, protocols, tutorials and training, and analysis tools, *Nucleic Acids Res.*, 2016, **44**, D1 D463–D1 D46470.
54. A. C. Dona, *et al.*, A guide to the identification of metabolites in NMR-based metabonomics/metabolomics experiments, *Comput. Struct. Biotechnol. J.*, 2016, **14**, 135–153.
55. L. M. Smith, *et al.*, Statistical Correlation and Projection Methods for Improved Information Recovery from Diffusion-Edited NMR Spectra of Biological Samples, *Anal. Chem.*, 2007, **79**, 5682–5689.
56. B. J. Blaise, *et al.*, Two-Dimensional Statistical Recoupling for the Identification of Perturbed Metabolic Networks from NMR Spectroscopy, *J. Proteome Res.*, 2010, **9**, 4513–4520.
57. D. J. Crockford, *et al.*, Statistical Heterospectroscopy, an Approach to the Integrated Analysis of NMR and UPLC-MS Data Sets: Application in Metabonomic Toxicology Studies, *Anal. Chem.*, 2006, **78**, 363–371.

58. J. K. Nicholson, *et al.*, Statistical Heterospectroscopy, an Approach to the Integrated Analysis of NMR and UPLC-MS Data Sets: Application in Metabonomic Toxicology Studies, *Anal. Chem.*, 2004, **9**(3), 363–371.
59. P. G. Takis, *et al.*, Deconvoluting interrelationships between concentrations and chemical shifts in urine provides a powerful analysis tool, *Nat. Commun.*, 2017, **8**, 1662.
60. L. Da-Wei, *et al.*, Reliable resonance assignments of selected residues of proteins with known structure based on empirical NMR chemical shift prediction of proteins with known structure based on empirical NMR chemical shift prediction, *J. Magn. Reson.*, 2015, **254**, 93–97.
61. J. Chong, *et al.*, Metaboanalyst 4.0: towards more transparent and integrative metabolomics analysis, *Nucleic Acids Res.*, 2018, **46**(W1), W486–W494.
62. E. Saccenti, *et al.*, Reflections on univariate and multivariate analysis of metabolomics data, *Metabolomics*, 2014, **10**(3), 361–371.
63. V. Ruiz-Rodado, *et al.*, ^{1}H NMR-Linked Metabolomics Analysis of Liver from a Mouse Model of NP-C1 Disease, *J. Proteome Res.*, 2016, **15**(10), 3511–3527.
64. A. Lemanska, *et al.*, Chemometric variance analysis of ^{1}H NMR metabolomics data on the effects of oral rinse on saliva, *Metabolomics*, 2012, **8**(S1), 64–80.
65. D.-A. Kwon, *et al.*, Assessment of green coffee bean metabolites dependent on coffee quality using a ^{1}H NMR-based metabolomics approach, *Food Res. Int.*, 2015, 175–182.
66. J. Camacho, *et al.*, Group-Wise Principal Component Analysis for Exploratory Data Analysis, *J. Comput. Graph. Stat.*, 2017, **26**(3), 501–512.
67. D. I. Broadhurst and D. B. Kell, Statistical strategies for avoiding false discoveries in metabolomics and related experiments, *Metabolomics*, 2006, **2**(4), 171–196.
68. J. Xia, D. J. Broadhurst, M. Wilson and D. Wishart, Translational biomarker discovery in clinical metabolomics: An introductory tutorial, *Metabolomics*, 2013, **9**(2), 280–299.
69. R. Bünger and R. T. Mallet, Metabolomics and ROC Analysis: A Promising Approach for Sepsis Diagnosis, *Crit. Care Med.*, 2016, **118**(24), 6072–6078.
70. S. F. Graham, *et al.*, Metabolic signatures of Huntington's disease (HD): 1H NMR analysis of the polar metabolome in post-mortem human brain, *Biochim. Biophys. Acta, Mol. Basis Dis.*, 2016, **1862**(9), 1675–1684.
71. E. Quansah, *et al.*, Methylphenidate alters monoaminergic and metabolic pathways in the cerebellum of adolescent rats, *Eur. Neuropsychopharmacol.*, 2018, **28**(4), 513–528.
72. A. Rinnan, F. Savorani and S. B. Engelsen, Simultaneous classification of multiple classes in NMR metabolomics and vibrational spectroscopy using interval-based classification methods: iECVA vs iPLS-DA, *Anal. Chim. Acta*, 2018, **1021**, 20–27.

73. E. López-Rituerto, *et al.*, Investigations of La Rioja Terroir for Wine Production Using ^{1}H NMR Metabolomics, *J. Agric. Food Chem.*, 2012, **60**, 3452–3461.
74. J. Xia and D. S. Wishart, MSEA: a web-based tool to identify biologically meaningful patterns in quantitative metabolomic data, *Nucleic Acids Res.*, 2010, **38**, W71–W77.
75. J. Xia and D. S. Wishart, MetPA: a web-based metabolomics tool for pathway analysis and visualization, *Bioinformatics*, 2010, **26**(18), 2342–2344.
76. H. N. B. Moseley, Error Analysis and Propagation in Metabolomics Data Analysis, *Comput. Struct. Biotechnol. J.*, 2013, **4**(5), e201301006.
77. E. Quansah, *et al.*, ^{1}H NMR-based metabolomics reveals neurochemical alterations in the brain of adolescent rats following acute methylphenidate administration, *Neurochem. Int.*, 2017, **108**, 109–120.
78. P. S. Gromski, *et al.*, A tutorial review: Metabolomics and partial least squares-discriminant analysis – a marriage of convenience or a shotgun wedding, *Anal. Chim. Acta*, 2015, **879**, 10–23.
79. S. Greenland, *et al.*, Statistical test, P values, confidence intervals and power: a guide to misinterpretations, *Eur. J. Epidemiol.*, 2016, **31**, 337–350.
80. E. Glaab, *et al.*, Integrative analysis of blood metabolomics and PET brain neuroimaging data for Parkinson's disease, *Neurobiol. Dis.*, 2019, **124**, 555–562.

CHAPTER 2

Recent Advances in Computational NMR Spectrum Prediction

ABRIL C. CASTRO*[a] AND MARCEL SWART*[b,c]

[a] Hylleraas Centre for Quantum Molecular Sciences, Department of Chemistry, University of Oslo, P.O. Box 1033 Blindern, 0315 Oslo, Norway; [b] Institut de Química Computacional i Catàlisi (IQCC) & Departament de Química, Universitat de Girona, Campus Montilivi, 17003 Girona, Spain; [c] ICREA, Pg. Lluís Companys 23, 08010 Barcelona, Spain
*Emails: abril.c.castro@gmail.com; marcel.swart@gmail.com

2.1 Introduction and Scope

Nuclear Magnetic Resonance (NMR) is an indispensable structural tool in the modern analytical arsenal of chemists and structural biologists (see Figure 2.1 for an example of an NMR spectrum). Although proton (^{1}H) and carbon (^{13}C) shielding constants hold a prominent place in organic chemistry, other magnetic nuclei such as ^{15}N, ^{19}F, ^{19}Si, or ^{31}P, but also heavier nuclei such as transition-metals are increasingly important in many areas of chemistry. Certainly, all NMR active nuclei are amenable to computational investigations that work in synergy with experimental NMR spectroscopic techniques for interpreting the observed NMR spectra, yielding new understanding difficult to achieve by computational or experimental approaches alone. Unfortunately, some methodological questions still need to be answered with regards to the scopes and limitations of quantum-chemistry methods.

Theoretical and Computational Chemistry Series No. 20
Computational Techniques for Analytical Chemistry and Bioanalysis
Edited by Philippe B. Wilson and Martin Grootveld

Published by the Royal Society of Chemistry, www.rsc.org

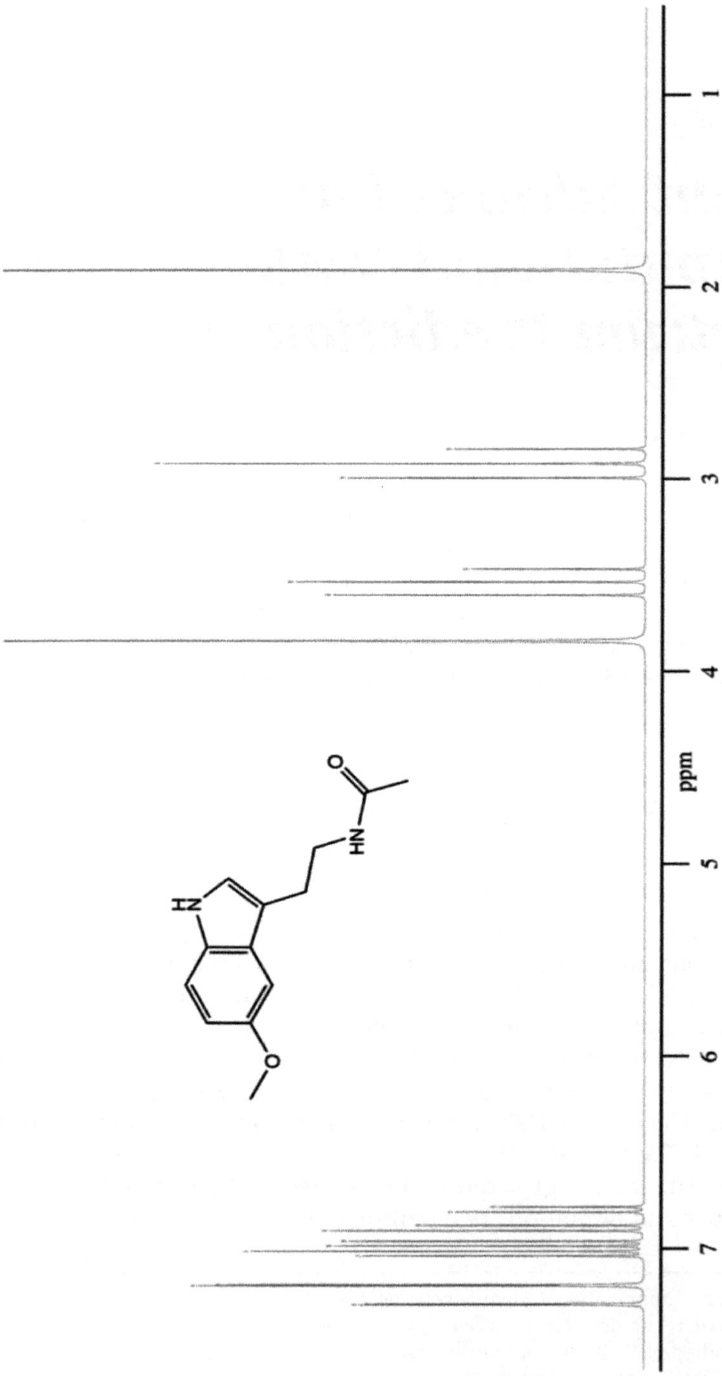

Figure 2.1 Experimental ^{1}H NMR spectra of melatonine, based on data from the Human Metabolome Database.[1]

The theoretical quantum-chemistry methods for the analysis of NMR experiments and the resulting spectra have a long history.[2–4] The detailed description of various wave-function and density functional theory methods for the calculation of parameters in the NMR spin Hamiltonian, that is, the nuclear magnetic shielding constants (or the chemical shifts δ on a relative scale) and the indirect spin–spin coupling constants, has been reviewed over the years. For instance, a comprehensive coverage of this field was captured in 2004 by a number of authors; leaders in the development and application of theoretical NMR techniques.[5] Hence, the accuracy of such calculated NMR parameters depends strongly on the quantum chemical calculation methods used. An excellent summary of the situation was reviewed by Helgaker, Jaszunski, and Ruud in 1999.[6] Accurate reproductions of experimentally observed NMR shift constants can nowadays be very abundant, mainly owing to the ability to carry out precise high-level calculations. At the present time, the size and complexity of molecules are not a limitation, and the modeling of NMR parameters has now become a routine task; on the other hand, there are numerous cases where the accurate prediction of NMR spectra has proven to be a challenging task and a protocol for these calculations is not always well established and understood. This applies particularly to difficult models for solution NMR spectra of systems containing heavy element(s), and/or molecules of an open-shell nature.

This chapter will first provide a broad overview of the currently available methodologies for the calculation of NMR parameters, discussing some of the questions and difficulties connected with their calculation. For more comprehensive descriptions of NMR theoretical treatments and other practical aspects of computational protocols, the reader is directed to excellent reviews by other authors.[7–12] The primary goal of this chapter is to highlight advances related to the development and analysis of implementations associated with NMR spectral prediction. Guidance and suggestions concerning the choice of appropriate tools for a given problem are provided, through illustrative applications of NMR computations in various areas of chemistry, limiting ourselves to those we consider representative of the field of computational NMR spectroscopy or relevant to the analysis of the outlined theoretical methods.

2.2 Calculation of NMR Chemical Shifts

A limited number of articles have demonstrated, by stepwise calculations, the importance of different factors influencing the accuracy of the total NMR chemical shift; among them are the reliability of density functionals, basis sets, gauge-independent canonical formalisms, relativistic approximations, solvent effects, and rovibrational corrections.[13–16] First of all, it is important to mention that computed magnetic properties are in general extremely sensitive to the geometry chosen: even small changes in bond lengths or angles may lead to significant deviations in the chemical shifts. Hence, reliable NMR chemical shifts can only be expected if these calculations are based on appropriate geometries. Secondly, it is interesting to note that concerning chemical shifts, the results of the calculations are absolute shieldings (σ ppm). For a good approximation, the chemical shift δ is habitually related to the absolute shielding constant σ by:

$$\delta = \sigma_{\text{ref}} - \sigma,$$

In which σ_{ref} is the absolute shielding constant of an appropriate reference system (*e.g.*, tetramethylsilane (TMS) in ^{13}C NMR). Therefore, chemical shifts can benefit from fortuitous cancellation of absolute shielding errors and lead to an artificially high agreement between calculated and experimental values, although the performance of some commonly used methods for shielding calculations could be rather poor.[17–19]

For small isolated systems, accurate results for the NMR chemical shifts may be achieved by using wave-function based *ab initio* approaches, such as second-order many-body perturbation (MP2),[20] third-order many-body perturbation (MP3),[21] coupled cluster singles and doubles (CCSD)[22] and perturbative triples CCSD(T),[23] and multiconfiguration self-consistent-field (MC-SCF),[24] to take into account correlation effects. However, such calculations are computationally demanding and impractical for larger molecules. A rather more popular approach is density functional theory (DFT), which has a significantly lower computational cost and offers an alternative way to treat the correlation effects more efficiently, although the proper treatment of magnetic perturbations using DFT is still under debate.[25–27] Currently, there are numerous articles in the literature performing accurate NMR calculations by using different methods derived from the basic DFT formalism and comparing with experimental values for particular sets of small molecules. For instance, several studies in a wide range of molecules showed a high accuracy of less than 1 ppm of difference for the ^{1}H nuclei between the theory and experimental result.[19,28,29] In contrast, for ^{13}C and ^{15}N these accuracies are somewhat less (3–10 ppm), depending on the method and basis set used.[30] On the other hand, the NMR chemical shift calculations of third- and higher-row nuclei are a challenging task for a meaningful practical value, and reliable results have only been obtained recently. A more exhaustive discussion about NMR shift calculations of heavy-element compounds can be found in Section 2.5.

Although the performance of individual density functional approximations differs dramatically and there appears to be no agreement on a single generally accurate functional, currently exist density functionals that are specifically optimized for NMR calculations. An example is the Keal–Tozer functional KT2, as well as KT1 and KT3, that have been specially designed to provide high-quality shielding constants for light main-group nuclei.[28,31] These three exchange-correlation functionals contain a modified exchange-correlation gradient correction, mimicking the behavior of the potential around the nuclei from a Brueckner coupled cluster, which is the key issue for high quality shieldings. Until now, the KT2 functional has been proven to give the most reliable data for NMR chemical shifts in several molecular structures/nuclei.[32–34]

When using standard application functionals, it is well known that hybrid generalized gradient approximation (GGA) functionals tend to provide a somewhat better agreement than their nonhybrid counterparts. For instance, Vícha *et al.* found that the standard 25% exact-exchange admixture in the Perdew–Burke–Ernzerhof hybrid (PBE0) functional seems to be sufficient for transition-metal complexes, if the correct geometry and the treatment of solvent effects is combined with a proper description of the relativity.[13] Furthermore, Xu and co-workers investigated the performance of a selective set of density functional methods (B3LYP, PBE0, BLYP, PBE, OLYP, and OPBE) and found that the OPBE exchange-correlation functional performs remarkably well for small one- and two-heavy-atom molecules containing first-row elements.[35]

It is important to note that most standard density functionals, as discussed above, do not depend on the external magnetic field and therefore produce exchange and correlation energies, which are unphysically constant in the presence of such magnetic fields. Thus, an alternative to introduce an explicit magnetic field dependence is the current-density functional theory (CDFT) approach, which is still actively being worked on and was recently extended to functionals based on the meta-generalized approximation (meta-GGA).[36] General meta-GGA functionals such as Tao–Perdew–Staroverov–Scuseria (TPSS), VS98, and Minnesota 2006 local functional (M06L),[37,38] have been shown to be particularly well suited for estimating NMR shielding constants in organic and inorganic chemistry, although this choice is also the subject of ongoing discussion.[39] CDFT results do not necessarily significantly improve the results obtained with common current-independent density functionals.[36,40,41] Recently, analytical calculations of chemical shielding tensors have been implemented for double-hybrid DFT (DHDFT),[42,43] showing that the double-hybrid DSD-PBEP86 functional is significantly improved compared to the meta-GGA density functionals such as M06L and TPSS for these calculations.[44]

On the other hand, special attention needs to be paid to the basis set employed. In general, the basis set needs a good representation of the orbitals near the nucleus. Several schemes have been developed to modify existing basis sets for this purpose. These usually involve decontracting basis functions to make the basis sets more flexible, and tight basis functions to describe the wave function core region. The most popular basis sets for the computation of magnetic properties are, for example, the Dunning's augmented correlation-consistent basis sets,[45] which include polarization functions, Jensen's aug-pcS-2,[46] Ahlrichs' triple-ζ quality basis sets,[47] and the IGLO basis sets.[48]

An alternative to preserve high accuracy at a much lower computational cost is to employ a locally dense basis set (LDBS) scheme, using a large high-quality basis set on (a) particular atom(s) of interest and a much smaller basis set elsewhere in the molecule. As exemplified in a number of recent publications,[49,50] this method has been used successfully for calculating NMR chemical shifts and spin-spin coupling constants within a general gauge-including atomic orbital (GIAO) approach. The latest approach is related to an additional complication in NMR shielding calculations, which is the gauge dependence problem,[51] a consequence of using a finite basis in the expansion of the wave function or electron density. Currently, the most widely accepted solution is the GIAO method, although some authors have used the alternative continuous set of gauge transformations (CSGT).[37]

2.3 Influence of Conformational Effects on NMR Shielding Constants

Recent advances in theoretical methods resulted in a marked breakthrough into computational NMR applications to structural studies of (bio)organic molecules including conformational analysis, configurational assignment, studies of intra- and intermolecular interactions, together with spectral assignment and structural elucidation. Many NMR spectroscopy studies have focused on the detection and characterization of hydrogen bond interactions, providing an insight into many fields of biological chemistry. Hydrogen bonds play, for instance, a key role in the working of the genetic code, protein folding, and also advanced drugs and materials design.[52–54] Therefore, several computational studies have been devoted to explaining these observations.[55–59] It is known that when a hydrogen atom is part of a H-bond, X–H···Y, in which X and Y are electronegative atoms, the proton experiences a further deshielding effect; that is, the proton feels a stronger magnetic field and its chemical shift is increased. The NMR signal of the bridging proton typically shifts downfield upon formation of the H-bond; that is, the shorter the H-bond, the larger the proton deshielding.

Several theoretical and experimental studies have attempted to analyse the effects of H-bonds on NMR spectral data. For instance, calculations have been employed to find a correlation between H-bond lengths and ^{1}H shieldings, suggesting that when the donor–acceptor distance decreases, ^{1}H shielding becomes smaller.[60] This deshielding (down-field shift) upon H-bond formation is mostly explained by the loss of electron density around the hydrogen nucleus.[61,62] However, a simple measurement of the shifts can lead to misleading interpretation as to the relative strengths on the H-bonds; the shape of this relationship is unclear and cannot be generalized. Various computational studies have suggested that these are two independent and uncorrelated effects;[63–66] we have shown, for instance, that the results of NMR experiments in DNA and RNA base pairs do not directly reflect the strength of the H-bonds.[64]

Recently, the importance of the Pauli repulsion interaction was demonstrated using the electron density around the hydrogen atom that is associated with the ^{1}H shielding. It was shown that the ^{1}H NMR shielding constants of H-bonds are determined by the depletion of the charge around the hydrogen atom, which stems from the fact that the electrons obey the Pauli exclusion principle.[67] In particular, the ^{1}H NMR shielding constant behaviour of the hydrogen-bonded protons was investigated as a function of the molecular structure in a series of guanine C8-substituted GC pair model compounds. To understand the shielding values in hydrogen-bond donors, the electron density distribution around the hydrogen atom was analysed with the Voronoi deformation density (VDD) atomic charges.[68] The change in VDD atomic charges, $\Delta Q(\mathrm{A})$, can be decomposed into a term associated with the rearrangement in electronic density owing to Pauli repulsive orbital interactions, $\Delta Q_{\mathrm{pauli}}(\mathrm{A})$, and a term associated with the bonding orbital interactions, $\Delta Q_{oi}(\mathrm{A})$. Notably, the $\Delta Q_{\mathrm{pauli}}(\mathrm{A})$ values can be used as descriptors of the lengths of hydrogen bonds, in the same way as ^{1}H shieldings are employed.

The effect of the conformational flexibility of DNA and especially RNA makes it difficult to use NMR for structural determination, as the nucleobase can rotate around the glycosidic bond χ in a nucleotide (see Figure 2.2). These rotations lead to variations of the chemical shifts (Figure 2.2), which make it difficult to pinpoint the conformation of the nucleobase directly. Through a combination of theory and experiment, three-dimensional structures of several RNA chains were investigated, and based on the computed chemical shift profile, values for the glycosidic torsion χ could be determined.[59]

2.4 Influence of Environmental Effects on NMR Shielding Constants

The perturbation of NMR parameters owing to solvent effects can be modeled using implicit or explicit solvent models in *ab initio* and DFT calculations. Whereas implicit solvent models are in general limited to accounting for electrostatic effects of the solvent environment (dielectric continuum), an explicit solvent model accounts also for short-range direct interactions between the solute and solvent, which can significantly alter the NMR chemical shifts of the atoms involved in these interactions.

An extensive review of dielectric continuum models up to the most recent developments and applications can be found in several ref. 69–71. Fundamentally, in a standard dielectric continuum model, the molecule of interest (solute) is assumed to reside in a *cavity* and is fully described using quantum chemistry. The solvent, however, rather than representing its charge distribution with electrons and nuclei explicitly, is incorporated as a dielectric continuum (represented by local electric fields that represent a statistical average over all solvent degrees of freedom at thermal equilibrium). In theoretical modeling, dielectric continuum methods generally improve the vacuum results and can be included in all types of quantum chemistry calculations (structure determinations, computational spectroscopy, *etc.*). Thus, many authors have used these models to calculate NMR parameters,[72–74] especially important are those of Ruud and co-workers.[75–77]

Recent studies have described new quantum-chemistry based protocols to improve NMR chemical shift predictions in solvent environments. Grimme and co-workers have developed a fully automated procedure for the quantum-chemical computation of spin-spin-coupled ^{1}H-NMR spectra for general, flexible molecules in solution.[78] Energies, shielding constants, and spin-spin couplings are computed at state-of-the-art DFT levels with continuum solvation. The procedure is based on four main steps, namely the generation of a conformer/rotamer ensemble (CRE), computation of relative free energies and

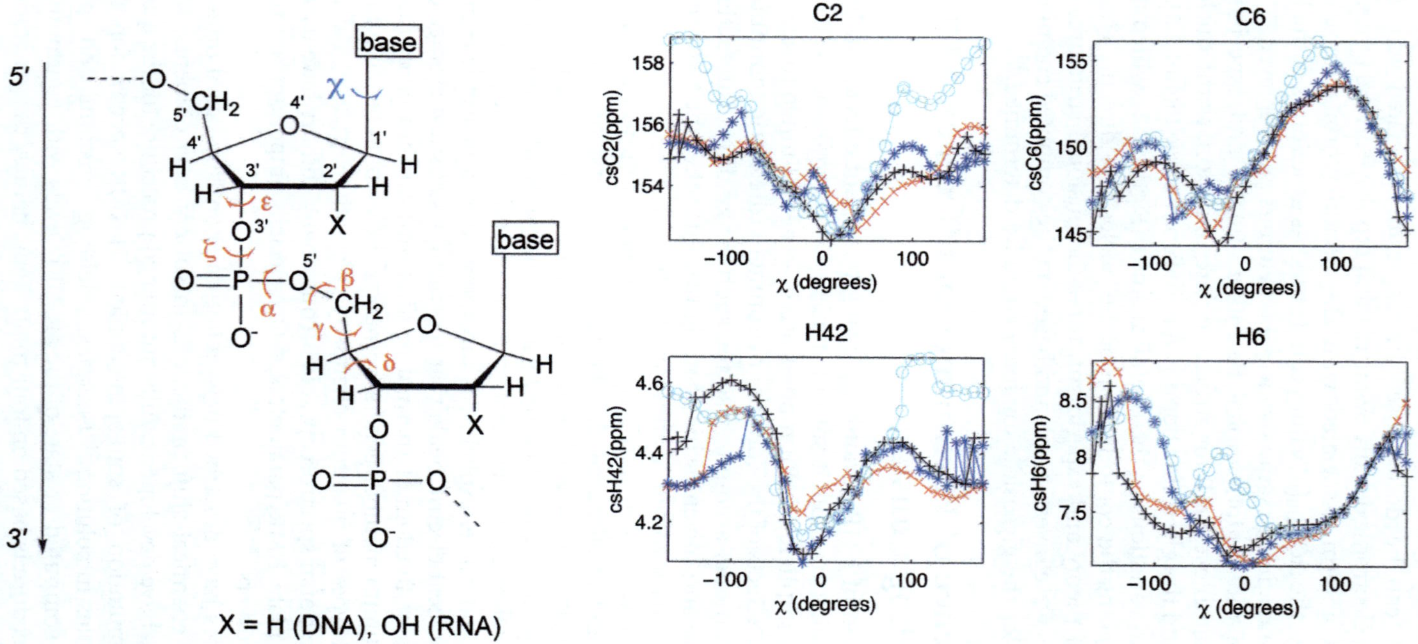

Figure 2.2 Computed ^{1}H and ^{13}C NMR chemical shifts for cytosine nucleotides as a function of the glycosidic torsion χ. Reproduced from ref. 59 with permission from John Wiley and Sons, Copyright © 2012 WILEY-VCH Verlag GmbH & Co. KGaA, Weinheim.

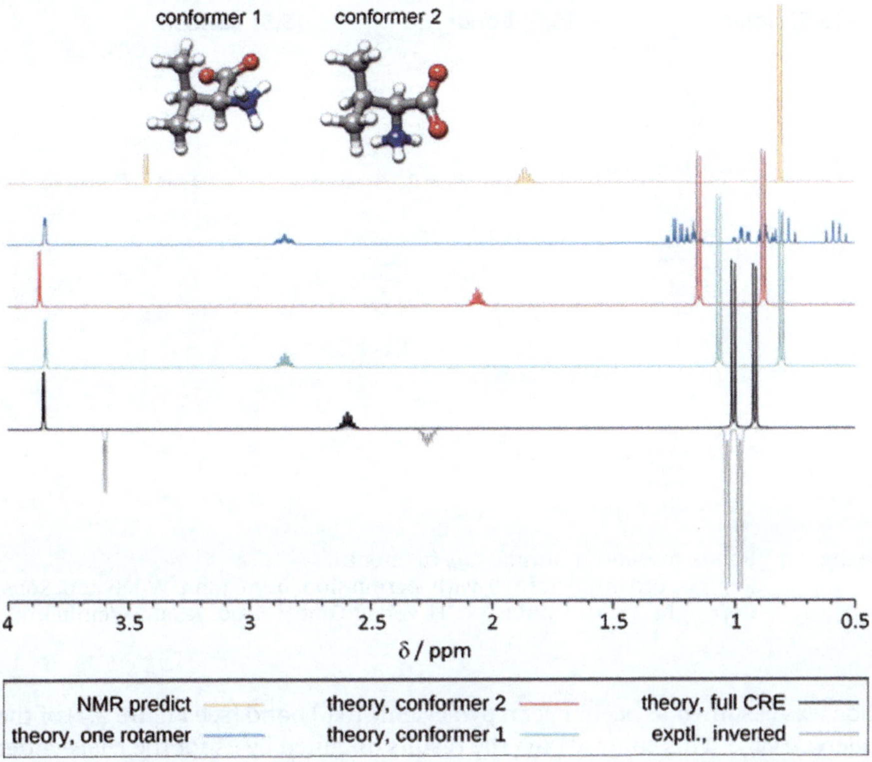

Figure 2.3 Experimental and calculated ^1H NMR spectra of valine (500 MHz, D_2O). Reproduced from ref. 78 with permission from John Wiley and Sons, Copyright © 2017 The Authors. Published by Wiley-VCH Verlag GmbH & Co. KGaA.

NMR parameters, and solution of the (fragmented) spin Hamiltonian for the spectrum, which can be directly compared to the experimental results. An example used to illustrate this approach is the amino acid valine shown in Figure 2.3. If we compare experimental and calculated spectra, not considering rotamers would yield totally erroneously simulated spectra owing to the substantially different chemical shifts of the two conformers. In this case, only the proper generation of the CRE average yields a simulated spectrum that is in good agreement with the experimental results, while the single rotamer spectrum is incomparable and contains additional signals *cf.* experiment.

A few (in)organic and transition-metal complexes were also presented in this study, and a very good, unprecedented agreement between the theoretical and experimental spectra was achieved. Computed chemical shifts can also assist structural assignments in complex systems such as fullerenes. For instance, a computational approach was used to determine the most stable Prato adduct of a C_{80} fullerene, which was previously assigned to a pyracylene adduct based on the highly symmetric NMR spectra. In particular, it was assigned to the D_{5h}-symmetric adduct of $Sc_3N@C_{80}$, in which the addition of the *N*-tritylazomethine

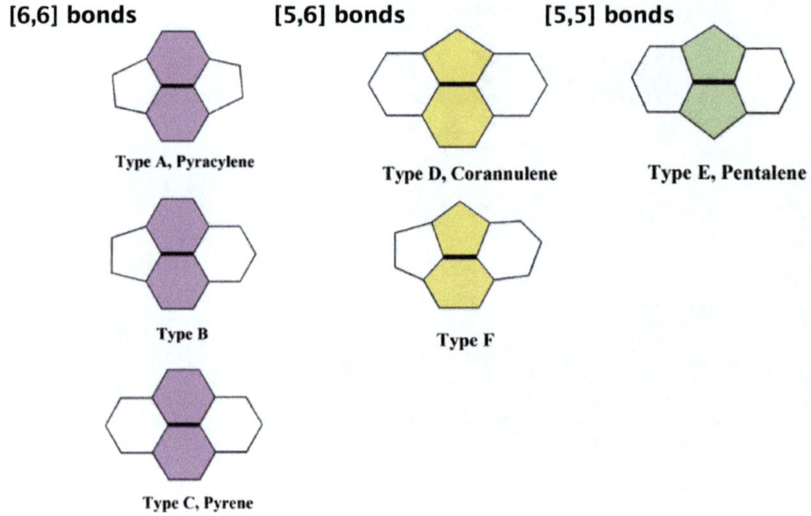

Figure 2.4 Bonds present in normal C_{2n} fullerenes.
Reproduced from ref. 79 with permission from John Wiley and Sons, Copyright © 2012 WILEY-VCH Verlag GmbH & Co. KGaA, Weinheim.

ylide was assumed to occur over a pyracylenic [6,6] bond (see Figure 2.4) of the fullerene cage. This contradicted the results obtained by us for the Diels–Alder reaction over $Sc_3N@C_{80}$, which had shown the most favorable adduct to result from addition on a [5,6] bond.[79]

A follow-up study[80] using the Prato reaction (see Figure 2.5) confirmed the same trend with the most stable adduct occurring for the reaction with a [5,6] bond (see Figure 2.4) of the fullerene. Notably, the computed ^{1}H NMR chemical shifts (H1-H4, see Figure 2.5) showed the most symmetric signals for both the pyracylene [6,6] and corannulene [5,6] adducts. However, the [5,6]-adduct values were much closer to those observed in the experiment, while the pyracylenic [6,6]-adduct showed chemical shifts that were completely off compared to the experimental values.

Relevant to the field of organic chemistry, Yesiltepe and co-workers developed the *in silico* Chemical Library Engine (ISiCLE) NMR chemical shift module[81] to accurately and automatically calculate NMR chemical shifts of small organic molecules through the use of multiple DFT methods, including implicit solvent effects *via* the conductor-like screening model (COSMO),[82] one of the many dielectric continuum models commonly implemented. ISiCLE also calculates the corresponding errors if experimental values are available. This protocol was used to successfully calculate ^{1}H and ^{13}C NMR chemical shifts for a set of 312 molecules under chloroform solvation using eight different levels of DFT theory and was then compared with experimental data, showing promise in the automation of chemical shift calculations of small organic molecules and, ultimately, the expansion of chemical shift libraries.[81]

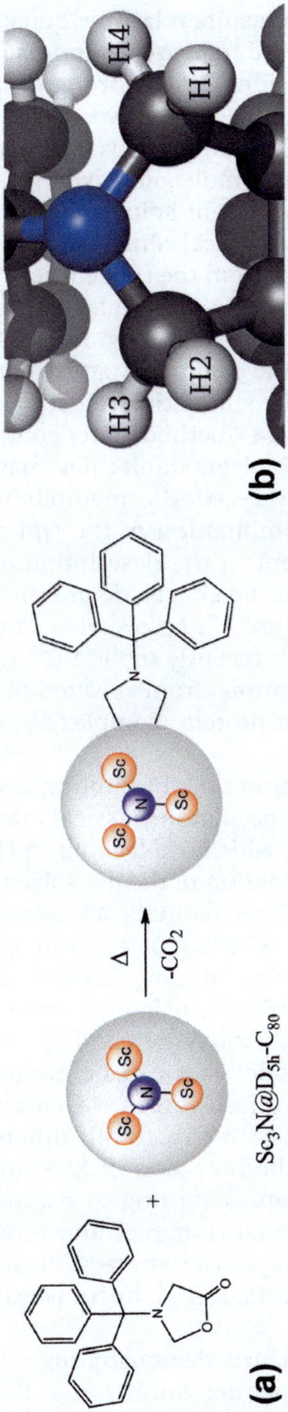

Figure 2.5 Reaction scheme for Prato reaction over $Sc_3N@D_{5h}$-C_{80} (left) and hydrogens involved in assignment of the adduct structure (right).
Reproduced from ref. 80 with permission from the Royal Society of Chemistry.

The continuum methods are the most widely used schemes as they are readily implemented and give results relatively quickly and automatically. However, they may be considered inadequate approaches in cases in which specific solute-solvent interactions have a strong influence on the spectra or conformations.[83,84] Thus, an alternative strategy for a proper determination of the NMR chemical shifts is based on a sequential protocol, in which a classical or *ab initio* molecular dynamics (MD) simulation is performed to determine the different solute-solvent configurations to be used in the following NMR chemical shift calculations averaged over a sufficient number of snapshots from the trajectories. Usually, it is possible to adopt a mixed quantum mechanics/molecular mechanics (QM/MM) description for the conformational sampling. In QM/MM hybrid methods, the central concept is to treat the chemically active part of the system using QM, whereas the majority of the simulation (mostly solvent) is treated with a MM level of theory. As MM methods are computationally less demanding, a large number of solvent molecules may be included in the system. Thus, application of the QM/MM methods to NMR shift calculations have shown that the combination of the QM and MM regions can indeed lead to an improvement in the description of the solvent effects, especially when hydrogen bonding effects are present between the nucleus of interest and the environment.[85–87] An overview of QM/MM applications to NMR spectroscopy has been recently reported.[88] Examples for a variety of nuclei include many applications from solvated molecules to biological systems, such as a pigment-protein complex,[89] or cross-linked DNA oligomers.[90]

Of course, even if the conformational sampling is correct and the most precise classical embedding is used, non-classical interactions between the solute and solvent molecules, which are lacking in QM/MM descriptions, could lead to an incorrect estimation of the net solvent effect and affect the structure of the solute.[91,92] This requires an extension of the QM description to the first solvation shell(s), or an evaluation of the conformational sampling by treating the bonding, as well as the solute-solvent interactions quantum mechanically, using *ab initio* molecular dynamics (AIMD) simulations. Computationally demanding AIMD simulations are limited by the size of the model that can be evaluated in a reasonable period of time, whereas specialized force-field parameters for MD simulations must be developed for each structurally different molecule or non-standard solvent. Currently, the inclusion of MD simulations to improve NMR chemical shift predictions is starting to rise owing to the rapid increase of powerful computational resources and robust software. Various studies have successfully employed this strategy, most of them dealing with accurate structural determinations and unequivocal assignment of the relative configuration.[93–97]

For instance, there is evidence demonstrating that realistic dynamic models are essential for a proper determination of the NMR parameters in various transition-metal complexes.[5] Good examples of such assessments

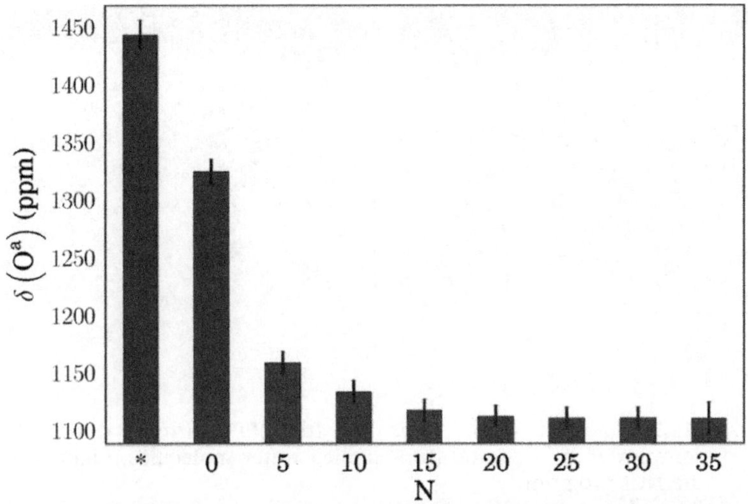

Figure 2.6 Dependence of $\delta(O^a)$ for UO_2^{2+} on the number of nearest water molecules included in the NMR calculations. Each bar represents an average over 64 evenly spaced snapshot geometries taken from the production CPMD run. Values on the horizontal axis indicate the number of explicit water molecules in addition to COSMO for bulk solvent effects. The "*" calculations omit both the COSMO and explicit solvent molecules. Reproduced from ref. 100 with permission from the Royal Society of Chemistry.

are the studies of Autschbach and co-workers for various Pt, Hg, U–O, and Tl–Pt bonded complexes,[98–100] as well as the calculations performed by Bühl and co-workers for a number of Fe, Mn, V, and Co complexes.[101–103] For instance, the ^{17}O NMR chemical shifts of the UO_2^{2+} complex computed from snapshots along the Car–Parrinello MD (CPMD) trajectory provide a good agreement with experimental data.[100] It was demonstrated that this complex has a large dependence on the presence of explicit and implicit solvation, caused by the direct coordination of water molecules to the uranium center. The calculations showed that the inclusion of implicit solvation causes a decrease in the chemical shift by about 120 ppm (see "*" and "0", in Figure 2.3). The inclusion of the first solvation shell obviously further decreases the chemical shift by 170 ppm (see "0" to "5", Figure 2.3), while including a third shell (past 20 water molecules) had a little effect on the ^{17}O NMR chemical shift (Figure 2.6).

Likewise, we examined the reliability of different levels of theory for the calculation of ^{31}P NMR chemical shifts in a series of *trans*-platinum(II) complexes.[34] We have shown that the calculation of the ^{31}P chemical shifts for these compounds remains a challenge for computational chemistry and that no simple protocol can be established for such studies. The identification of the species observed experimentally failed when *static* ^{31}P NMR calculations using both two- and four-component relativistic corrections

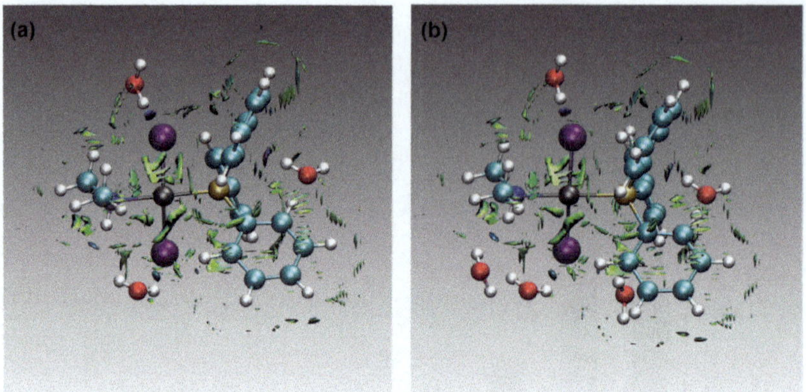

Figure 2.7 AIMD snapshots of the trans-[$PtCl_2$(dma)PPh_3] (dma = dimethylamine) complex with (a) 3, and (b) 5 explicit water molecules, identified using the NCI program.
Reproduced from ref. 34 with permission from the Royal Society of Chemistry.

were used. Notably, we found that solvation influences the ^{31}P chemical shift *via* structural parameters, dynamics, and electronic solvent-solute interactions. The latter cannot be adequately modeled with a continuum model and it is necessary to perform dynamic averaging on the calculated ^{31}P NMR shifts to obtain a satisfactory accuracy. The explicit solvation process was also explored with the addition of explicit water molecules, identified based on the non-covalent interaction (NCI) regions using the NCI program (see Figure 2.7). In this case, the presence of explicit water molecules has some effect on the ^{31}P NMR shift, showing that the explicit treatment of the solvent needs to be carefully monitored to ensure that convergence upon inclusion has been achieved.

2.5 Relativistic Effects on NMR Chemical Shifts

Electrons in heavy-element compounds move with such velocity that the effects of relativity become essential for the accurate prediction of the chemical and physical properties. Thus, it is not surprising that the relativistic effects have a particularly strong impact on the NMR parameters. For instance, the calculated relativistic effects in the vicinity of a heavy metal are generally very sensitive to: (i) the character of metal-ligand bonding, requiring the use of correct and accurate structures; (ii) the inclusion of environmental effects (*e.g.*, solvents); and (iii) reliable methods for treating the relativistic effects.[13,104–106] Hence, in the field of heavy-element compounds, calculation of the NMR chemical shifts is a challenging task that requires elaborate computational models.

The relativistic effects can be treated in several ways at various levels of theory. We refer to some of the many excellent reviews focusing on relativistic quantum chemistry for details.[51,107–110] These effects are usually separated into scalar (SR) and spin-orbit (SO) coupling relativistic interactions, in which the SR are relativistic effects at the one-electron level and the SO are relativistic effects that affect the electron-electron interactions in many-electron systems. All-electron relativistic quantum chemical calculations can be carried out in a relativistic two- (2c) or four-component (4c) framework and with a Hamiltonian including only SR or both SR and SO terms. The quasi-relativistic 2c methods are an attractive alternative if 4c methodology is not available or is considered to be computationally very demanding. For instance, it is known that in general the two-component SO zeroth-order regular (SO-ZORA) approximation (as implemented in the Amsterdam density functional program, ADF) performs very well for NMR observables[111] and does not appear to be a source of large errors for computations of NMR chemical shifts (although we recently observed some drawbacks when compared to 4c methods[34]). Nevertheless, the inclusion of the relativistic effects is more complete with a full 4c relativistic Dirac–Kohn–Sham (DKS) Hamiltonian (as implemented, for example, in the ReSpect[112] and DIRAC[113] program packages). Recently, these methods have reached a level of maturity that make them useful tools for modelling and understanding chemically interesting systems.[114] The best example of this without a doubt is the dirhodium carbene system[115] by Berry and co-workers, who asked Autschbach to compute the NMR properties of an elusive short-lived intermediate involved in a series of cyclopropanation and C–H insertion reactions. With this information in hand, they knew which range of the spectrum to look at for the presence of this intermediate, which was eventually observed. The predicted NMR parameters were substantially larger than anticipated, and hence, without prior computational input the intermediate would have remained elusive.

Additionally, the relativistic contributions on the NMR parameters can be categorized into HAHA (heavy-atom on heavy-atom itself) and HALA (heavy-atom on the light-atom) effects. The HAHA effect mainly disturbs the NMR observables of the same nucleus in which the relativistic effect originates, which is important for NMR calculations of heavy nuclei, such as ^{195}Pt and ^{199}Hg, ^{205}Tl, or ^{207}Pb.[116–119] On the other hand, the HALA effect involves a heavy atom (HA) that induces relativistic *shielding* or *deshielding* at a light-atom (LA) nearby. Whereas the HAHA effect is strongly influenced by both the SR and SO coupling, the HALA effect tends to be dominated by SO coupling. The latest effect is the denominated SO-HALA effect, and the induced shifts, are called SO-HALA NMR chemical shifts, δ^{so} (LA). Extended discussions on the SO-HALA effects in NMR can be found in previously published literature.[120]

The relativistic HALA effects in heavy-atom compounds are particularly interesting from the chemical point of view because of their influence on the bonding and electronic structure. Owing to the SO-HALA effects, trends in NMR shifts are difficult to predict and understand, and various studies have extensively focused on analysing the role of these effects in diamagnetic and paramagnetic heavy-atom compounds.[121–123] As different factors influence the size of the SO-HALA effects, these studies have been generally tackled with the aid of the orbital analysis, as mentioned in the following chapter.

2.6 Molecular Orbital Analysis for Understanding Chemical Shieldings

Interest in understanding and visualizing the relationships between molecular symmetry, electronic structure, and NMR shielding constants has increased through the years.[124–129] The NMR chemical shift values of atoms directly bonded to a metal centre provide detailed information about the electronic structure and are powerful reporters of the location, orientation, and relative energy of the frontier molecular orbitals (MOs) that control reactivity.[130] This framework for examining the relationship between MOs and shielding constants can be of predictive value of a compound's stability and reactivity, and has been discussed recently in several interesting cases.[131–134]

The molecular properties responsible for the generation of the NMR spectra were first identified and analysed in terms of perturbation theory by Ramsey in 1950. According to the nonrelativistic Ramsey equation, the NMR shielding can be formally split into the diamagnetic (σ_{dia}) and paramagnetic (σ_{para}) contributions. The diamagnetic term contributes to the *shielding* of the nucleus and describes the screening of paired electrons on the electronic ground state in the molecule. In contrast, the paramagnetic part contributes to the *deshielding* of the nucleus and arises from the ability of the external magnetic field to force the electrons to circulate through the molecules by making use of virtual orbitals.[135–137] The paramagnetic term often involves the field-induced mixing between filled MOs and empty excited states.[130] Although all excited states are involved in the mixing, frontier MOs, such as the highest occupied MO (HOMO) and lowest unoccupied MO (LUMO), and other near-frontier orbitals contribute most to the mixing. As a result of these MOs and thus the atoms on which the unpaired electrons reside, the paramagnetic term is typically the main source of the chemical shift variations in the nuclei.[5]

Moreover, the relativistic contribution to the NMR shielding constant is typically modulated by the SO coupling,[6,12,138,139] which is included in the paramagnetic term ($\sigma_{para+SO}$). This partitioning has been considered in numerous frameworks such as the natural chemical shielding (NCS) analysis.[135,137] The essential features of the NCS analysis can be explained as discussed in a recent paper on various metal alkylidene complexes.[140] For ethylene, the corresponding orbitals that participate in the paramagnetic term are the σ_{CC}, π_{CC}, as well as the two σ_{CH} bonds and their associated antibonds. In particular, the frontier or near-frontier orbitals, σ_{CC} and π_{CC}, and associated antibonds, σ^*_{CC} and π^*_{CC}, are associated with the most deshielded component of the shielding tensor. These orbitals are coupled *via* the angular momentum L_x, parallel to X_c (see Figure 2.8).

The question of why compounds with occupied frontier π MOs cause SO-HALA shielding at the LA nuclei, while the frontier σ MOs causes deshielding, was recently answered by Vícha and co-workers,[141] who described a new and simple method for understanding the often enigmatic trends in HALA NMR chemical shifts. The ^{1}H NMR chemical shifts for hydrides of the sixth-period (Cs-At) were analysed to study the nature of the SO-HALA effect and its

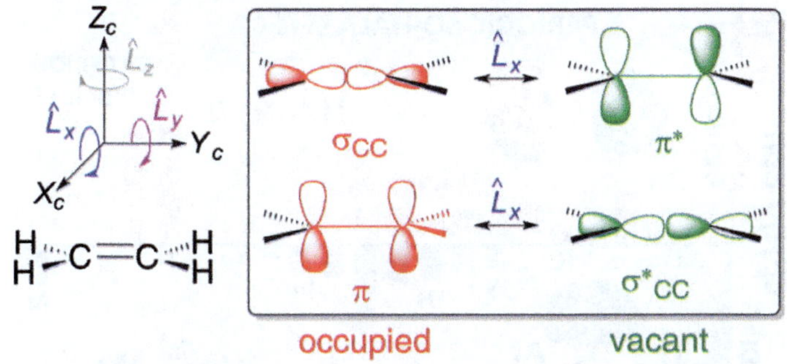

Figure 2.8 Schematic localized orbitals contributing to the shielding of the carbon in ethylene along X_c.
Reproduced from ref. 140 with permission from American Chemical Society, Copyright 2016.

relationship to the electronic structure of the HA. Two general mechanisms are related to the electronic structure at the HA, which control the sign and size of the δ^{so} (LA). In summary, empty HA valence shells induce relativistic *deshielding* at the LA nuclei, while partially occupied HA valence shells induce relativistic *shielding*. In particular, the LA nucleus is relativistically shielded in the $5d^2$–$5d^8$ and $6p^4$ HA hydrides and deshielded in the $4f^0$, $5d^0$, $6s^0$, and $6p^0$ HA hydrides. Notably, this work explains the periodic trends in the ^{1}H-NMR chemical shifts along the sixth-period hydrides (see Figure 2.9).[141]

Various computational studies have focused on the SO-HALA effects induced by neighbouring heavy transition metals on the ^{13}C chemical shifts.[142] In these cases, the CS tensors and associated ^{13}C chemical shifts (δ_{iso} and the principal tensor components $\delta_{11} \geq \delta_{22} \geq \delta_{23}$) are directly related to the frontier molecular orbitals that control reactivity. The polar/covalent character of the HA-LA bond can be judged from the size of δ^{So} (LA),[143,144] and this concept can be further extended to address the stability and reactivity of the HA complexes. For example, studies of NMR shifts of selected d^n transition metal complexes have led to an understanding of some trends in the complexes with *d* electrons.[145] This is particularly interesting for d^0 olefin metathesis catalysts, in which reactivity can be predicted through the chemical shift contribution from δ^{So} (LA).[140]

The analysis of the SO-HALA effects and structure/bonding in NMR spectroscopy can also be extended to the electron paramagnetic resonance (EPR) electronic *g*-tensor and hyperfine coupling tensor (*A*-tensor). For example, the effect of SO interactions originating from HA on the NMR shielding at a neighbouring LA, σ^{SO}, has been pointed out recently in a series of d^6 complexes of iridium.[134] The trends in σ^{SO} observed in this study can be fully explained neither in terms of the s-character of the HA-LA bonding nor by trends in the energy differences between occupied and virtual MOs. It was demonstrated that the σ^{SO} of the LA correlates with the HA d-character of bonding in a similar way to some EPR parameters. Another study analysed

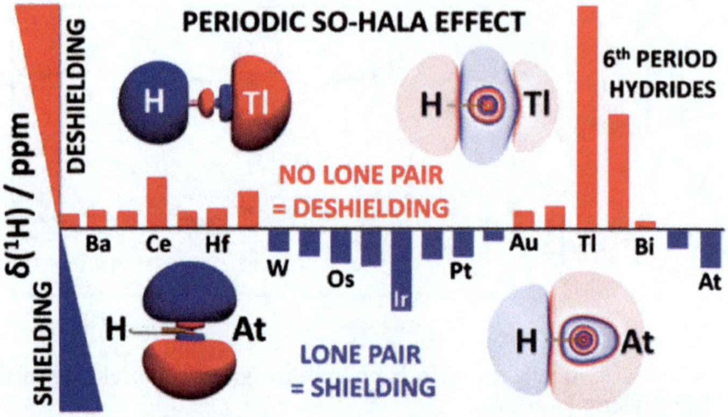

Figure 2.9 Schematic representation of the periodic SO-HALA effect reported by Vícha *et al.*
Reproduced from ref. 141 with permission from American Chemical Society, Copyright 2018.

the paramagnetic contributions to NMR shielding constants and their modulation by relativistic SO effects in a series of transition-metal complexes of Pt(II), Au(II), Au(III), and Hg(II),[146] revealing how the paramagnetic NMR shielding and SO effects relate to the characteristics of the metal-ligand bond. Therefore, the known qualitative relationships between the electronic structure and EPR parameters can be applied to reproduce, predict, and understand the SO-induced contributions to NMR shielding constants of light atoms in HA compounds.

2.7 Electron Paramagnetic Resonance Spectroscopy

Experimentally, the paramagnetic (open-shell) molecular systems can be characterized by both EPR and NMR spectroscopies. In particular, EPR spectroscopy is a powerful source of long-range structural data and provides valuable information about small electron spin densities and hyperfine couplings. Collecting and interpreting the NMR spectra of paramagnetic molecules presents technical challenges and can be difficult. Nevertheless, this process can be substantially facilitated by computational methods to complement the experimental data.[147,148]

The paramagnetic contributions to the NMR shifts for systems in the doublet electronic state can be reconstructed from the electronic *g*-tensors and hyperfine coupling tensors (*A*-tensors).[149–151] The *g* represents the global response property of the molecular system and is related to the magnetic susceptibility tensor, whereas *A* is specific to any nucleus in the system and gives details about the chemical bonds and electron spin structure distributed over the entire molecule. In parallel, the nucleus-electron(s) hyperfine interaction dominates the NMR shielding of the individual atomic nuclei and induces a fast nuclear-spin relaxation that can result in significant broadening of the NMR signals.[152,153] This hyperfine contribution is explicitly temperature-dependent, as determined by the Boltzmann distribution over the Zeeman-split states and can be calculated using knowledge of the electronic *g*- and *A*-tensors.[154,155]

In the same manner as for diamagnetic transition-metal complexes, analysis of the magnetic response parameters for open-shell transition-metal complexes represents a challenge for computational chemistry because of the SR and SO coupling effects.[156–158] Important contributions in this field are, among others, the studies of Marek and co-workers in paramagnetic ruthenium and iridium complexes,[159–163] and Autschbach and co-workers in various transition-metal complexes.[164–166] Thus, the relationship between the electronic spin structure of open-shell systems and the relativistic spin-orbit effects on the EPR and hyperfine NMR parameters has been carefully analysed in a series of paramagnetic Ru(III) and diamagnetic Rh(III) complexes with 2-substituted β-diketones.[159] In particular, the long-range hyperfine NMR effects were investigated systematically by experimental ^{1}H and ^{13}C NMR spectroscopy together with fully relativistic DFT calculations. The substituents were shown to alter the electronic and spin structures of the ligand moieties, resulting in large hyperfine effects on the ligand NMR resonances. Moon and Patchovskii[154] developed a method of prediction for NMR shieldings of paramagnetic systems for a doublet state, which was later generalized for arbitrary multiplicity by Kaupp and co-workers.[167]

Electron paramagnetic resonance spectroscopy has also played an important role in the heme protein field for collating and interpreting NMR spectra of paramagnetic heme proteins such as cytochromes *c*, globins, nitrophorins, cytochrome P450, the hemophore HasA, and heme oxygenases.[168] In particular for heme cytochromes *c*, the NMR and EPR spectra have been shown to be sensitive to the extent of heme nonplanar *ruffling* distortion and to provide insights into the effect of *ruffling* on the electronic structure. The Bren group has combined NMR spectroscopy and DFT calculations to correlate this heme *ruffling* distortion with the ^{1}H and ^{13}C hyperfine NMR shifts and proposed a mechanism for how *ruffling* affects heme redox properties.[169,170]

The final example given is by Otten and co-workers who studied a pseudo-tetrahedral bis-(formazanate) iron(II) complex, in which the ligand might be involved in a non-innocent redox process.[171] They used variable temperature NMR to show how the spin state of the complex changes from low-spin (diamagnetic) to high-spin (paramagnetic), and hence showed spin-crossover behaviour (see Figure 2.10).

2.8 Conclusions

Nuclear magnetic resonance spectroscopy is a well-established technique, which is used routinely in both experiments and computational chemistry. The combination of the two is often used to assign peaks in the spectra to specific atoms, leading to an enhanced understanding of the electronic structure of the species involved. Furthermore, a deep understanding of the NMR nuclear shielding (and the corresponding chemical shift) can be of predictive value of a compound's stability and reactivity, making their physical interpretation an

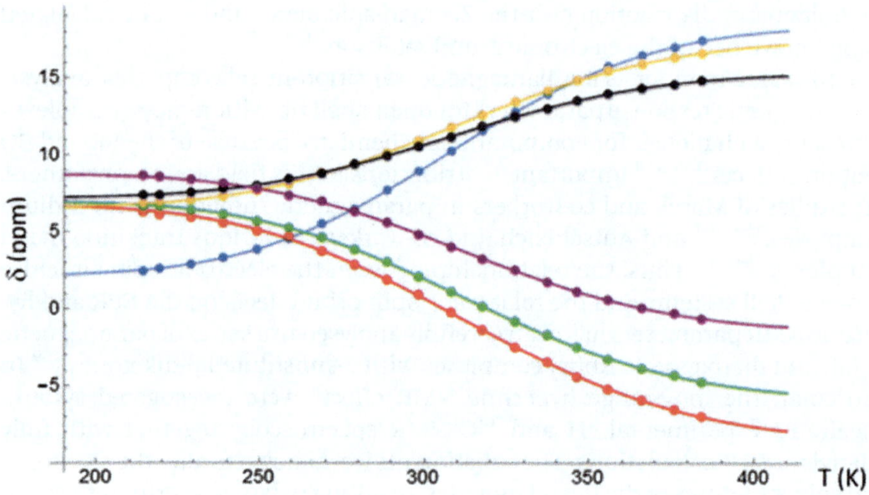

Figure 2.10 Temperature dependence of the ^{1}H NMR signals of Fe(II)-complex with formazanate ligands in toluene-d_8 solution.
Reproduced from ref. 171, https://pubs.acs.org/doi/10.1021/jacs.6b01552, with permission from American Chemical Society, Copyright 2016.

invaluable tool for the development and the understanding of mechanisms and reactivity. The advances attained in quantum-chemistry methods in terms of the quantum-chemical method, basis set, inclusion of solvent, relativity, and dynamics, have now reached the point at which computational NMR spectroscopy can be applied to predict and assist structural assignments in complex systems, but also to unravel important chemistry problems for synthesis and catalysis.

Acknowledgements

ACC acknowledges support from the Norwegian Research Council through the CoE Hylleraas Centre for Quantum Molecular Sciences (project 262695) and the European Union's Framework Programme for Research and Innovation Horizon 2020 (2014–2020) under the Marie Sklodowska-Curie Grant Agreement No. 794563 (ReaDy-NMR). MS thanks the Ministerio de Economía y Competitividad (MINECO, CTQ2015-70851-ERC, and CTQ2017-87392-P), GenCat (XRQTC network), Software for Chemistry and Materials (SCM), and European Fund for Regional Development (FEDER, UNGI10-4E-801) for financial support.

References

1. D. S. Wishart, D. Tzur, C. Knox, R. Eisner, A. Chi Guo, N. Young, D. Cheng, K. Jewell, D. Arndt, S. Sawhney, C. Fung, L. Nikolai, M. Lewis, M.-A. Coutouly, I. Forsythe, P. Tang, S. Shrivastava, K. Jeroncic,

P. Stothard, G. Amegbey, D. Block, D. D. Hau, J. Wagner, J. Miniaci, M. Clements, M. Gebremedhin, N. Guo, Y. Zhang, G. E. Duggan, G. D. MacInnis, A. M. Weljie, R. Dowlatabadi, F. Bamforth, D. Clive, R. Greiner, L. Li, T. Marrie, B. D. Sykes, H. J. Vogel and L. Querengesser, *Nucleic Acids Res.*, 2007, **35**, D521–D526.
2. I. Alkorta and J. Elguero, *Computational Spectroscopy: Methods, Experiments and Applications*, Wiley-VCH Verlag GmbH & Co. KGaA, Weinheim, 2010.
3. A. C. de Dios and C. J. Jameson, in *Annual Reports on NMR Spectroscopy*, ed. G. A. Webb, Academic Press, vol. 77, 2012, pp. 1–80.
4. I. Alkorta and J. Elguero, *Struct. Chem.*, 2003, **14**, 377–389.
5. *Calculation of NMR and EPR Parameters: Theory and Applications*, ed. M. Kaupp, M. Bühl and V. G. Malkin, Wiley-VCH, Weinheim, 2004.
6. T. Helgaker, M. Jaszuński and K. Ruud, *Chem. Rev.*, 1999, **99**, 293–352.
7. P. J. Wilson, *Annual Reports on NMR Spectroscopy*, Academic Press, vol. 49, 2003, pp. 117–168.
8. C. J. Jameson and A. C. De Dios, *Nuclear Magnetic Resonance*, The Royal Society of Chemistry, vol. 44, 2015, pp. 46–75.
9. T. Helgaker, M. Jaszuński and M. Pecul, *Prog. Nucl. Magn. Reson. Spectrosc.*, 2008, **53**, 249–268.
10. C. J. Jameson, *Encycl. Anal. Chem.*, 2014, DOI: 10.1002/9780470027318.a6109.pub2.
11. L. B. Krivdin, *Prog. Nucl. Magn. Reson. Spectrosc.*, 2017, **102–103**, 98–119.
12. J. Vaara, *Phys. Chem. Chem. Phys.*, 2007, **9**, 5399–5418.
13. J. Vícha, J. Novotný, M. Straka, M. Repisky, K. Ruud, S. Komorovsky and R. Marek, *Phys. Chem. Chem. Phys.*, 2015, **17**, 24944–24955.
14. J. Vícha, M. Patzschke and R. Marek, *Phys. Chem. Chem. Phys.*, 2013, **15**, 7740–7754.
15. S. K. Latypov, F. M. Polyancev, D. G. Yakhvarov and O. G. Sinyashin, *Phys. Chem. Chem. Phys.*, 2015, **17**, 6976–6987.
16. L. Jeremias, J. Novotný, M. Repisky, S. Komorovsky and R. Marek, *Interplay of Through-Bond Hyperfine and Substituent Effects on the NMR Chemical Shifts in Ru(III) Complexes*, 2018.
17. J. C. Facelli, *Prog. Nucl. Magn. Reson. Spectrosc.*, 2011, **58**, 176–201.
18. T. H. Sefzik, D. Turco, R. J. Iuliucci and J. C. Facelli, *J. Phys. Chem. A*, 2005, **109**, 1180–1187.
19. M. J. Allen, T. W. Keal and D. J. Tozer, *Chem. Phys. Lett.*, 2003, **380**, 7 0–77.
20. J. Gauss, *J. Chem. Phys.*, 1993, **99**, 3629–3643.
21. S. A. M. Cybulski and D. M. Bishop, *J. Chem. Phys.*, 1997, **106**, 4082–4090.
22. J. Gauss and J. F. Stanton, *J. Chem. Phys.*, 1995, **103**, 3561–3577.
23. A. A. Auer, J. Gauss and J. F. Stanton, *J. Chem. Phys.*, 2003, **118**, 10407–10417.
24. C. van Wüllen and W. Kutzelnigg, *J. Chem. Phys.*, 1996, **104**, 2330–2340.

25. V. G. Malkin, O. L. Malkina, L. A. Eriksson and D. R. Salahub, in *Theoretical and Computational Chemistry*, ed. J. M. Seminario and P. Politzer, Elsevier, vol. 2, 1995, pp. 273–347.
26. S. Reimann, U. Ekström, S. Stopkowicz, A. M. Teale, A. Borgoo and T. Helgaker, *Phys. Chem. Chem. Phys.*, 2015, **17**, 18834–18842.
27. S. Reimann, A. Borgoo, J. Austad, E. I. Tellgren, A. M. Teale, T. Helgaker and S. Stopkowicz, *Mol. Phys.*, 2019, **117**, 97–109.
28. T. W. Keal and D. J. Tozer, *J. Chem. Phys.*, 2003, **119**, 3015–3024.
29. J. Poater, E. V. Lenthe and E. J. Baerends, *J. Chem. Phys.*, 2003, **118**, 8584–8593.
30. L. Armangué, M. Solà and M. Swart, *J. Phys. Chem. A*, 2011, **115**, 1250–1256.
31. T. W. Keal and D. J. Tozer, *J. Chem. Phys.*, 2004, **121**, 5654–5660.
32. S. V. Fedorov, Y. Y. Rusakov and L. B. Krivdin, *Magn. Reson. Chem.*, 2014, **52**, 699–710.
33. Y. Y. Rusakov, I. L. Rusakova and L. B. Krivdin, *Magn. Reson. Chem.*, 2018, **56**, 1061–1073.
34. A. C. Castro, H. Fliegl, M. Cascella, T. Helgaker, M. Repisky, S. Komorovsky, M. A. Medrano, A. Gomez Quiroga and M. Swart, *Dalton Trans*, 2019, **48**(23), 8076–8083.
35. Y. Zhang, A. Wu, X. Xu and Y. Yan, *Chem. Phys. Lett.*, 2006, **421**, 383–388.
36. J. W. Furness, J. Verbeke, E. I. Tellgren, S. Stopkowicz, U. Ekström, T. Helgaker and A. M. Teale, *J. Chem. Theory Comput.*, 2015, **11**, 4169–4181.
37. Y. Zhao and D. G. Truhlar, *J. Phys. Chem. A*, 2008, **112**, 6794–6799.
38. G. L. Stoychev, A. A. Auer, R. Izsák and F. Neese, *J. Chem. Theory Comput.*, 2018, **14**, 619–637.
39. E. I. Tellgren, S. Kvaal, E. Sagvolden, U. Ekström, A. M. Teale and T. Helgaker, *Phys. Rev. A*, 2012, **86**, 062506.
40. A. M. Lee, N. C. Handy and S. M. Colwell, *J. Chem. Phys.*, 1995, **103**, 10095–10109.
41. E. I. Tellgren, A. M. Teale, J. W. Furness, K. K. Lange, U. Ekström and T. Helgaker, *J. Chem. Phys.*, 2014, **140**, 034101.
42. S. Grimme, *J. Chem. Phys.*, 2006, **124**, 034108.
43. G. Lars and G. Stefan, Double-hybrid density functionals, *WIREs Comput. Mol. Sci.*, 2014, **4**, 576–600.
44. G. L. Stoychev, A. A. Auer and F. Neese, *J. Chem. Theory Comput.*, 2018, **14**, 4756–4771.
45. T. H. Dunning Jr., *J. Chem. Phys*, 1989, 90.
46. F. Jensen, *J. Chem. Theory Comput.*, 2008, **4**, 719–727.
47. A. Schäfer, C. Huber and R. Ahlrichs, *J. Chem. Phys.*, 1994, **100**, 5829–5835.
48. W. Kutzelnigg, U. Fleischer and M. Schindler, *Deuterium and Shift Calculation*, Springer, Berlin, Heidelberg, DOI: 10.1007/978-3-642-75932-1_3, 1990.

49. I. L. Rusakova, Y. Y. Rusakov and L. B. Krivdin, *J. Comput. Chem.*, 2016, **37**, 1367–1372.
50. S. V. Fedorov, Y. Y. Rusakov and L. B. Krivdin, *Russ. J. Org. Chem.*, 2017, **53**, 643–651.
51. J. Autschbach and T. Ziegler, *Relativistic Computation of NMR Shieldings and Spin-Spin Coupling Constants*, Wiley, Chichester, 2007.
52. G. A. Jeffrey and W. Saenger, *Hydrogen Bonding in Biological Structures*, Springer-Verlag, Berlin, 1991.
53. G. R. Desiraju and T. Steiner, *The Weak Hydrogen Bond*, University Press, Oxford, U.K., 1999.
54. C. L. Perrin and J. B. Nielson, *Annu. Rev. Phys. Chem.*, 1997, **48**, 511–544.
55. F. A. A. Mulder and M. Filatov, *Chem. Soc. Rev.*, 2010, **39**, 578–590.
56. J. Vícha, R. Marek and M. Straka, *Inorg. Chem.*, 2016, **55**, 10302–10309.
57. M. Dračínský, in *Annual Reports on NMR Spectroscopy*, ed. G. A. Webb, Academic Press, vol. 90, 2017, pp. 1–40.
58. M. Barfield, A. J. Dingley, J. Feigon and S. Grzesiek, *J. Am. Chem. Soc.*, 2001, **123**, 4014–4022.
59. J. M. Fonville, M. Swart, Z. Vokáčová, V. Sychrovský, J. E. Šponer, J. Šponer, C. W. Hilbers, F. M. Bickelhaupt and S. S. Wijmenga, *Chem. – A Eur. J.*, 2012, **18**, 12372–12387.
60. M. G. Siskos, A. G. Tzakos and I. P. Gerothanassis, *Org. Biomol. Chem.*, 2015, **13**, 8852–8868.
61. M. Babinský, K. Bouzková, M. Pipíška, L. Novosadová and R. Marek, *J. Phys. Chem. A*, 2013, **117**, 497–503.
62. E. D. Becker, Hydrogen Bonding, in *Encyclopedia of Nuclear Magnetic Resonance*, ed. D. M. Grant and R. K. Harris, John Wiley, New York, 1996, pp. 2409–2415.
63. M. Swart, C. Fonseca Guerra and F. M. Bickelhaupt, *J. Am. Chem. Soc.*, 2004, **126**, 16718–16719.
64. A. C. Castro, M. Swart and C. F. Guerra, *Phys. Chem. Chem. Phys.*, 2017, **19**, 13496–13502.
65. A. Pérez, J. Sponer, P. Jurecka, P. Hobza, F. J. Luque and M. Orozco, *Chem. Eur. J.*, 2005, **11**, 5062–5066.
66. P. Vidossich, S. Piana, A. Miani and P. Carloni, *J. Am. Chem. Soc.*, 2006, **128**, 7215–7221.
67. M. N. C. Zarycz and C. F. Guerra, *J. Phys. Chem. Lett.*, 2018, **9**, 3720–3724.
68. C. Fonseca Guerra, J.-W. Handgraaf, E. J. Baerends and F. M. Bickelhaupt, *J. Comput. Chem.*, 2004, **25**, 189–210.
69. C. J. Cramer and D. G. Truhlar, *Chem. Rev.*, 1999, **99**, 2161–2200.
70. R. F. Ribeiro, A. V. Marenich, C. J. Cramer and D. G. Truhlar, *J. Phys. Chem. B*, 2011, **115**, 14556–14562.
71. J. Tomasi, *Theor. Chem. Acc.*, 2004, **112**, 184–203.
72. J. Kaminský, M. Buděšínský, S. Taubert, P. Bouř and M. Straka, *Phys. Chem. Chem. Phys.*, 2013, **15**, 9223–9230.

73. B. Maryasin and H. Zipse, *Phys. Chem. Chem. Phys.*, 2011, **13**, 5150–5158.
74. J. Kongsted and B. Mennucci, *J. Phys. Chem. A*, 2007, **111**, 9890–9900.
75. J. Kongsted and K. Ruud, *Chem. Phys. Lett.*, 2008, **451**, 226–232.
76. M. Repisky, S. Komorovsky, P. Hrobarik, L. Frediani and K. Ruud, *Mol. Phys.*, 2017, **115**, 214–227.
77. R. D. Remigio, M. Repisky, S. Komorovsky, P. Hrobarik, L. Frediani and K. Ruud, *Mol. Phys.*, 2017, **115**, 214–227.
78. S. Grimme, C. Bannwarth, S. Dohm, A. Hansen, J. Pisarek, P. Pracht, J. Seibert and F. Neese, *Angew. Chem., Int. Ed.*, 2017, **56**, 14763–14769.
79. S. Osuna, R. Valencia, A. Rodriguez-Fortea, M. Swart, M. Sola and J. M. Poblet, *Chemistry*, 2012, **18**, 8944–8956.
80. S. Osuna, A. Rodríguez-Fortea, J. M. Poblet, M. Solà and M. Swart, *Chem. Commun.*, 2012, **48**, 2486–2488.
81. Y. Yesiltepe, J. R. Nuñez, S. M. Colby, D. G. Thomas, M. I. Borkum, P. N. Reardon, N. M. Washton, T. O. Metz, J. G. Teeguarden, N. Govind and R. S. Renslow, *J. Cheminformatics*, 2018, **10**, 52.
82. A. Klamt and G. Schüürmann, *J. Chem. Soc. Perk. Trans. 2*, 1993, 799–805.
83. M. Dračínský and P. Bouř, *J. Chem. Theory Comput.*, 2010, **6**, 288–299.
84. C. C. Roggatz, M. Lorch and D. M. Benoit, *J. Chem. Theory Comput.*, 2018, **14**, 2684–2695.
85. C. Steinmann, J. M. H. Olsen and J. Kongsted, *J. Chem. Theory Comput.*, 2014, **10**, 981–988.
86. J. Kongsted, C. B. Nielsen, K. V. Mikkelsen, O. Christiansen and K. Ruud, *J. Chem. Phys.*, 2007, **126**, 034510.
87. K. Aidas, A. Møgelhøj, C. B. Nielsen, K. V. Mikkelsen, K. Ruud, O. Christiansen and J. Kongsted, *J. Phys. Chem. A*, 2007, **111**, 4199–4210.
88. U. N. Morzan, D. J. Alonso de Armiño, N. O. Foglia, F. Ramírez, M. C. González Lebrero, D. A. Scherlis and D. A. Estrin, *Chem. Rev.*, 2018, **118**, 4071–4113.
89. S. Caprasecca, L. Cupellini, S. Jurinovich, D. Loco, F. Lipparini and B. Mennucci, *Theor. Chem. Acc.*, 2018, **137**, 84.
90. E. Pauwels, D. Claeys, J. C. Martins, M. Waroquier, G. Bifulco, V. V. Speybroeck and A. Madder, *RSC Adv.*, 2013, **3**, 3925–3938.
91. Q. Cui and M. Karplus, *J. Phys. Chem. B*, 2000, **104**, 3721–3743.
92. B. Mennucci, J. M. Martínez and J. Tomasi, *J. Phys. Chem. A*, 2001, **105**, 7287–7296.
93. M. Dračínský, H. M. Möller and T. E. Exner, *J. Chem. Theory Comput.*, 2013, **9**, 3806–3815.
94. N. Özcan, J. Mareš, D. Sundholm and J. Vaara, *Phys. Chem. Chem. Phys.*, 2014, **16**, 22309–22320.
95. T. S. Pennanen, P. Lantto, A. J. Sillanpää and J. Vaara, *J. Phys. Chem. A*, 2007, **111**, 182–192.
96. J. Přecechtělová, P. Novák, M. L. Munzarová, M. Kaupp and V. Sklenář, *J. Am. Chem. Soc.*, 2010, **132**, 17139–17148.

97. J. Přecechtělová, M. L. Munzarová, J. Vaara, J. Novotný, M. Dračínský and V. Sklenář, *J. Chem. Theory Comput.*, 2013, **9**, 1641–1656.
98. L. A. Truflandier, K. Sutter and J. Autschbach, *Inorg. Chem.*, 2011, **50**, 1723–1732.
99. L. C. Ducati, A. Marchenko and J. Autschbach, *Inorg. Chem.*, 2016, **55**, 12011–12023.
100. A. Marchenko, L. A. Truflandier and J. Autschbach, *Inorg. Chem.*, 2017, **56**, 7384–7396.
101. M. Bühl, F. T. Mauschick, F. Terstegen and B. Wrackmeyer, *Angew. Chem., Int. Ed.*, 2002, **41**, 2312–2315.
102. M. Bühl, R. Schurhammer and P. Imhof, *J. Am. Chem. Soc.*, 2004, **126**, 3310–3320.
103. M. Bühl, S. Grigoleit, H. Kabrede and F. T. Mauschick, *Chem. – A Eur. J.*, 2006, **12**, 477–488.
104. T. Pawlak, M. L. Munzarová, L. Pazderski and R. Marek, *J. Chem. Theory Comput.*, 2011, **7**, 3909–3923.
105. J. Roukala, A. F. Maldonado, J. Vaara, G. A. Aucar and P. Lantto, *Phys. Chem. Chem. Phys.*, 2011, **13**, 21016–21025.
106. M. Pascual-Borràs, X. López and J. M. Poblet, *Phys. Chem. Chem. Phys.*, 2015, **17**, 8723–8731.
107. J. Autschbach, *Encycl. Anal. Chem.*, 2014, DOI: 10.1002/9780470027318.a9173.
108. M. Kaupp, in *Theoretical and Computational Chemistry*, ed. P. Schwerdtfeger, Elsevier, vol. 14, 2004, pp. 552–597.
109. J. Autschbach and S. Zheng, in *Annual Reports on NMR Spectroscopy*, ed. G. A. Webb, Academic Press, vol. 67, 2009, pp. 1–95.
110. M. Reiher and A. Wolf, *Relativistic Quantum Chemistry: The Fundamental Theory of Molecular Science*, Wiley, 2nd edn, 2015.
111. J. Kaminský, J. Vícha, P. Bouř and M. Straka, *J. Phys. Chem. A*, 2017, **121**, 3128–3135.
112. ReSpect 5.1.0 (2019), relativistic spectroscopy DFT program of authors M. Repisky, S. Komorovsky, V. G. Malkin, O. L. Malkina, M. Kaupp, K. Ruud, with contributions from R. Bast, R. Di Remigio, U. Ekstrom, M. Kadek, S. Knecht, L. Konecny, E. Malkin, I. Malkin Ondik, http://www.respectprogram.org.
113. *DIRAC, a relativistic ab initio electronic structure program*, Release DIRAC18, 2018, written by T. Saue, L. Visscher, H. J. A. Jensen, and R. Bast, with contributions from V. Bakken, K. G. Dyall, S. Dubillard, U. Ekström, E. Eliav, T. Enevoldsen, E. Faßhauer, T. Fleig, O. Fossgaard, A. S. P. Gomes, E. D. Hedegård, T. Helgaker, J. Henriksson, M. Iliaš, C. R. Jacob, S. Knecht, S. Komorovský, O. Kullie, J. K. Lærdahl, C. V. Larsen, Y. S. Lee, H. S. Nataraj, M. K. Nayak, P. Norman, G. Olejniczak, J. Olsen, J. M. H. Olsen, Y. C. Park, J. K. Pedersen, M. Pernpointner, R. Di Remigio, K. Ruud, P. Sałek, B. Schimmelpfennig, A. Shee, J. Sikkema, A. J. Thorvaldsen, J. Thyssen, J. van Stralen, S. Villaume, O. Visser, T. Winther, and S. Yamamoto,

available at DOI: 10.5281/zenodo.2253986, see also http://www.diracprogram.org.
114. J. Vícha, R. Marek and M. Straka, *Inorg. Chem.*, 2016, **55**, 1770–1781.
115. K. P. Kornecki, J. F. Briones, V. Boyarskikh, F. Fullilove, J. Autschbach, K. E. Schrote, K. M. Lancaster, H. M. L. Davies and J. F. Berry, *Science*, 2013, **342**, 351–354.
116. J. Roukala, S. T. Orr, J. V. Hanna, J. Vaara, A. V. Ivanov, O. N. Antzutkin and P. Lantto, *J. Phys. Chem. A*, 2016, **120**, 8326–8338.
117. B. Adrjan, W. Makulski, K. Jackowski, T. B. Demissie, K. Ruud, A. Antušek and M. Jaszuński, *Phys. Chem. Chem. Phys.*, 2016, **18**, 16483–16490.
118. T. B. Demissie, B. D. Garabato, K. Ruud and P. M. Kozlowski, *Angew. Chem., Int. Ed.*, 2016, **55**, 11503–11506.
119. M. Jaszuński, T. B. Demissie and K. Ruud, *J. Phys. Chem. A*, 2014, **118**, 9588–9595.
120. M. Repisky, S. Komorovsky, R. Bast and K. Ruud, *Gas Phase NMR*, The Royal Society of Chemistry, 2016, pp. 267–303.
121. Y. Y. Rusakov and I. L. Rusakova, *Int. J. Quantum Chem*, 2019, **119**, e25809.
122. A. H. Greif, P. Hrobárik, J. Autschbach and M. Kaupp, *Phys. Chem. Chem. Phys.*, 2016, **18**, 30462–30474.
123. Y. Y. Rusakov, I. L. Rusakova and L. B. Krivdin, *Int. J. Quantum Chem*, 2016, **116**, 1404–1412.
124. M. Barfield and P. Fagerness, *J. Am. Chem. Soc.*, 1997, **119**, 8699–8711.
125. K. B. Wiberg, J. D. Hammer, K. W. Zilm and J. R. Cheeseman, *J. Org. Chem.*, 1999, **64**, 6394–6400.
126. J. Bohmann and T. C. Farrar, *J. Phys. Chem.*, 1996, **100**, 2646–2651.
127. D. Auer, M. Kaupp and C. Strohmann, *Organometallics*, 2005, **24**, 6331–6337.
128. D. Auer, C. Strohmann, A. V. Arbuznikov and M. Kaupp, *Organometallics*, 2003, **22**, 2442–2449.
129. D. Auer, M. Kaupp and C. Strohmann, *Organometallics*, 2004, **23**, 3647–3655.
130. C. M. Widdifield and R. W. Schurko, *Concepts Magn. Reson., Part A*, 2009, **34A**, 91–123.
131. S. Standara, K. Bouzková, M. Straka, Z. Zacharová, M. Hocek, J. Marek and R. Marek, *Phys. Chem. Chem. Phys.*, 2011, **13**, 15854–15864.
132. J. Toušek, M. Straka, V. Sklenář and R. Marek, *J. Phys. Chem. A*, 2013, **117**, 661–669.
133. S. V. C. Vummaleti, D. J. Nelson, A. Poater, A. Gómez-Suárez, D. B. Cordes, A. M. Z. Slawin, S. P. Nolan and L. Cavallo, *Chem. Sci.*, 2015, **6**, 1895–1904.
134. J. Vícha, M. Straka, M. L. Munzarová and R. Marek, *J. Chem. Theory Comput.*, 2014, **10**, 1489–1499.
135. J. A. Bohmann, F. Weinhold and T. C. Farrar, *J. Chem. Phys.*, 1997, **107**, 1173–1184.

136. J. Autschbach and S. Zheng, *Magn. Reson. Chem*, 2008, **46**, S45–S55.
137. J. Autschbach, *J. Chem. Phys.*, 2008, **128**, 164112.
138. P. Pyykkö, A. Görling and N. Rösch, *Mol. Phys.*, 1987, **61**, 195–205.
139. J. Autschbach and T. Ziegler, Encyclopedia of Nuclear Magnetic Resonance, *Advances in NMR*, John Wiley and Sons, Chichester, vol. 9, 2002, pp. 306–323.
140. S. Halbert, C. Copéret, C. Raynaud and O. Eisenstein, *J. Am. Chem. Soc.*, 2016, **138**, 2261–2272.
141. J. Vícha, S. Komorovsky, M. Repisky, R. Marek and M. Straka, *J. Chem. Theory Comput.*, 2018, **14**, 3025–3039.
142. C. P. Gordon, K. Yamamoto, K. Searles, S. Shirase, R. A. Andersen, O. Eisenstein and C. Copéret, *Chem. Sci.*, 2018, **9**, 1912–1918.
143. J. Vícha, C. Foroutan-Nejad, T. Pawlak, M. L. Munzarová, M. Straka and R. Marek, *J. Chem. Theory Comput.*, 2015, **11**, 1509–1517.
144. D. E. Smiles, G. Wu, P. Hrobárik and T. W. Hayton, *J. Am. Chem. Soc.*, 2016, **138**, 814–825.
145. Z.-L. Xue, T. M. Cook and A. C. Lamb, *J. Organomet. Chem.*, 2017, **852**, 74–93.
146. J. Novotný, J. Vícha, P. L. Bora, M. Repisky, M. Straka, S. Komorovsky and R. Marek, *J. Chem. Theory Comput.*, 2017, **13**, 3586–3601.
147. J. Vaara, in *Science and Technology of Atomic, Molecular, Condensed Matter & Biological Systems*, ed. R. H. Contreras, Elsevier, vol. 3, 2013, pp. 41–67.
148. S. Komorovsky, M. Repisky, K. Ruud, O. L. Malkina and V. G. Malkin, *J. Phys. Chem. A*, 2013, **117**, 14209–14219.
149. F. E. Mabbs and D. Collison, *Electron Paramagnetic Resonance of d Transition Metal Compounds*, Elsevier Science, Amsterdam, vol. 16, 1992, pp. 338–441.
150. M. C. R. Symons, *Chemical and Biochemical Aspects of Electron Spin Resonance Spectroscopy*, Van Nostrand Reinhold Inc., U.S., New York, 1978.
151. Z. Rinkevicius, K. J. de Almeida and O. Vahtras, *J. Chem. Phys.*, 2008, **129**, 064109.
152. G. N. L. Mar, W. D. Horrocks and R. H. Holm, *NMR of Paramagnetic Molecules: Principles and Applications*, Elsevier, New York, 2013.
153. I. Bertini, C. Luchinat and G. Parigi, Solution NMR of Paramagnetic Molecules: Applications to Metallobiomolecules and Models, *Current Methods in Inorganic Chemistry*, Elsevier Science, Amsterdam, 2016.
154. S. Moon and S. Patchkovskii, First-Principles Calculations of Paramagnetic NMR Shifts, in *Calculation of NMR and EPR Parameters*, ed. M. Kaupp, M. Bühl and V. Malkin, Wiley-VCH Verlag, 2004, pp. 325–338.
155. J. Autschbach, Chapter One- NMR Calculations for Paramagnetic Molecules and Metal Complexes, in *Annual Reports in Computational Chemistry*, ed. D. A. Dixon, Elsevier, vol. 11, 2015, pp. 3–36.

156. J. Autschbach, in *Annual Reports in Computational Chemistry*, ed. D. A. Dixon, Elsevier, vol. 11, 2015, pp. 3–36.
157. S. A. Rouf, J. Mareš and J. Vaara, *J. Chem. Theory Comput.*, 2017, **13**, 3731–3745.
158. S. A. Rouf, J. Mareš and J. Vaara, *J. Chem. Theory Comput.*, 2015, **11**, 1683–1691.
159. L. Jeremias, J. Novotný, M. Repisky, S. Komorovsky and R. Marek, *Inorg. Chem.*, 2018, **57**, 8748–8759.
160. J. Novotný, D. Přichystal, M. Sojka, S. Komorovsky, M. Nečas and R. Marek, *Inorg. Chem.*, 2018, **57**, 641–652.
161. J. Novotný, M. Sojka, S. Komorovsky, M. Nečas and R. Marek, *J. Am. Chem. Soc.*, 2016, **138**, 8432–8445.
162. P. L. Bora, J. Novotný, K. Ruud, S. Komorovsky and R. Marek, *J. Chem. Theory Comput.*, 2019, **15**, 201–214.
163. J. Chyba, M. Novák, P. Munzarová, J. Novotný and R. Marek, *Inorg. Chem.*, 2018, **57**, 8735–8747.
164. B. Pritchard and J. Autschbach, *Inorg. Chem.*, 2012, **51**, 8340–8351.
165. F. Aquino, B. Pritchard and J. Autschbach, *J. Chem. Theory Comput.*, 2012, **8**, 598–609.
166. F. Gendron and J. Autschbach, *J. Chem. Theory Comput.*, 2016, **12**, 5309–5321.
167. P. Hrobárik, R. Reviakine, A. V. Arbuznikov, O. L. Malkina, V. G. Malkin, F. H. Köhler and M. Kaupp, *J. Chem. Phys.*, 2007, **126**, 024107.
168. L. B. Kara, *Curr. Inorg. Chem.*, 2012, **2**, 273–291.
169. M. D. Liptak, X. Wen and K. L. Bren, *J. Am. Chem. Soc.*, 2010, **132**, 9753–9763.
170. J. G. Kleingardner, S. E. J. Bowman and K. L. Bren, *Inorg. Chem.*, 2013, **52**, 12933–12946.
171. R. Travieso-Puente, J. O. P. Broekman, M.-C. Chang, S. Demeshko, F. Meyer and E. Otten, *J. Am. Chem. Soc.*, 2016, **138**, 5503–5506.

CHAPTER 3

Computational Vibrational Spectroscopy: A Contemporary Perspective

DIEGO J. ALONSO DE ARMIÑO,* MARIANO C. GONZÁLEZ LEBRERO, DAMIÁN A. SCHERLIS AND DARÍO A. ESTRIN*

Departamento de Química Inorgánica, Analítica y Química Física/INQUIMAE-CONICET, Facultad de Ciencias Exactas y Naturales, Universidad de Buenos Aires, Ciudad Universitaria, Pab. II, Buenos Aires (C1428EHA), Argentina
*Emails: diegoarmino@gmail.com; dario@qi.fcen.uba.ar

3.1 Introduction

Vibrational spectroscopy is without a doubt among the most important and versatile experimental techniques available today in the chemical sciences. From infrared (IR) spectroscopy to all possible variants of Raman spectroscopy, including resonant Raman, coherent anti-Stokes Raman spectroscopy (CARS), stimulated Raman spectroscopy (SRS), surface-enhanced Raman spectroscopy (SERS), and of course the more recent bidimensional techniques, which sometimes even combine vibrational with electronic spectroscopy. All of these techniques probe the structure of matter as a function of the structure of its vibrational energy levels, as well as the interactions between them. Modern experimental techniques have achieved an outstanding level of accuracy, for example in the gas phase, in rare-gas matrix isolation spectroscopy or,

Theoretical and Computational Chemistry Series No. 20
Computational Techniques for Analytical Chemistry and Bioanalysis
Edited by Philippe B. Wilson and Martin Grootveld

Published by the Royal Society of Chemistry, www.rsc.org

more recently, in superfluid helium nanodroplets, which in many cases call for high precision theoretical methods to aid in the interpretation of experimental results. Moreover, rotational and vibrational techniques are central to the study of organic molecules in the interstellar medium or in the atmosphere of other celestial bodies. The Cassini–Huygens mission, for example, probed the atmosphere of Titan (one of Saturn's moons) and demostrated that simple molecules in the gas phase can produce complex organic molecules, particularly pre-biotic molecules,[1] and although many molecular species have been detected, there are still many more waiting for identification, as many signals obtained with Cassini's composite infrared spectrometer (CIRS) have not yet been assigned.[2] Vibrational spectroscopy is also a paramount tool in the study of complex systems such as those found in material and biological sciences. The modelling of these kinds of systems is very challenging, as the model in question must not only be able to reproduce the vibrational structure of the solute in great detail, but also its interactions with a rich and complicated environment.

In this chapter, we present a review of the theoretical techniques used to deal with a wide variety of scenarios in the context of the modelization of the vibrational properties of molecular systems. In Section 3.2.1 we start with a brief introduction to vibrational structure theory at its most basic level, (*i.e.* harmonic approximation) as well as intensity calculations for Raman and infrared spectroscopies. Section 3.3 deals with advanced vibrational structural methods for high accuracy anharmonic calculation of vibrational spectra. This section encompases the theoretical introduction to the basic theory of the diverse theoretical approximations to the solutions of the nuclear Schrödinger equation, as well as different representations of the potential energy surface, and the kinetic energy operator. The section ends with a review of the latest developments in the field and their applications. Section 3.4 is concerned with the simulation of vibrational spectroscopies in complex environments, with a special emphasis on hybrid quantum mechanical-molecular mechanical methodologies, together with a review of selected applications. Special attention is devoted to simulation methods for SERS and related techniques. Finally we present our concluding remarks in Section 3.5.

3.2 Fundamentals of Computational Vibrational Spectroscopy

3.2.1 The Harmonic Approximation

The standard approach to the excitation frequencies and normal modes in electronic structure simulations of vibrational spectroscopy involves the so called harmonic approximation. If M is the number of atoms and q_i are the internal spatial degrees of freedom of the molecule (which amount to

$3M-6=n$ in non-linear structures), then the potential energy $V(q_1, \ldots q_n)$ can be expanded as a Taylor series around the set of coordinates q_i^0 corresponding to the minimum or optimized geometry:

$$V(q_1,\cdots,q_n)=V(q_1^0,\cdots,q_n^0)+\sum_i^n \left.\frac{\partial V}{\partial q_i}\right|_{q_0} \cdot q_i+\frac{1}{2}\sum_i^n\sum_j^n \left.\frac{\partial^2 V}{\partial q_i \partial q_j}\right|_{q_0} \cdot q_i \cdot q_j+\cdots \quad (3.1)$$

Keeping the expansion to the second order (harmonic), introducing the Hessian matrix elements $V_{ij}=\left.\frac{\partial^2 V}{\partial q_i \partial q_j}\right|_{q_0}$, and noting that $\left.\frac{\partial V}{\partial q_i}\right|_{q_0}=0$, the last equation can be rewritten as:

$$V(\mathbf{q})=\frac{1}{2}\sum_i^n\sum_j^n V_{ij}\cdot q_i \cdot q_j, \quad (3.2)$$

In which $\mathbf{q}$ represents the vector containing the internal coordinates $V(q_1, q_2 \ldots q_n)$. Or, in matrix notation:

$$V(\mathbf{q})=\frac{1}{2}\mathbf{q}^\dagger \mathbf{V}\mathbf{q}. \quad (3.3)$$

As $\mathbf{V}$ is symmetric it can be diagonalized through a unitary transformation, $\mathbf{U}^{-1}\mathbf{V}\mathbf{U}=\mathbf{v}$, $\mathbf{v}_{ij}=\mathbf{v}_i\delta_{ij}$, so that eqn (3.3) assumes the following form:

$$V(\mathbf{q})=\frac{1}{2}\mathbf{Q}^\dagger \mathbf{v}\mathbf{Q} = \frac{1}{2}\sum_{i=1}^n Q_i^2 v_i \quad (3.4)$$

In which $\mathbf{Q}^\dagger=\mathbf{q}^\dagger\mathbf{U}=(Q_1Q_2\ldots Q_n)$ and $\mathbf{Q}=\mathbf{U}^{-1}\mathbf{q}=(Q_1Q_2\ldots Q_n)^\dagger$. The transformed coordinates Q_i are the normal modes of the molecule, and the eigenvalues $\mathbf{v}_i$ are the harmonic vibrational frequencies associated with these modes. This procedure requires calculation of the Hessian matrix, that is, the second derivatives of the Hamiltonian with respect to the nuclear coordinates. This can always be realized numerically in the equilibrium structure, although analytical expressions have been devised for the most common electronic structure methods. Additionally, anharmonicity can be introduced into the potential representation using a fourth order Taylor series (a quartic force field)[3,4] or resorting to grids,[5] in which case more advanced methods are necessary to compute the vibrational wavefunctions and energies. Such methods can be based on variational (vibrational self-consistent field (VSCF), vibrational configuration interaction (VCI)),[6,7] perturbative (2nd order vibrational Møller-Plesset perturbation (VMP2))[8] or coupled pair (vibrational coupled clusters (VCC))[9] theories and these will be discussed in detail in Section 3.3.

3.2.2 Intensities Calculations

The light–matter interaction can be expressed in the following classical Hamiltonian.

$$H = \frac{1}{2m}\left(\mathbf{p} - \frac{e}{c}\mathbf{A}\right)^2 \tag{3.5}$$

In which $\mathbf{A}$ is the vector potential that acts as a modification to the effective momentum. Introducing quantization and considering an internal molecular potential V_i,

$$\hat{H} = -\frac{\hbar^2}{2m} + \nabla^2 + \left[V_i + \frac{i\hbar e}{2mc}(\nabla \cdot \mathbf{A} + \mathbf{A} \cdot \nabla) + \frac{e^2}{2mc^2}\mathbf{A} \cdot \mathbf{A}\right] = \hat{H}_0 + V. \tag{3.6}$$

By introducing the weak-field approximation we can neglect the third term inside the square brackets and using the gauge condition $\nabla \cdot \mathbf{A} = 0$, the second term is also simplified.

$$V = -\frac{e}{mc}\mathbf{A} \cdot \mathbf{p} = -\frac{e}{mc}A_0 \cos(\mathbf{k} \cdot \mathbf{r} - \omega t)\epsilon \cdot \mathbf{p} = -\frac{e}{2mc}A_0(e^{-i\omega t} + e^{i\omega t})\epsilon \cdot \mathbf{p} \tag{3.7}$$

In which $\mathbf{p}$ is the quantum mechanical momentum operator $\mathbf{p} = -\frac{i\hbar}{2}\nabla$, and light is considered to be an electromagnetic wave with the electric field in the direction of the unit vector ϵ propagating in the direction $\mathbf{k}$ with frequency ω.

The so-called Fermi's Golden Rule obtained from the application of quantum-mechanical perturbation theory[10] gives us the probability for a transition between the initial ($|i\rangle$) and final ($|f\rangle$) quantum states as:

$$P_{i \to f} = \frac{2\pi}{\hbar}|V_{fi}|^2 \delta(E_f - E_i \pm \hbar\omega). \tag{3.8}$$

This equation expresses the transition probability from an initial to a final state, i and f respectively, per unit time. Replacing the interaction potential matrix element V_{if} and also the long wavelength approximation and the dipole approximation we arrive at the following expression:

$$P_{i \to f} = \frac{\pi A_0^2}{2\hbar^2}\epsilon \cdot |\langle f|\mu|i\rangle|^2 \left[\delta(\omega_f - \omega_i - \omega) + \delta(\omega_f - \omega_i + \omega)\right]. \tag{3.9}$$

This is the time-independent expression of the transition probability between two quantum states according to Fermi's Golden Rule.

In the vibrational regime, we can express the dipole moment μ as a Taylor series in the normal coordinates Q_i of the molecule.

$$\mu = \mu_0 + \sum_i \left(\frac{\partial \mu}{\partial Q_i}\right)_0 Q_i + \frac{1}{2!}\sum_{ij}\left(\frac{\partial^2 \mu}{\partial Q_i Q_j}\right)_0 Q_i Q_j + \ldots \tag{3.10}$$

In which all derivatives are evaluated in the equilibrium molecular geometry. By only considering transitions between states inside the same electronic potential energy function (no electronic transitions) we can obtain expressions for the infrared intensities by building the dipole moment transition matrix elements:

$$\mu_{if} = \langle \nu_i | \mu | \nu_f \rangle = \mu_0 \langle \nu_i | \nu_f \rangle + \sum_s \left(\frac{\partial \mu}{\partial Q_s} \right)_0 \langle \nu_i | Q_s | \nu_f \rangle + \ldots \quad (3.11)$$

In which the first term vanishes owing to the orthonormality of the vibrational states $|\nu_i\rangle$ and $|\nu_f\rangle$. Therefore in order to obtain the IR absorption intensities for vibrational transitions we need to calculate the first derivatives of the dipole moment with respect to the normal coordinates of the molecule. If the second and higher order terms of the previous Taylor series are neglected, intensities are considered to be "electrically harmonic". Under these approximations, transitions to overtone and combination states (states with two or more quanta in one or more vibrational modes, respectively) are not considered. Electrical anharmonicity refers to the inclusion of second and/or higher terms in the Taylor series of eqn (3.10), which does include the possibility of transitions between the fundamental and higher excited states. Electrical anharmonicity should not be confused with mechanical anharmonicity, which is the inclusion of the third and higher terms into the Taylor series expansion of the potential energy operator, rather than in the light–matter interaction operator.

In IR spectroscopy matter absorbs a photon owing to the coupling between the electric field oscillation and the change in the molecular dipole moment caused by the nuclear motion, which is reflected in the derivatives of the dipole moment with respect to the normal coordinates in the last equation. Raman spectroscopy, on the other hand, is not based on the absorption of light, but on innelastic scattering. The theory of Raman scattering was first developed by Kramers, Heisenberg and Dirac[11] in 1925. However, their theory was much more complicated to implement than that for IR absorption. Placzek was the first to realize an approximation to the Raman scattering calculation, usually known as classical polarizability theory.[12] Placzek's theory describes the Raman scattering phenomenon as being due to radiation emitted by an induced dipole which is in turn a product of the interaction between an incident beam on the molecular system. The relationship between the induced dipole and the incident light is:

$$\mu^{\text{ind}} = \alpha(Q) \cdot \varepsilon(t) \quad (3.12)$$

In which $\varepsilon(t)$ is the electric field of incident light and $\alpha(Q)$ is the polarizability tensor, a function only of the normal coordinates under Placzek's approximation. The light–matter interaction potential would be:

$$V(t) = -\,\varepsilon(t) \cdot \mu^{\text{ind}} = -\varepsilon'(t) \cdot \alpha(Q) \cdot \varepsilon(t) \quad (3.13)$$

Here $\varepsilon'(t)$ is the scattered field. Now both fields can be represented as plane-wave oscillations so the previous equation can be rewritten as:

$$V(t) = U(e^{i(\omega+\omega')t} + e^{-i(\omega+\omega')t} - e^{i(\omega-\omega')t} - e^{-i(\omega-\omega')t}) \tag{3.14}$$

$$U = \frac{\omega\omega' A_0 A_0'}{4c^2}(\epsilon' \cdot \alpha \cdot \epsilon) \tag{3.15}$$

In which the primes again mark the scattered field. In eqn (3.14) only the third term is responsible for Stokes Raman scattering, the remaining terms produce anti-Stokes Raman and two-photon emission and absorption. If we replace these in Fermi's Golden rule:

$$P_{i\to f} = \frac{2\pi\omega'^2\omega^2 A_0'^2 A_0^2}{16\hbar c^4}\rho|\epsilon' \cdot \alpha_{fi} \cdot \epsilon|^2 \tag{3.16}$$

$$\alpha_{fi} = \langle f|\hat{\alpha}|i\rangle \tag{3.17}$$

In which f and i refer to the final and initial quantum states and ρ is the photon density of states. From this expression it is simple to derive another one for the Stokes Raman differential cross-section:

$$\frac{\mathrm{d}\sigma}{\mathrm{d}\Omega} = (N+1)\frac{\omega^3\omega'}{c^4}|\epsilon' \cdot \alpha_{fi} \cdot \epsilon|^2. \tag{3.18}$$

Now, expressing the initial and final states as a product of the rotational, vibrational and electronic states, and performing a classical isotropic average over rotational states,[13] expanding the polarizability tensor components in a Taylor series over nuclear normal coordinates, using the harmonic oscillator wavefunctions as a representation of vibrational states and finally performing a Boltzmann average over all possible transitions, it is possible to obtain the following expression:

$$\frac{\mathrm{d}\sigma}{\mathrm{d}\Omega} = \frac{\pi^2}{\varepsilon_0}(\nu_{in} - \nu_p)^4 \frac{h}{8\pi^2 c\nu_p}\left(\frac{45a_p'^2 + 7\gamma_p'^2}{45}\frac{1}{1 - \exp(-hc\nu_p/k_B T)}\right) \tag{3.19}$$

$$a_p' = \frac{1}{3}\{\langle\bar{\alpha}_{xx}'\rangle_p + \langle\bar{\alpha}_{yy}'\rangle_p + \langle\bar{\alpha}_{zz}'\rangle_p\} \tag{3.20}$$

$$\begin{aligned}\gamma_p'^2 = \frac{1}{2}\{&[\langle\bar{\alpha}_{xx}'\rangle_p - \langle\bar{\alpha}_{yy}'\rangle_p]^2 + [\langle\bar{\alpha}_{yy}'\rangle_p - \langle\bar{\alpha}_{zz}'\rangle_p]^2 + [\langle\bar{\alpha}_{zz}'\rangle_p - \langle\bar{\alpha}_{xx}'\rangle_p]^2 \\ &+ 6[\langle\bar{\alpha}_{xy}'\rangle_p^2 + \langle\bar{\alpha}_{yz}'\rangle_p^2 + \langle\bar{\alpha}_{zx}'\rangle_p^2]\}\end{aligned} \tag{3.21}$$

$$\langle\bar{\alpha}_{r's'}'\rangle_p = \left(\frac{\partial\bar{\alpha}_{r's'}}{\partial Q_p}\right)_o. \tag{3.22}$$

Here ν_{in} and ν_p correspond to the wavenumbers of the incident and of the vibrational transition for the normal mode p, respectively, and a_p' and γ_p' are derivatives of the isotropic and anisotropic polarizabilities.

From these equations it transpires that the key quantities are the derivatives of the polarizabilities with respect to the nuclear normal coordinates. However, calculation of the polarizability and its derivatives is not such a simple matter[11,14] and the most rigorous expressions (from Kramers, Heisemberg and Dirac, and therefore known as the KHD equations) are, in general, not applicable to all but the simplest of cases. For off-resonance Raman it is possible to use a very simplified expression. Using the Hellmann–Feymann theorem it is possible to derive the following expression for the polarizability tensor:

$$\alpha_{rs} = \frac{\partial^2 H}{\partial \mathscr{E}_r \partial \mathscr{E}_s} \tag{3.23}$$

With this formula the derivatives with respect to the normal coordinates are a much simpler problem. It is possible to derivate eqn (3.23) with respect to normal coordinates, or derivate the gradient with respect to the field. This later method was first proposed by Komornicki and McIver[15] and is the standard way to calculate derivatives of the polarizability. This methodology is usually known as the finite field gradient method.

In the near resonance regime, however, this approximation is not valid, nor is Placzek's approximation. Albrecht *et al.*[16] derived approximated expressions for the dynamic polarizability using a sum-over-states approach. Albrecht's expression, although significantly more expensive than Placzek's, in computational terms, allows the calculation of Raman differential cross-sections in the resonance regime, although it does have the drawback of a rather narrow applicability to simple cases owing to its computational cost. Arguably the most commonly used approximation for resonance Raman intensities is enclosed into what is called "short-time approximations", of which the simplest realization is the excited state gradient method. Using this methodology it is possible to obtain the relative intensities of the resonance Raman bands corresponding to normal modes i and j using the following simple formula.

$$\frac{I_i}{I_j} = \left(\frac{\partial H^{ex}}{\partial Q_i}\right)^2 \bigg/ \left(\frac{\partial H^{ex}}{\partial Q_j}\right)^2 \tag{3.24}$$

In which H^{ex} are the Frank–Condon electronic transition energies, Q_i and Q_j are normal coordinates for modes i and j, and I_i and I_j are the intensities of the fundamental bands corresponding to the said modes. This approximation has a series of limitations. It is considered valid when only one excited state contributes significantly to the resonance Raman intensity, it neglects Duschinsky rotations and Herzberg–Teller effects[17] and of course it is restricted to fundamental transitions. Despite its limitations, it has been shown to be reliable in a suprisingly broad range of applications.[18–20]

A more recent approach, also based on short time dynamics, was reported by Jensen and Schatz[21,22] and relies on linear response time-dependent density functional theory (LR-TDDFT). This scheme uses geometrical derivatives of the dynamic polarizability. This method has the important advantage of not being limited to systems in which only one excited state is the sole contributor to the resonant Raman intensities. A real-time time-dependent density functional theory (RT-TDDFT) adaptation was presented by Chen *et al.*[23] for calculation of the dyamic polarizabilities and then by Thomas *et al.*[24] for resonant Raman intensities. The use of RT-TDDFT has the advantage of improved scaling with the system size compared to the linear-response theory based methods.

The previous development is time-independent. It is also possible to translate these equations into the time-dependent framework. Assuming the state populations follow the Boltzmann distribution, and considering the energy flux of the radiation, $S = \frac{c}{8\pi} n E_0^2$ (with c representing the speed of light and n the refraction index), eqn (3.9) leads to the absorption lineshape in an isotropic fluid:[10,24]

$$I(\omega) = \frac{1}{2}\int \mathrm{d}t\, e^{-i\omega t} \langle \vec{\mu}(0) \cdot \vec{\mu}(t) \rangle \tag{3.25}$$

Here, the brackets denote an equilibrium ensemble average, which can be obtained from a molecular dynamics simulation:

$$\langle \mu(0) \cdot \mu(t) \rangle \approx \sum_{i}^{N} \frac{t}{N - \dfrac{t_i}{\Delta t + 1}} \dot{\mu}(t_i)\dot{\mu}(t_i + t) \tag{3.26}$$

In which Δt is the integration time-step. The Fourier transform of the time-correlation function in eqn (3.25) reproduces the experimental IR spectrum, with the proper relative intensities of the bands. In a similar way, it can be shown that the Raman spectrum results from the Fourier transform of the autocorrelation of the molecular polarizability. The Fourier transform of the autocorrelation of all particle velocities, known as the power spectra, contains all the vibrational frequencies of the system, both IR and Raman.[24,25] Alternatively, vibrational frequencies of a given mode or modes can be computed from the Fourier transform of the time-correlation functions of the positions or velocities of the atoms associated with that particular mode, or modes.

3.3 Beyond the Harmonic Approximation: Advanced Techniques in Vibrational Spectroscopy

3.3.1 Motivation

The most commonly used methodology for the calculation of quantum energies and states is based on the harmonic approximation (HA). This method, although simple and widely available in most computational chemistry software packages, has some rather important limitations. Even small to medium-sized molecules often show strong anharmonicity effects in their vibrational spectra, in which case, the HA collapses.

Examples of these anomalous behaviors can be found in low frequency modes with large amplitude movements. These types of complications can already be found in systems as small as methanol,[26] and have long intrigued spectroscopists, being the subject of numerous experimental[26] and theoretical[27–29] studies. Another fundamentally anharmonic phenomenon occurs when a molecule possesses two conformers connected by a low energy barrier, in such a way that the molecule can pass from one minimum to another, sometimes even by tunneling. Such is the case, for example, for the inversion of ammonia.[30] This molecule has a trigonal pyramid structure with two possible conformers, produced by the inversion of the pyramid. Under these conditions the vibrational wave functions of both minima (originally degenerate) interact and mix, giving rise to linear symmetric and antisymmetric combinations with an energy below and above the vibrational energies of each conformer separately. This is evidenced in the gaseous phase vibrational spectra as a splitting of approximately 8 cm^{-1} of the vibrational bands.[31,32] Another species that shows a similar behavior is H_3O^+, with an inversion split of 55.4 cm^{-1}.[33] However, the inversion split is not limited to small molecules. Davies *et al.*[34] for example, demonstrated a redox switch mechanism to control the inversion rate of aziridines. Interest in this type of phenomena has led to the development, in recent years, of various methodologies for the calculation of inversion splitting.[35–38] Anharmonicity also plays a key role in the redistribution of intramolecular vibrational energy (in a harmonic system vibrational modes cannot exchange energy), a phenomenon that is of central importance in studies of mode or bond-selective reactivity, a field that has acquired notoriety in recent years, for example, in relation to the production of molecular hydrogen by reaction of water and methane in the presence of nickel as a catalyst. In this sense, a combination of experimental and theoretical techniques allowed researchers to establish that the speed of dissociation of water vapor at a high temperature is dominated by the excited vibrational states highly populated at the temperature of the process.[39] The effects of vibrational excitation on chemical reactivity can be understood thanks to the recently proposed sudden vector projection model (SVP), which proposes that the reaction occurs at the instantaneous limit and that the degree of enhancement of the reactivity by a given vibrational mode reflects the degree of coupling between that mode and the reaction coordinate in the transition state,[40,41] in which it is evident that a purely harmonic representation is insufficient. A phenomenon that is a direct result of anharmonicity and that is very commonly observed in vibrational spectroscopy is the Fermi resonance. Fermi resonances are the result of strong couplings between near-degenerate vibrational states that interact and mix, resulting in an energy separation accompanied by an exchange of intensities. This, of course leads to the appearance of a greater number of bands in the experimental spectrum compared to the 3N-6 (3N-5 in the case of linear molecules) expected fundamental bands. The most common scenario for the appearance of Fermi resonances is that a vibrational state involved in a fundamental transition resonates with an overtone (type I) or a combination state (type II).

These bands are often very useful because they have a strong sensitivity to structural factors of the molecule, as well as to the chemical environment. For example, the well-known Fermi doublet of tryptophan at 1340/1360 cm^{-1} (W7), is widely used in UV resonance Raman (UVRR) experiments, as it has a great sensitivity to the polarity of the chemical environment immediately to the chain side of this amino acid. Likewise, tyrosine has a Fermi doublet at 830/850 cm^{-1} sensitive to hydrogen bonds with the –OH phenolic group, as well as to its protonation state,[42] and has been widely used in the analysis of protein systems by Raman spectroscopy.[43] However, the empirical rules that are commonly used for interpretation were proposed on the basis of a set of model systems and should be applied with great caution in complex systems such as proteins. Despite its importance, and its routine use, there is no way to simulate this type of band using existing methods in computer chemistry packages available to the general public, which are based, almost without exception, on the harmonic approach. That is why there is interest in methods that allow an accurate modeling of this type of transition.[44–46] Moreover, even in the case of relatively well-behaved vibrational states under the harmonic approach, there is a great wealth of dynamic information that is impossible to exploit under the constraints it imposes. For example, coupling between vibrational states, vibrational energy relaxation, or intramolecular energy transfer, is of great interest in light of the many experimental techniques recently developed, such as coherent two-dimensional vibrational spectra, currently a very active field of study.[47] In some other cases, systems with several minima connected by low energy barriers, can have highly anharmonic potentials that lead to strongly delocalised wave functions, and that can only be modeled by means of more advanced methods. In short, the calculation of energies of anharmonic vibrational states represents today an important challenge, and an active field of research.

3.3.2 The Watson Hamiltonian

Let us remember that the vibrational degrees of freedom of a non-linear molecule can be described by a set of $M = 3N - 6$ coordinates, which can be rectilinear normal coordinates or a set of more general curvilinear coordinates (bond lengths, angles, dihedrals, *etc.*). In contrast to the electronic problem, in which the electrons are indistinguishable, the vibrational degrees of freedom are distinguishable. This difference accounts for the electronic kinetic energy operator having a very simple expression:

$$T_e = \sum_i \frac{1}{2} \nabla_i^2. \tag{3.27}$$

The exact expression for the kinetic energy operator (reported by Watson[48]), is much more complicated, even in rectilinear normal coordinates, as it includes rovibrational coupling:

$$T = -\frac{1}{2}\sum_k \frac{\partial^2}{\partial Q_k^2} + \frac{1}{2}\sum_{a,b}(J_a - \pi_a)\mu_{ab}(J_b - \pi_b) - \frac{1}{8}\sum_a \mu_{aa} \tag{3.28}$$

In which J_a and π_a are the total and vibrational angular momenta, respectively, and μ_b is the inverse of the effective moment of inertia. With this expression it is easy to see that even if $J_a = 0$ there is still a coupling term left containing the inverse moment of inertia. Although this term can usually be discarded, it can produce noticeable coupling in some cases, especially in small molecules in which the inverse moment of inertia can be significant. In more general curvilinear coordinates, the kinetic energy operator can be significantly more involved, to the point that many methods have been devised for the automatic generation of this operator using numerical methodologies.[49–55] Therefore, we can see that in solving the nuclear problem outside the harmonic approximation, we can have significant coupling terms arising in the kinetic energy operator, whereas in the electronic structure theory we had none.

Having said this, however, the most critical aspect of the vibrational problem within the framework of the Born–Oppenheimer approximation is generally the representation of the potential energy surface (PES), as it is in most cases the main source for coupling between vibrational degrees of freedom. In the case of the electronic potential operator, the origin of the electron-electron interactions is the Coulomb two-body potential for which an exact expression is known. In contrast, the vibrational potential is a many-body expression which couples essentially all degrees of freedom, and for which there is no exact expression. This means that on top of the more complex coupling, the vibrational potential must be approximated in some way or another. The definition of the approximation for the PES is a decisive factor for the precision and efficiency of any high level vibrational structure calculation, therefore it comes as no surprise that it is one of the most active research themes in this field.

Quartic force fields (QFFs), that is, fourth order Taylor expansions of the potential with respect to the normal coordinates of the system, currently represent the best trade-off between the accuracy and computational cost for the representation of molecular potential energy surfaces (PESs) for anharmonic vibrational structure applications.

$$V(Q) = V_0 + \frac{1}{2}\sum_{i=1}^{M} f_{ii}Q_i^2 + \frac{1}{6}\sum_{ijk}^{M} f_{ijk}Q_iQ_jQ_k + \frac{1}{24}\sum_{ijkl}^{M} f_{ijkl}Q_iQ_jQ_kQ_l. \tag{3.29}$$

Here, Q_i is the cartesian normal coordinates for simplicity, and f_i, f_{ijk} and f_{ijkl} are the second, third, and fourth order force constants. Once the representation of the PES has been defined, a method is needed to solve the vibrational Schrödinger equation. It should be noted that in contrast to the harmonic approximation, for which an analytical solution exists for the Schrödinger equation, when we enter into the anharmonic realm we are forced into numerical methods to determine the solution, in much the same way as with electronic structure theory. In fact, just as in electronic structure theory, we can classify vibrational structure methods as variational (including a vibrational version of Hartree–Fock), perturbational or coupled clusters theory.

By far, the most widely used method is the VSCF, introduced decades ago by Bowman.[56–58] It is a variational method based on the mean field approach, analogous in most of its aspects to the Hartree–Fock method for electronic structure, and is today the most powerful approach for the study of anharmonic vibrational states of polyatomic systems.[59] It has been applied to a large variety of molecules, from isolated small systems[60,61] to complete proteins including solvent.[62,63] Similar to the Hartree and Fock method for the electronic structure, the VSCF wave function only incorporates the correlation between normal modes in an averaged manner, a limitation that is inherent to the mean field approximation. To introduce the intermodal correlation explicitly, it is necessary to resort to post-VSCF methods, of which there are three variants: perturbative (VMP2),[64] variational (VCI)[57] and those based on the theory of coupled pairs (VCC).[9,65]

In what remains of this section we will present a more detailed overview of some of the mean methodologies in use.

3.3.3 Vibrational Self-consistent Field

The vibrational self-consistent field is arguably the most popular and venerable of all high-level vibrational structure methods. It was first introduced by Bowman in his seminal paper of 1978,[56] and has been the subject of considerable study over more than 40 years since its proposal. It is also the starting point for most higher-level methodologies that take mode-mode correlation into account explicitly, in much the same way as its electronic structure counterpart. In this section we will start with an overview of the most relevant theoretical aspects of the method and then review some of the later advances in the matter. After separation of the translational and rotational degrees of freedom and also considering all rovibrational coupling terms negligible, we obtain the following nuclear Schrödinger equation:

$$\left(-\frac{1}{2}\sum_{i=1}^{M}\frac{\partial^2}{\partial Q_i^2}+V(\mathbf{Q})\right)\psi_n(\mathbf{Q})=E_n\psi_n(\mathbf{Q}) \tag{3.30}$$

In which $\mathbf{Q}=\{Q_1, Q_2,\ldots,Q_M\}$. The simplest way to represent the vibrational wavefunction is as a product of one-dimensional functions, (*i.e.* a Hartree product). The one-dimensional functions forming the Hartree product are known as "modals" in analogy to the orbitals in electronic structure theory:

$$\psi_n(\mathbf{Q})=\sum_{i=1}^{M}\phi_{i,n}(Q_i) \tag{3.31}$$

In which $\psi_n(\mathbf{Q})$ is the full vibrational wavefunction, and $\{\phi_{i,n}\}$ are the modals, each of which is a function of only one mass-weighted normal coordinate Q_i. With this test wavefunction and the variational principle, we can derive the VSCF equations for each modal.

$$[H^0(Q_i)+\nu_i(Q_i)]\phi_i(Q_i)=\varepsilon_i\phi_i(Q_i) \tag{3.32}$$

In which ($i = 1, 2, \ldots, M$), $H^0(Q_i)$ is the core Hamiltonian for mode i and consists of the kinetic energy (assuming no kinetic coupling) and the diagonal potential energy (*i.e.* the part of the potential not including mode-mode coupling),

$$H^0(Q_i) = -\frac{1}{2}\sum_{i=1}^{M}\frac{\partial^2}{\partial Q_i^2} + V_d(\mathbf{Q}_i) \tag{3.33}$$

and $\nu_i(Q_i)$ is the effective potential, defined as,

$$\nu_i(Q_i) = \left\langle \prod_{j \neq i}^{M} \phi_j(Q_j) \,|\, V_c(\mathbf{Q}) \,|\, \prod_{j \neq i}^{M} \phi_j(Q_j) \right\rangle. \tag{3.34}$$

Note that in this last expression the product of modals does not contain the modal for the normal mode i, therefore the "effective one-dimensional potential" for mode i is obtained as the average of the interactions between mode i and all the rest of the modals. Also, because the effective potential contains the solutions (the modals) this system of equations must be solved iteratively until self-consistency, and therefore the name. In practice the calculations begin with an initial guess, which can be obtained, for example, by solving the diagonal Hamiltonian (neglecting the effective potential) and solving the corresponding (generalized) eigensystem. Once an initial guess has been obtained, we can use the modals to compute $\nu_i(Q_i)$ for all i, and start again, repeating the procedure until energies and wavefuctions are considered converged.

Notice that we have not yet considered how the modals themselves are to be represented. There are several ways to do this, of course, but a very simple one (which is implemented in our computer code, QUMVIA[4]) is to use distributed Gaussian functions (DGFs), developed by Hamilton and Light.[66] DGFs are a non-orthogonal set, which is why we previously considered solving a generalized eigensystem instead of a normal one when calculating the initial guess. This extra complication when compared to an orthonormal set, such as harmonic oscillator functions, is compensated in flexibility and provides a very good description of the VSCF modals.[66] The basis set consists of a Gaussian function centered in predefined points along a certain normal coordinate, which are commonly taken to be Gauss–Hermite quadrature points. Each Gaussian function situated at position $Q_{i,u}$ of a normal coordinate Q_i can be expressed as:

$$\gamma_\mu(Q_i) = \left(\frac{2A_{i,\mu}}{\pi}\right)^{1/4} \exp[-A_{i,\mu}(Q_i - Q_{i,\mu})^2] \tag{3.35}$$

In which $A_{i,\mu}$ is a constant that determines the width of each Gaussian function and is fixed by the separation between adjacent quadrature points. For more details the reader may want to read Hamilton and Light's original paper.[66] In this ansatz the modals are represented as a linear combination of DGFs:

$$\phi_i(Q_i) = \sum_{\mu=1}^{N_g} C_{i,\mu}\gamma_\mu^{(i)}(Q_i). \tag{3.36}$$

Figure 3.1 represents a set of sixteen Gaussian functions in their respective Gauss–Hermite quadrature points along the normal coordinate Q_i. Now, if we replace eqn (3.36) into the VSCF system of eqn (3.32), after a little algebra we arrive at the following integrals that we must solve:

$$S_{\mu\nu} = \langle \gamma_\mu | \gamma_{\mu\nu} \rangle \tag{3.37}$$

$$T_{\mu\nu} = \langle \gamma_\mu | \hat{T}_\nu | \gamma_\nu \rangle \tag{3.38}$$

$$V_{\mu\nu} = \langle \gamma_\mu | \hat{V} | \gamma_\nu \rangle \tag{3.39}$$

Here $S_{\mu\nu}$, $T_{\mu\nu}$ and $V_{\mu\nu}$ are the overlap, kinetic energy and potential energy integrals. These integrals can be solved either numerically or analytically. In our own code we implemented the second alternative. For more details the interested reader may read the work by Hamilton[66] and also Neff and Rauhut's more recent work on the matter.[67] Here, we see one of the advantages of expressing the potential as a Taylor series. All integrals involving the potential can be expressed as a sum of products of one-mode integrals of the form:

$$Q^n_{\mu\nu} = \langle \gamma_\mu | \hat{Q}^n | \gamma_\nu \rangle \tag{3.40}$$

In which Q is a normal coordinate and $n = 1, 2, 3, 4$ is simply an exponent coming from the Taylor series expression. Therefore, no multidimensional

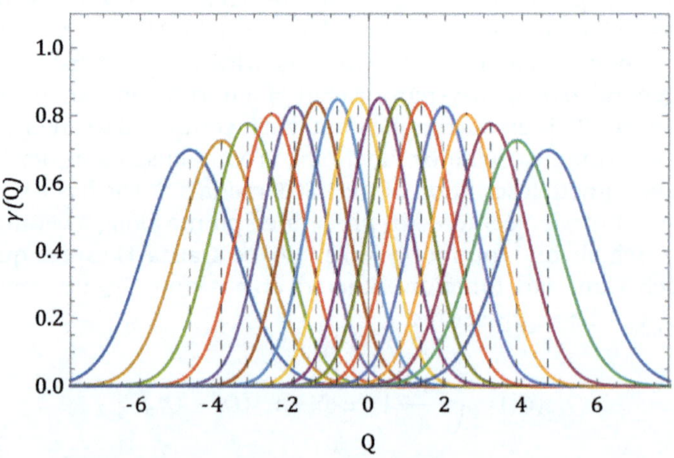

Figure 3.1 Sixteen distributed Gaussian functions centered on their respective Gauss–Hermite quadrature points. Discontinuous lines indicate the center of each Gaussian function.

integration is necessary. Moreover, as all modes can be viewed as distinguishable particles, no exchange integrals are necessary. Needless to say, this simplifies the matter of the VSCF calculation substantially.

Once all the integrals are computed, the Fock-like matrix is built and the generalized eigenvalue problem solved, we obtain the eigenvectors, C_i (coeficients for the modals) and eigenvalues ϵ_i. Now the VSCF energy is not, as the reader may have anticipated, simply the sum of eigenvalues, but has a slightly more involved expression, which is as follows:

$$E_{VSCF} = \sum_{i=1}^{M} \varepsilon_i + \sum_{n=2}^{4} (n-1)\langle \psi_{VSCF} | V^{(n)} | \psi_{VSCF} \rangle. \tag{3.41}$$

The VSCF energy is the sum of the eigenvalues of the modals forming the Hartree product of the VSCF wavefunction, ψ_{VSCF} plus an extra term that corrects for the double counting in the mode-mode interactions and can be thought of as a perturbative first order correction to the energy.

Therefore, having discussed the peculiarities of the vibrational self-consistent field methodology, the interested reader may resort to the cited literature for more details or even Szabo's excellent introduction to electronic structure theory,[68] given that what is left to discuss for the vibrational version of self-consistent field method is very similar to its electronic counterpart. We will now discuss some of the more recent developments regarding this methodology.

3.3.4 Vibrational Configuration Interaction

Owing to the inherent limitations of the mean-field approximation, VSCF wave functions only introduce correlation in an averaged manner. To account for it in a more explicit way different methods are needed that improve upon VSCF. These more sophisticated methodologies are broadly called post-VSCF. In this section we will begin with an introduction to one of the conceptually simplest methodologies among these, known as VCI, first introduced by Bowman *et al.*, Ratner *et al.* and Thompson and Truhlar.[69–71] The wave functions and correlated vibrational energies are obtained by diagonalizing the complete Hamiltonian on the basis of virtual VSCF states.[72,73] In principle, VCI has the potential to produce an exact solution, however, in practice the method is limited by the size of the basis set, so the VCI energies are actually an upper bound for the exact ones, which is to be expected from a variational method. Recall that if we start with a number of g gaussian functions as a basis, the VSCF method produces a set of g modals (one fundamental and g-1 excited ones) for each of the M normal modes of the molecule. The VCI basis set $\{\psi_n\}$ is the direct product of these sets:

$$\{\psi_n\} = \otimes_{i=1}^{M} \{\phi_i^{\nu_i}\}. \tag{3.42}$$

Here ν_i are the vibrational quantum numbers for the modal i. The number of elements in the set $\{\psi_n\}$ is equal to $N_{VCI} = (1+g)^M$. Therefore, considering a non-linear triatomic molecule with three vibrational normal modes and using sixteen gaussians as the basis set for a VSCF calculation, the VCI basis set would contain a total of 4913 basis functions. For a non-linear four atom molecule the size of the basis set rises to 24 million. One more atom, and we get 10^{11}. It is clear then that the exponential scaling of the size of the basis set means that it is indispensable to truncate the basis set in some way.

In general terms we can express the VCI wavefunction as a linear combination of VSCF configurations $\{\psi_n\}$, this is:

$$|\Phi_{VCI}\rangle = \sum_{v} C_v |\psi_v\rangle. \tag{3.43}$$

The sum is over all possible VSCF configuations, and **v** is a vector of M components that specifies all the vibrational quantum numbers (*i.e.* one for each normal mode):

$$|\psi_v\rangle = \prod_{i=1}^{M} \phi_i^{v_i}. \tag{3.44}$$

At this point it is more useful to represent each configuration relative to the fundamental state $|0\rangle$ so that $|r_i\rangle$ is a configuration excited in r quanta in normal mode i with all the remaining modes in the ground state. Likewise, $|r_a s_b\rangle$ is an excited configuration with r quanta in the mode a, and s quanta in the mode b with all other modes in the ground state, and so on. We can now factor the VCI wave function according to the number of excited modes with respect to the ground state:

$$|\Phi_{VCI}\rangle = C_0|0\rangle + \sum_{ar} C_a^r |r_a\rangle + \sum_{\substack{a<b \\ r,s>0}} C_{ab}^{rs} |r_a s_b\rangle + \sum_{\substack{a<b<c \\ r,s,t>0}} C_{abc}^{rst} |r_a s_b t_c\rangle + \cdots \tag{3.45}$$

With:

$$|0\rangle = \prod_{i=1}^{M} |\phi_i^0\rangle \tag{3.46}$$

and the VSCF wavefunctions, also called configurations, $|r_a\rangle$,$|r_a s_b\rangle$ and $|r_a s_b t_c\rangle$ corresponding to VSCF states excited in one, two and three modes simultaneously, respectively. These configurations are normally reffered to as singles, doubles and triples, respectively.

Once the basis set is defined it is necessary to build the VCI matrix, for which the matrix elements are defined as:

$$H_{RS} = \langle \psi_R | H | \psi_S \rangle = \sum_{i=1}^{M} \epsilon_i^{R_i} \delta_{R_i S_i} + \langle \psi_R | \Delta V | \psi_S \rangle \tag{3.47}$$

In which R and S are two quantum number tuples, $\delta_{R_iS_i}$ is a Kronecker delta, and ΔV is the difference potential between the real Hamiltonian and the VSCF effective potential.

$$\Delta V = V_c - \sum_{i=1}^{M} v_i. \tag{3.48}$$

The main bottleneck for VCI methods is the exponential growth of the VCI basis with respect to the size of the system. Therefore, the set of VSCF bases is normally truncated in single, double or triple excitations or de-excitations (in the case of using excited states as a reference). It has been shown, however, that the rejection of excitations of three and four modes leads to unacceptable errors.[74] Therefore, whenever possible, quadruple excitations should be included in the VCI basis set. That said, VCI matrices are generally diagonal-dominant and sparse, so most of their non-diagonal elements are zero or nearly so. We have implemented approaches to take advantage of this structure.[4] These approaches aim to accelerate the construction of the VCI matrix, as well as to reduce the memory requirements for the storage and diagonalization of the Hamiltonian matrix. Taking advantage of the orthogonality of the VSCF virtual states, it is possible to derive a series of Condon–Slater (RCS) rules analogous to those applied to electronic structure methods. These rules help to predict which elements of the Hamiltonian matrix are identically zero, and greatly simplify the expressions for the calculation of the rest of the matrix elements.

3.3.5 Vibrational Perturbation Theory

Although VCI methods are exact (in principle) and conceptually simple, the problems related to the basis set size scaling with the number of atoms in the systems results in the implementation being quite complicated in practice. Perturbative methods, and second order perturbative methods in particular, are significantly more computationally efficient and relatively simpler to implement, therefore they can be applied to medium-to-large systems with relative ease. Moreover, contrary to VCI, vibrational perturbation theory (VPT) does not have its size-consistency limitations, but as a counterpart it is not variational, therefore there is no guarantee that the resulting energy will be a higher bound to the exact one.

Perturbation theory treats the Hamiltonian of the system as a sum of a zero-order Hamiltonian (H^0), for which a complete set of wavefunctions and energies is known, and a perturbation (V).

$$H = H^{(0)} + V. \tag{3.49}$$

The exact energy can then be expressed as an infinite sum of contributions of increasing complexity, which contain the eigenvalues of (H^0)

and products of the matrix elements of the perturbation in the basis of the eigenvalues of (H^0). Terms that involve products of n said matrix elements are grouped into what is called nth-order perturbation energy. It is assumed that the pertubation V is small in magnitude, such that the perturbed wavefunction will be similar to the unperturbed one and therefore the perturbation expansion will converge quickly. One reason for the accuracy of perturbation theory is that it produces the exact eigenvalues of the Morse potential if a quartic force field is used, so in effect, vibrational perturbation theory produces a "morsification" of the potential[75–79] which leads to an effective inclusion of almost exact higher order terms. This feature explains why in some cases perturbation theory can produce results which are in better agreement with the experimental results than its variational counterparts.

There are, of course, many formulations of VPT, such as Van Vleck,[80] or Bloch's projector formalism.[81] For the sake of simplicity here we will only consider Rayleigh–Schrödinger methods. If one chooses the VSCF Hamiltonian as the zero-order, the perturbation becomes the mode-mode correlation energy, and the formalism is usually called Møller–Plesset perturbation theory (MPPT). MPPT was pioneered by Norris, Jung and Gerber in the 90s,[82,83] and has been applied to a great variety of systems since, usually up to second order in the energy (VMP2). Although higher order perturbative methods have been devised, it has become clear that the perturbation series are inherently divergent,[84–86] even in the absence of near degeneracies, therefore in most cases VMP2 is the most accurate result of the perturbation series. The matter is usually complicated when near degenerate vibrational states appear, in particular Fermi resonances produce a divergence in the energy of the states involved, therefore the method is not suited for the treatment of such states. The divergence in such cases is due to the strong coupling as a result of the near-degeneracy, which invalidates the original assumption that the perturbation should be weak in magnitude. Nevertheless, many degeneracy-tolerant perturbational methods have been devised and applied to a variety of systems.[64,85,87,88]

Defining F_i as the VSCF Hamiltonian (eqn (3.32)),

$$\hat{H}^{(0)} = \hat{F}_i = [\hat{T}_I + \hat{V}_{d_i} + \hat{v}_i] \tag{3.50}$$

$$\hat{H} = \hat{F} + \Delta\hat{V} \tag{3.51}$$

$$\hat{F} = \sum_{i=1}^{M} \hat{F}_i \tag{3.52}$$

$$\Delta\hat{V} = \hat{V}_c - \sum_{i=1}^{M} \hat{v}_i \tag{3.53}$$

therefore the perturbation is expressed as $\Delta\hat{V}$ and is the difference between the exact coupling potential $\hat{V}_c$ and the VSCF mean-field potential $\hat{v}_i$. Following the formalism by Knowles *et al.*[89] in electronic structure theory, we can avoid explicit formulas for each order of the series. The VMP expansion can be expressed as:

$$E = \sum_{n=0}^{\infty} E^{(n)} \tag{3.54}$$

$$|\psi\rangle = \sum_{n=0}^{\infty} |\psi^{(n)}\rangle. \tag{3.55}$$

Applying intermediate normalization,

$$\langle\psi^{(0)}|\psi^{(n)}\rangle = 0 \tag{3.56}$$

we can introduce eqn (3.55) into the time-independent Schrödinger equation. Collecting same-order terms, we obtain:

$$(\hat{H}^{(0)} - E^{(0)})\,|\psi^{(n)}\rangle = -\,\hat{U}|\psi^{(n-1)}\rangle + \sum_{k=1}^{n} E^{(k)}|\psi^{(n-k)}\rangle. \tag{3.57}$$

Defining a transformed wavefunction $|\rho^{(k)}\rangle$

$$|\rho^{(k)}\rangle = -\,\hat{U}|\psi^{(k-1)}\rangle \tag{3.58}$$

and a projector $\hat{P}$ into the space orthogonal to $\psi^{(0)} = 1 - |\psi^{(0)}\rangle\langle\psi^{(0)}|$, we can express the perturbed energies and wavefunctions up to order n, as follows:

$$|\psi^{(n)}\rangle = P(\hat{H}^{(0)} - E^{(0)})^{-1}P\left[-|\rho^{(n)}\rangle + \sum_{k=1}^{n-1} E^{(k)}|\psi^{(n-k)}\rangle\right] \tag{3.59}$$

$$E^{(n)} = \langle\psi^{(0)}|\rho^{(n)}\rangle \tag{3.60}$$

We note here that it is not necessary to use a VSCF Hamiltonian and zero-order. In fact there are many implementations which apply perturbation theory using the harmonic oscillator as a zero-order Hamiltonian.[90]

3.3.6 Vibrational Coupled Clusters

Coupled clusters theory is certainly among the most successful computational tools for the calculation of high accuracy vibrational frequencies, and has become in recent years the gold standard, both in electronic structure, as well as vibrational structure theory.

Coupled clusters theory is generally presented in terms of the second quantization formalism, an extended discussion of which is outside the scope of the present work. However, the interested reader may consult any electronic structure introductory text, such as Szabo[68] or more specifically the work by Christiansen who developed a second quantization formulation specifically for vibrational theory.[91] In short, all quantum mechanical operators are expressed in terms of the creation and annihilation operators. The creation operator ($a_{s^m}^{m\,\dagger}$) puts a quanta in mode m into the state with index s^m, while the corresponding annihilation operator $a^m_{s^m}$ removes said quanta. These operators satisfy the following commutation relationships:

$$[a^m_{r^m}, a^{m'\dagger}_{s^{m'}}] = \delta_{m,m'}\delta_{r^m,s^m} \tag{3.61}$$

$$[a^m_{r^m}, a^{m'}_{s^{m'}}] = [a^{m\dagger}_{r^m}, a^{m'\dagger}_{s^{m'}}] = 0. \tag{3.62}$$

A Hartree product state (such as a VSCF configuration) can be represented as a product of creation operators acting on the "vacuum state":

$$|\psi\rangle = \prod_{m=1}^{M} a^{m\dagger}_{r^m}|vac\rangle. \tag{3.63}$$

The Hamiltonian (and any other quantum mechanical operator) can be represented in terms of creation and annihilation operators:

$$\hat{H} = \sum_{t=1}^{N_t}\sum_{m=1}^{M} h^m_t \tag{3.64}$$

In which

$$h^m_t = \sum_{r^m, s^m} h^{m,t}_{r^m s^m} a^{m\dagger}_{r^m} a^m_{s^m} \tag{3.65}$$

$$h^{m,t}_{r^m s^m} = \langle \varphi^m_{r^m}(Q_m)|H^{m,t}|\phi^m_{s^m}(Q_m)\rangle \tag{3.66}$$

and $|\phi^m_{r^m}\rangle$ are the one-mode basis functions (modals in the VSCF formalism) corresponding to mode m and excited in the r^m quanta. Note that the Hamiltonian here is expressed as a sum over products of one mode Hamiltonian operators h^m_t which is the preferred representation when using a quartic force field for the potential. Also $H^{m,t}$ is simply the first quantization one-mode operator.

The vibrational coupled clusters ansatz can be expressed in terms of the exponential operator:

$$|\psi_{\nu cc}\rangle = \exp(T)|\Phi_i\rangle \tag{3.67}$$

In which $|\Phi_i\rangle$ is the reference VSCF wavefunction, T is the so-called cluster operator

$$T = \sum_{\mu} t_\mu \tau_\mu, \tag{3.68}$$

where t_μ are the so-called cluster amplitudes and τ_μ are excitation operators. The cluster operator can be expressed in terms of single, double, triple, and so on, excitations in a similar way to in VCI theory:

$$T = \sum_{i=1}^{M}\sum_{\mu} t_{\mu_i}\tau_{\mu_i}. \tag{3.69}$$

Here, the symbol μ_i carries all the information required to specify the transition in question and the index i indicates the excitation level. The single-excitation operator can be expressed as:

$$T_1 = \sum_{\mu_1} t_{\mu_1} \tau_{\mu_1} = \sum_{m_1}^{M} \sum_{a^{m_1}} t_{a^{m_1}}^{m_1} a_{a^{m_1}}^{m_1\,\dagger} a_{i^{m_1}}^{m_1} \quad (3.70)$$

and similar expressions for the double, triple, and so on, cluster operators.

Now, replacing eqn (3.67) into the Schrödinger equation and multiplying by $\exp(-T)$,

$$\exp(-T)\hat{H}\,\exp(T)|\Phi_i\rangle = E|\Phi_i\rangle, \quad (3.71)$$

and projecting onto the reference state means that the VCC equations for the energies and wavefunction amplitudes are obtained:

$$E_{\mathrm{VCC}} = \langle\Phi_i|\exp(-T)\hat{H}\,\exp(T)|\Phi_i\rangle = 0 \quad (3.72)$$

$$e_\mu = \langle\mu|\exp(-T)\hat{H}\,\exp(T)|\Phi_i\rangle = 0 \quad (3.73)$$

In the limit of including all possible excitations, VCC is identical to VCI. This scheme has important advantages if the excitation level is truncated. To begin with, the VCC method is, contrary to VCI, size-consistent, which translates into a quicker convergence of the method relative to the excitation level.

3.3.7 Latest Developments and Applications

In the following section we will present a selection of the latest developments in vibrational structure calculations, as well as some of their applications. The presented work does not pretend to be comprehensive, nor exhaustive, and inevitably will reflect the biases and preferences of the authors. We present it in the hope that the reader may find it interesting and informative.

Since its beginnings in the 1970s at the hands of Bowman,[56] the field of vibrational structure calculations has seen an enormeous expansion. A significant a effort has been made in the development of more efficient and accurate methodologies and the quality and quantity of work seems to rise by the day. This is manifested by the significant number of recent reviews and perspectives on the subject. Hirata and Hermes[8] wrote an excellent review piece on the diagramatic theories of vibrational structure with a special enphasis on self-consistent and perturbational methods. Ove Christiansen[92] prepared a very recent review centered around coupled clusters methodologies discussing ways to introduce tensor decomposition approaches into it. Puzzarini and Barone very recently reviewed methodologies to obtain highly accurate molecular structures from rovibrational structure calculations.[93] Barone *et al.*[94] also reviewed strategies to compute the physicochemical and spectroscopic (particularly rovibrational) properties of prebiotic molecules of astrochemical interest taking into account ways to include different chemical environments into the analysis, such as solvation, surface chemistry and intra and inter-molecular interactions. Continuing with the astrochemical theme, Fortenberry recently made a contribution regarding the study of interstellar anions by means of quantum chemical anharmonic vibrational calculations[95] and another one in 2017 with a broader scope, also on astrochemistry.[96] Bloino, Baiardi and Biczysko[97] focused on cost effective methods, suitable for the simulation of electronic or vibrational spectra of medium to large molecular systems, ranging from half a dozen to more than a 100 atoms. Qu and Bowman[75] presented a very interesting analysis on the question of ease of interpretation *versus* methodological complexity in anharmonic vibrational structure theory. A very interesting review was also prepared by Gerber and Roy,[98] centered around VSCF and post-VSCF methods with applications in medium to large molecular systems. Of course we should also mention the excellent review by Bowman, Carrington and Meyer centered mainly on VSCF/VCI methods and including some discussion of multi-configurational time-dependent Hartree methods.[59] Therefore the field is certainly very active. In what follows we will try to present the reader with a summary of the latest developments in the field as viewed by ourselves.

Probably the greatest bottleneck in any anharmonic vibrational structure calculation is the calculation of the PES. Given the high computational cost usually attached to this part of the calculation, the steep scaling with regards to system size, and the fact that no matter how good or accurate an approximation of the Schrödinger equation can be, the results will only be as accurate as the potential used in the Hamiltonian. Therefore, it is understandable that much of the work in the past few decades has been devoted to the development of new and more efficient ways of modeling the PES.

Numerical grids represent a very accurate alternative[99] in which the PES can be represented by a collection of energy values collected in N^M points of the conformational space, in which N is a predefined number of points per degree of vibrational freedom and M is the number of normal vibrational modes. We can see that the number of points needed to build the grid, and therefore the number of evaluations of the electronic Schrödinger equation, grows exponentially with the size of the system. As an example, for the water molecule, with three normal modes and using 16 points by normal mode, $16^3 = 4096$ energy evaluations are necessary, therefore this brute force method touches its limit of applicability in triatomic molecules. In order to avoid this serious limitation, a great variety of strategies have been designed over the years. The "curse of dimensionality" can be mitigated somewhat by the use of an idea first proposed by Bowman that has appeared time and again in the literature under different names: N-mode coupling representations,[100] many-body expansion,[101] mode-coupling expansion,[102] high dimensional model representaion,[103] and others.[104] This familly of schemes allows a finer control of the coupling of the different modes in the system, by expressing the full potential as a sum of the lower dimensional functions.

More recently, Strobusch *et al.* proposed a very interesting strategy called adaptive sparse grid expansions[105–107] which consists of an algorithm that optimizes the number and position of grid points relative to energy precision. Grids may be used directly, or converted into an analytical expression either by fitting[108] or interpolation[109] procedures. For example, very high accuracies have been obtained by reproducing kernel Hilbert space interpolation coupled to a grid representation for triatomic reactive systems.[110]

Grid methods, even the most efficient examples of them, are still limited to small systems, therefore alternatives are needed for bigger molecules. QFFs, this is, fourth order Taylor expansions of the potential with respect to the normal coordinates of the system, currently represent the best trade-off of the accuracy and computational cost for the representation of molecular PES for anharmonic vibrational structure applications. N-mode coupling and similar schemes can also be easily applied to QFFs meaning that $V(Q)$ can be expressed as the simple sum of two, three, and four-mode coupling terms.

$$V(Q) = V^{(1)} + V^{(2)} + V^{(3)} + V^{(4)} \tag{3.74}$$

$$V^{(1)} = \sum_{i=1}^{M}\left(\frac{1}{2}f_{ii}Q_i^2 + \frac{1}{6}f_{iii}Q_i^3 + \frac{1}{24}f_{iiii}Q_i^4\right) \tag{3.75}$$

$$V^{(2)} = \sum_{i<j}^{M}\left(\frac{1}{2}f_{ijj}Q_iQ_j^2 + \frac{1}{2}f_{iij}Q_i^2Q_j + \frac{1}{6}f_{ijjj}Q_iQ_j^3 + \frac{1}{6}f_{iiij}Q_i^3Q_j + \frac{1}{4}f_{iijj}Q_i^2Q_j^2\right) \tag{3.76}$$

$$V^{(3)} = \sum_{i<j<k}^{M} f_{ijk}Q_iQ_jQ_k + \frac{1}{2}\sum_{i\neq j<k}^{M}\left(f_{iijk}Q_i^2Q_jQ_k + f_{ijjk}Q_iQ_j^2Q_k + f_{ijkk}Q_iQ_jQ_k^2 +\right) \tag{3.77}$$

$$V^{(4)} = \sum_{i<j<k<l}^{M} f_{ijkl}Q_iQ_jQ_kQ_l \tag{3.78}$$

The efficiency of the quartic force field (QFF) scheme comes from the possibility of calculating the expansion coefficients by numerical differentiation of the potential, analytical gradients, or Hessians, if available. Consider that the number of electronic structure evaluations necessary to obtain a complete QFF can decrease substantially if we calculate the force constants using numerical differentiation of gradients (obtained analytically within the electronic structure code) instead of fully numerical differentiation of the energies. The same is true if we use analytical Hessians instead of gradients or energies. In this later case, however, one must consider the actual efficiency of the analytical Hessian calculation. As Rauhut and Ramakrishnan demonstrated[111] after a certain number of atoms it is usually more efficient to use gradients rather than Hessians, as the computational time is actually less when using the former, rather than the later.

In this way, a complete QFF can be obtained at a small fraction of the cost that a potential grid would require. QFFs usually show a slow convergence of the Taylor series, which translates into a noticeable loss of accuracy, which in many cases may be insufficient even for the calculation of fundamental vibrational transitions, introducing errors in the order of tens of wavenumbers in the vibrational energies compared to grid methods.[77,90,99]

These large errors are in many cases a result of so called "variational holes", which occur when the gradient of the QFF is of the wrong sign at large molecular distortions, producing unphysically low energy regions. Proton stretching modes in particular are very challenging for QFFs. In the case of symmetric modes, for example, the potential energy tends to have an asymptotic limit as the stretch coordinate is displaced towards positive or negative infinity, depending on the phase of the normal mode, whereas the Taylor series in normal coordinates will always tend towards positive or negative infinite energies in the limit of infinite displacements of the coordinate.

A fourth order expansion often falls short of an accurate representation of the potential along these modes. Using higher order polynomials is usually out of the question, because of their cost, and numerical accuracy considerations. Modified Shepard Interpolation (MSI) schemes were proposed to improve the performance of the QFFs,[112–114] and have since been successfully applied to many molecular systems.[115–118] Although significant improvements were obtained for vibrational frequencies relative to simple QFF schemes, this method requires some prior knowledge of the PES to place additional reference points,[119] and although some methods for automatic selection of those points have been proposed,[120] the scheme still implies a significant increment in computational cost.

A very interesting way of improving QFFs was implemented by Dateo *et al.*[121] and later by Fortenberry *et al.*[3,122–124] based on the early work by Watson,[48] Meyer *et al.*,[125] and Carter and Handy,[126] in the context of internal valence coordinates. The scheme consists of substituting single bond stretching coordinates with Morse coordinates, which present the correct asymptotic behavior, obtaining excellent fits for the vibrational frequencies to within a few cm^{-1} of the experiment. What makes this approach so attractive is the fact that unlike, for example, MSI, there is no increase in the computational cost relative to that of a simple QFF and it is also much simpler to implement and use. Building on these ideas, Burcl *et al.*,[77] presented a similar scheme specially adapted for its use in the context of cartesian normal coordinates. In this case, Morse coordinate substitution can only be performed on symmetric normal coordinates, which may lead to an imbalance in the representation of the stretching modes, as Morse coordinates do not present the correct limiting behavior for substitution into antisymmetric stretch normal coordinates. For this reason, Burcl *et al.* proposed the use of Gaussian coordinates for these, following a previous paper by Carter and Handy[127] proposing the use of hyperbolic tangent coordinates for the same purpose.

Gaussian and tanh coordinates, however, should be better adapted for cases when the potential along the normal coordinate to be substituted shows even parity and a dissociative asymptotic limit. In cases in which this is not so, the coordinate substitution may still provide a poor convergence radii for the potential. Moreover, the integrals involving Gaussian or tanh coordinates for the solution of the variational vibrational structure equations must be solved numerically, which complicates its implementation in codes that make use of analytic integrals, such as our own.

The VSCF method has also seen considerable improvement since its initial proposal in the 1970s by Bowman,[56] Carney *et al.*,[128] Cohen *et al.*,[129] and Gerber and Ratner.[130] Several authors have proposed several modifications and enhancements over the years.[131–134] One instance in which this methodology often fails is in floppy systems with low frequency, large amplitude motions, such as the case with –CH_3 torsional motions. These large amplitude motions (LAMs) are defficiently described by normal mode coordinates owing to large couplings between these modes and higher frequency ones in the potential. Some early work by Horn and Gerber[135] experimented with hyperspherical and elipsoidal coordinates obtaining good results for the description of several small systems.[135–138] Bowman, Carter and Handy proposed the application of reaction path Hamiltonian theory as a strategy to deal with LAMs.[139,140] Wang and Bowman also proposed a localized coordinates scheme for floppy systems.[141–143] Also Cheng *et al.* implemented a localized coordinates scheme specially designed for large molecules.[144] Truhlar *et al.*,[145] were the first to propose vibrational coordinates optimization, and more recently Yagi *et al.* proposed another algorithm with the same main idea.[146] There has also been several attempts to implement curvilinear internal coordinates. This system of coordinates, however, although it produces much better results by significantly decoupling low frequency modes from higher ones, does have the problem of much greater mathematical complexity, posing serious difficulties in finding a general implementation, capable of being applied to any kind of molecular system. Most of these difficulties come from the kinetic energy operator. When curvilinear internal coordinates are used the kinetic energy operator is no longer diagonal, and significant coupling can be found in it, moreover, its precise mathematical form must be determined for each case. There has been several approaches used to obtain general analytical expressions for the kinetic energy operator,[147–152] as well as others more specific for a certain coordinate system.[49–54] More recently, Sielk *et al.*[55] proposed a numerical method to obtain a general kinetic energy operator, an idea first put forward by Laane *et al.* many years ago,[153] and later applied by Scribano, Lauvegnat and Benoit in a VSCF/VCI methodology,[154] and recently even to a multi-configurational time-dependent Hartree method[155] and applied to several cases.[156–159] The methodologies and applications were described in detail in a recent review by Gatti *et al.*[160] Gerber *et al.* introduced one of the first applications of perturbation theory into VSCF methodology, VSCF-PT2, also known as correlation corrected VSCF (cc-VSCF).[161–164] One particular problem that severely impairs the general applicability of perturbative methods is the strong divergence of the perturbative series in cases of strong coupling, such as when degeneracies or near degeneracies appear between two states, as is normally the case in all but the smallest of molecules, therefore clearly this is no small problem. Several methods have been proposed to deal with this. Matsunaga *et al.*[165] proposed an algorithm based on the use of a limited vibrational configuration interaction over the degenerate subspace in order to lift the degeneracy and then perform perturbation theory on the resulting mixed states. This of course also requires a configuration selection scheme capable of detecting all resonant states and including them in the degenerate space to be treated with VCI. One important limitation of the algorithm presented by Matsunaga *et al.* is that the selected degenerate space consists only of the VSCF configuration in which only one mode is excited (single excitations). This means the algorithm is only capable of treating 1 : 1 resonances leaving out other kinds, which are of common occurence. Yagi, Hirata and Hirao[64] proposed a new quasi-degenerate perturbation theory method with an improved configuration selection algorithm by including the condition that all states that differ in excitation in more than four quanta are de facto considered to not interact. This simple condition makes it possible to apply this methodology efficiently and accurately. They applied their methods to benzene, carbon dioxide and formaldehyde and revealed an excellent agreement with the experimental results, especially with resonant bands. More recently, Yagi and Otaki[166] presented an update of their method including an optimized coordinate scheme, and applied it to trans-1,3-butadiene. Yagi and Thomsen[167] recently applied this methodology to the study of Eigen and Zundel forms of protonated water clusters. They found that the Eigen clusters reproduce recent experimental bands very well, while their Zundel cluster simulations do not. Daněček and Bour also proposed a similar quasi-degenerate perturbation theory approach with some slight modifications.[168] Matito *et al.*[169] proposed a vibrational auto-adjusting perturbation theory. Finally, Barone *et al.*[170] recently proposed a general degeneracy-tolerant, rovibrational algorithm based on the Watson Hamiltonian which they call general vibrational second-order perturbation theory (GVP2), which they applied extensively (see below).

In much the same way that vibrational perturbation theory may be improved by the use of variational approaches (in this case, in order to lift degeneracies), variational approaches such as VCI may benefit from vibrational perturbation theory as well. As we exposed before, VCI methods are not very broadly applicable owing to the high speed with which the basis set grows with the system size. However, the VCI matrix is in general sparse and diagonal-dominant, which means that most of its elements are zero or nearly so. Therefore, one strategy to mitigate the unfavorable scaling of the method is by using a screening scheme so as to only include in the VCI matrix in those states that have a large contribution to the energy (and wavefunction). One way to do just this is to use perturbation theory in order to access the magnitude of the coupling between a new state and those already included in the matrix. Handy and Carter were the first to propose a configuration selection scheme for VCI[171] and implemented it into their code

MULTIMODE, and later applied it to the study of $H_3O_2^-$ and its isotopologue $D_3O_2^-$. Later, Scribano and Benoit[172] presented an iterative approach which they named vibrational configuration interaction with perturbation selected interactions (VCIPSI), and tested it on methane and benzoic acid molecules. Very recently, Garnier *et al.* presented an algorithm called adaptive VCI (AVCI)[173,174] with the particularity that in it a complete VCI matrix is never constructed, which alleviates the memory bottleneck that plagues most of the VCI implementations. These last approaches showed promising results in terms of size reduction, but still require the user to *a posteriori* determine a big matrix block designed to improve the accuracy of the solutions. The same group later[175] presented an improved version of their scheme called dual vibration configuration interaction (DVCI), in which they implemented a new factorization of the Hamiltonian that substantially reduces the number of operations that the algorithm requires compared to AVCI. Other interesting contributions to these kind of strategies were made by Sibaev *et al.*[176] the state-specific configuration-selective vibrational configuration interaction (cs-VCI) by Neff and Rauhut,[67] and the VCI-P code by Pouchan *et al.*[177] Other interesting methods designed to reduce the cost while minimizing the loss of accuracy involved restricting the maximum excitation level below the convergence limit,[7,178,179] restricting the extent of mode-coupling in excited states, particularly for larger molecules[177,180] and mode tailoring of the VCI basis.[171]

With regards to VCC methods there has been substantial developments in recent years. For example, as with other methods there have been implementations of coordinate optimization schemes for VCC.[9] In addition, there have been implementations with well defined polynomial computational scaling at all levels,[181] and the approximate inclusion of three[182] and four-mode couplings by use of perturbation theory (VCC[2pt3,4]).[183] Of course, VCC methods are always rather expensive in terms of computer resources, therefore alternatives have been devised in order to reduce, as much as possible, the requirements of these methods. In this regard, linear response theory has been applied to VCC by Christiansen's group.[184–186]

A great deal of work in the field has been dedicated to astrochemistry. In recent years a great deal of astronomical complex organic molecules have been identified in space in diverse contexts. Although previously thought to not be important, they are now proving to be ubiquitous and therefore there has been a surge of activity directed towards the identification and analysis of these species. Thanks to advances in many different techniques and NASA's successful Kepler exoplanet transit mission, thousands of diverse planets outside of our solar system have been discovered. The soon to be launched James Webb Space Telescope (JWST) will be very helpful in the identification of biosignature gases in the atmosphere of Earth-like planets and prebiotic molecule signatures in Titan-like atmospheres, by observing their absorption during transits. Although the search for key-target molecules in exoplanet atmospheres can be carried out using JWST transit spectroscopy in the IR region (0.6–29 μm wavelength range), opportunities for their detection in protostellar cores, protoplanetary disks, and on Titan are also offered by interferometric high spectral and spatial resolution observations using the Atacama large millimeter/submillimeter array. In this respect, one important challenge is the need for a very high precision, not only in vibrational, but also in rovibrational and rotational spectra simulation[93] which makes the use of advanced quantum anharmonic rovibrational methods a requirement. In particular the group lead by Barone and Puzzarini have been prolific in this field. One particular goal for these high-precision methods that have made an important contribution is the quest for pre-biotic molecules. That is, molecules that are precursors for complex biological molecules and form naturally in the interstellar, exoplanetary or otherwise extraterrestrial medium.[187] In a pair of recent publications, Ali *et al.* pointed out, using quantum chemistry methods, how the molecular structure and reactivity of some of the molecules detected by the Cassini mission in Saturn's moon, Titan, can lead to the formation of aromatic species and therefore possibly of RNA or DNA precursors.[188,189] Some of these molecules, in particular protonated and unprotonated oxirane (to probe for oxygen chemistry in space),[190,191] the cyclopropenyl cation and its cyclic derivative the methyl-cyclopropenyl cation, which are key precursor species of the molecular complexity on Titan, ortho- and *peri*-fused tri-cyclic aromatic rings, for example the phenalenyl cation ($C_{13}H_9{}^+$) and anion ($C_{13}H_9{}^-$), which are important intermediates in molecular growth, and uracil, can be spectroscopically characterized using second order vibrational perturbation theory and a coupled clusters-level potential with the aim of guiding astronomical line searches in the infrared and/or millimeter/submillimeter-wave ranges, as well as directing future laboratory measurements. These contributions show that high-level quantum-chemical computations with an adequate treatment of electron correlation effects, extrapolation of the basis-set limit, and inclusion of the core correlation are able to quantitatively predict spectroscopic parameters, with an accuracy of no less than 0.01% and up to 0.001% in some cases.[192–194]

There is also interest in building libraries of very high precision rovibrational spectra to aid the modelization of the atmosphere of celestial bodies. One important initiative in this respect is ExoMol, a database of molecular line lists that can be used for spectral characterisation and simulation, and as input to atmospheric models of exoplanets, brown dwarfs and cool stars, and other models including those for combustion and sunspots. The TROVE group, lead by Sergey Yurchenko has been very active and prolific in this line of work.[195–204] They developed and implemented a new methodology to produce symmetry adapted basis functions for solving rovibrational Schrödinger equations. The method is based on the properties of the Hamiltonian operator commuting with the complete set of symmetry operators.[205,206] Owens *et al.*[207] presented a very interesting paper studying the interesting phenomenon of rotationally induced chirality. This phenomenon consists of the emergence of chirality in achiral molecules in extremely highly excited rotational states. In their paper, the authors delineate a strategy to create the so-called "molecular superrotor" quantum states experimentally and also methods to study them theoretically. These quantum states are interesting because they show novel behavior, not only can they show rotationally induced chirality, but also interesting dynamics in scattering and spectroscopy.[208–211]

3.4 Computational Vibrational Spectroscopy in Complex Environments

3.4.1 Multiscale Methods for General Vibrational Spectroscopy

The calculation of vibrational spectra of complex chemical environments is quite clearly, complex. In fact it would be almost impossible were it not for the fact that the chemical species of interest does not encompass the whole of the system, but usually only a spatially limited region. This makes such systems excellent candidates for hybrid quantum mechanical molecular mechanical (QM/MM) simulations, also known as multi-scale schemes, in which a big system is partitioned into a small region modelled in great detail (usually quantum-mechanically) and the rest of the system using a low accuracy model, such as an empirical force field (the MM part). The history of vibrational spectra simulation using QM/MM techniques spans about two decades, and usually falls into two categories: (i) static or time-independent; and (ii) time-dependent formalism. In the former the system is first subject to a full or partial optimization and then the Hessian matrix is computed and diagonalized to obtain normal modes and transition frequencies. In the later one, several time-dependent simulations are performed and the relevant autocorrelation function is computed and the Fourier transformed in order to obtain the required spectrum.

In the case of static methods, there are some drawbacks. First, in condensed phases the spectrum should be the result of an ensemble average of multiple minima, so a conformational search is required in order to obtain converged results. This can be done using either molecular dynamics techniques, in which the system is evolved through time using Newton's laws, or Monte Carlo techniques, in which the conformational space is sampled stocastically. Second, the computational cost of the Hessian calculation normally scales with the square of the system size, which means that one should be careful to only optimize and/or include in the Hessian as many atoms as is strictly necessary in order to obtain meaningful results and no more. However, doing this is not as simple as it sounds. There are two methods for the calculation of vibrational frequencies from partially optimized structures. The partial Hessian vibrational analysis (PHVA) devised by Li and Jensen[212] assigns infinite masses to those atoms not included into the geometry optimized region, which essentially zeroes-out their contribution into the Hessian matrix and therefore, effectively uncouples their motion to the vibrational motions of the molecular system of interest. Later, Head and co-workers improved the model by allowing some coupling between the optimized and non-optimized regions.[213–215] On the other hand, the mobile block Hessian (MBH) scheme by Ghysels *et al.*[216] in which instead of giving infinite masses to the non-optimized region, it is allowed to move as a rigid body with respect to the optimized region, and therefore the name. The MBH method consists of partitioning the system into blocks of atoms which move as rigid bodies (only rotational and translational motions are allowed). The MBH method also has a few variants.[217,218] Along these lines, Durand and Tama developed a similar method called rotation-translation blocks (RTB),[219,220] although it has the problem of being incapable of producing physical frequencies in non-equilibrium points.

Regarding time-dependent methods, the spectra can be obtained from the time-autocorrelation function of the relevant dynamical variable: the dipole moment for the IR spectra, polarizability for Raman spectroscopy and even nuclear velocities if only the vibrational frequencies are of interest. One drawback of time-dependent methodologies is that the assignment of the vibrational bands is not straightforward. One could, in principle, project the trajectory into the vibrational normal modes obtained by other means, or some approximation thereof. In small molecules, different bands may be assigned simply by projecting the trajectory into relevant internal coordinates and then computing the autocorrelation function of the velocities thereof,[221] however, this approach is of little use in bigger and more complex systems owing to the fact that any internal coordinate is bound to be involved in several normal modes. In these cases one could still use more precisely defined normal coordinates for the projection, but the problem remains owing to the inherent anharmonicity of the MD simulations, which means that no matter how precisely defined a normal mode may be, even in a vacuum a trajectory projected onto one will always include the intensities of several bands owing to vibrational mode coupling. Of course the problem is much more serious in condensed phases in which the concept of a normal mode is not even very well defined. However, some attempts have been made to systematize the band assignment.[222] Moreover, the use of time-dependent techniques does not lift the requisite of performing a conformational search. The high computational cost of QM/MM simulations only allows for relatively short trajectories, therefore in many cases it may be necessary to make a conformational search by other means, such as fully classical MD simulations previous to the spectra calculation. Also, given the fact that vibrational spectra are usually readily obtained both experimentally and theoretically, it has been used in the past simply as a way to validate a QM/MM scheme.[223–231] Along these lines Carr and Parrinello[232] studied simulations of water, ethanol and a Schiff base using DFT with planewaves and empirical forcefields. They compared static and time-dependent approaches for the calculation of vibrational spectra and found them to be in very good agreement. They also used the study to analyze the effect of intramolecular link atoms on vibrational frequencies as a measure of the perturbation these schemes introduce into the molecular system.[223]

Cui and Karplus studied the vibrational spectra of formamide in water using a QM/MM model[233] using a QM/MM analytical second-derivative

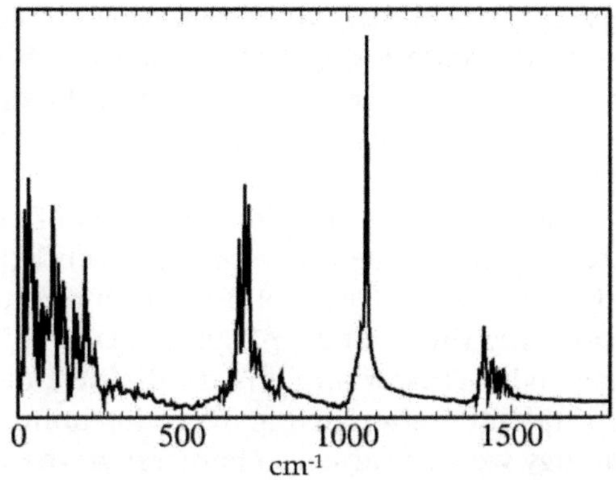

Figure 3.2 Vibrational density of states for nitrate in bulk water obtained by Fourier transform of the time autocorrelation function of atomic velocities. The signal at about 1500 cm^{-1} is broadened owing to symmetry breaking of the nitrate molecule in aqueous solution, this is in good accordance with the resonant Raman spectroscopy experimental results.

implementation. They studied the effect of intramolecular partitioning *via* link atoms on vibrational frequencies, the transition state structure of triosephosphate isomerase and the active site of myoglobin. They concluded that the neglect of the MM part in the Hessian matrix did not produce significant differences in the the vibrational frequencies and IR intensities.

Rovira *et al.*[234] studied the effect of the protonation of a nearby histidine on the vibrational frequencies of CO bound to a heme group. Our group has contributed several pioneer works to the field. Gonzalez Lebrero *et al.* studied the symmetry breaking of the nitrate ion in aqueous solution as observed in vibrational spectroscopy. A representative spectrum obtained in bulk solution is presented in Figure 3.2. The simulations allowed us to interpret the phenomenon in terms of solvation patterns using a polarizable force field together with a standard TIP4P water model.[235] Later Gonzalez Lebrero, Perissinotti and Estrin presented a study of the solvated peroxynitrite ion.[221] Both studies used the time-dependent approach based on time correlation functions in order to obtain the vibrational power spectra. Band assignment was performed by simple projection of the dynamics onto approximate normal modes. Other contributions from our group include the characterization of lithium hydrides in ether,[236] and a study of the reaction between nitroxyl and oxygen in water.[237]

Recently, the simulation of bidimensional vibrational spectra has seen some successful implementations.[238–240] These new methodologies are very much welcomed as they can provide information not only on the vibrational modes and frequencies of vibration of the molecules of interest, but also about the coupling of different parts of the system and relaxation times in condensed phases.

Adaptive QM/MM has also been applied to vibrational spectroscopy. In these schemes the MM and QM parts are allowed to exchange molecules, allowing, to some extent, researchers to better study the effect of solvation shells without unwanted wandering of QM atoms away from the region of interest.[241] One important inconvenience of this scheme is its elevated computational cost, which permits only lightweight quantum mechanical methods to be used.

Despite the significant potential, implementations of vibrational analysis for materials are only rarely seen. Our group recently presented an exception with a study of the water-TiO_2 solid-liquid interphase using a QM/MM implementation using the Quantum-Espresso DFT software interfaced to our own in-house code. This example is, to the best of our knowledge, the only QM/MM implementation using periodic boundary conditions. Using this methodology we were able to characterize the vibrational spectrum of interfacial H_2O at the water/anatase (101) interface. We observed and enhanced the importance of water librations and bending modes with a shift of the stretching bands to lower frequencies with respect to the gas phase.[242]

Mroginski, Hildebrandt and collaborators applied QM/MM techniques to the study of the phycocianobilin (PCB) cofactor of phycocianin.[243–246] These studies showed the importance of sampling the effect of the environment in order to obtain a higher accuracy on the vibrational spectra. In their studies, the authors were able to isolate chromophore-protein interaction-sensitive vibrational modes, which led them to revise previous conclusions based on simpler models.

Resonance Raman spectrosocpy constitutes an excellent probe for biomolecular systems owing to its high selectivity and sensitivity. It is able to provide meaningful information about systems with an unknown molecular structure, especially if coupled with accurate theoretical methods as demonstrated by Horch *et al.*,[247] who were able to study [NiFe]-hydrogenase by means of a theoretical and experimental approach based on resonance Raman spectroscopy and QM/MM simulations of the spectra. The relevance of this enzyme is due to its ability to catalyze the reversible cleavage of hydrogen and it is therefore a valuable model system for emision-free energy conversion processes. They obtained valuable structural information about the unknown structure of the H-bound intermediate in the catalytic cycle. Later, further studies of 65 resonance Raman samples of the same protein in different oxidation states were used to confirm the existence of –OH coordination to one of the Fe atoms of a nerby [4Fe-3S] cluster with the help of QM/MM simulations.[248]

3.4.2 Hybrid Models for Surface-enhanced Raman Spectroscopy

Surface enhanced Raman spectroscopy is a technique in which the Raman signal of a molecule absorbed on a surface is strongly amplified, sometimes up to a factor of 10^{14}, which makes it sensitive to single molecules. The phenomenon was first observed by Fleischmann *et al.*[249] in the 1970s while observing pyridine adsorbed onto a roughened silver electrode. It was later shown that the phenomenon had a scattering cross-section that was a million times greater than normal Raman.[250] Later, in the 1990s, Nie and Emory[251] and Kneipp *et al.*[252,253] suggested it was possible to obtain single-molecule SERS, producing a surge in activity in this field that has not dried up even today, with a broad range of applications.[254–257] Although the fundamental theoretical framework of the technique has been the cause of heated debate among experts for many years,[258,259] today it is widely agreed upon that most of the observed SERS enhancement is fairly well explained by plasmonic theory,[260–269] in what is commonly known as the "electromagnetic enhancement mechanism", which states that SERS is the result of the enhancement of electromagnetic fields in the immediate vicinity of nanomaterials as a result of excited plasmons.

The enhancement effect is not exclusive to Raman spectroscopy, though. It has been observed in absorption and fluorescence phenomena. Raman scattering has particularly benefitted, however, owing to the fact that both the incident, as well as the scattered light beams, are enhanced by the plasmonic effects, which in turn produces enhancement factors scaling with the 4th power of the local fields, in comparison with only a quadratic scaling for absorption and fluorescence. Single molecule SERS occurs when two or more nearby nanostructures cooperate to produce an enhanced field in a particular region of the material, known as a "hot spot", this is caused by the coupling between plasmonic resonances. One example of such hot spots is the gap between two closely located nanospheres, in which the space in the junction between both structures harbours an electric field orders of magnitude higher than any other place on the surface of any of the spheres.[263]

A former aspiring candidate to explain the SERS phenomena is the mechanism known as "chemical enhancement", which can be envisioned as the result of chemical adsorption of molecular species into the nanostructured material, resulting in alteration of the molecular electronic structure and therefore, its resonance properties. Today, it is a mostly settled matter that although chemical enhancement is not the prime cause of the phenomena, it does contribute significantly to SERS by modulating the enhancement factors, changing the enhanced band structure and also providing a wealth of information about the details of the molecule-material interactions. There are three contributions to this mechanism:[259] complex formation,[270] charge transfer resonances[271] and electron-hole effects.[272]

It is clear then that SERS is a very complex phenomenon, involving heterogeneous systems, interphases, plasmonic resonances and finally surface chemistry, which is, not at all suprisingly, quite challenging to model theoretically. This has hardly discouraged the theoretical comunity, though, whom have seen clear opportunities for the natural application of QM/MM methodology ideas. Of course such complex and extensive systems require some sacrifices to be made in the modelization. Usually there are two groups of approaches, which either focus on the nanostructured material or on the adsorbed molecule. In this work we focused only on the later. Molecule-centered models usually represent the molecule at the quantum mechanical level of theory, while the nano-structure material is represented in a much more simplified form, usually based on classical electrostatic theory. Chemical enhancement effects can be coarsely introduced by the inclusion of small metallic clusters into the QM region.

One early example of SERS theoretical models was presented by Croni and Tomasi.[273–275] They represented the nanostructured material as a fusion of spheres characterized by a frequency-dependent dielectric constant, while the molecule was represented by standard electronic structure methods. Another approach consists of treating the nanomaterial using classical discrete dipole approximation and many-body Green functions for the molecular system, a method introduced by Masiello *et al.*[276,277] A methodology combining the CAM-B3LYP DFT functional for pyridine chemically adsorbed to a small Ag cluster, and a classical atomistic force field for the rest of the nanoparticle was presented by Arcisauskaite *et al.*[278] Morton and Jensen presented a series of very interesting works combining TDDFT with long-range interaction corrections to be able to study long range charge transfer excitations for the molecular part of the system and a classical atomistic discrete interaction model for the nanoparticle.[279–284] They developed a scheme which they called quantum mechanics/capacitance-molecular mechanics, in which they used a capacitance-polarization method which introduced the novelty of taking into account capacitance properties, as well as the more usual point charges and polarization properties for the classical atomistic representation of the nanomaterial. For the QM part they used linear response TDDFT,[285] quadratic response TDDFT,[286] and complex polarization propagator TDDFT[287] methodologies. They recently presented a very interesting application of their technique for gold nanoclusters functionalized with organic compounds.[288] A very noteworthy scheme is based on RT-TDDFT for the molecule and finite difference time dependent (FDTD) methods for the nanomaterial.[23,289–292] This methodology features the very remarkable inclusion of resonance Raman effects within the short-time approximation. FDTD theory was also used by Mullin *et al.* alongside linear response TDDFT theory for the molecular part of the system.[293,294] A recent implementation making use of the dressed tensor formalism for the nanomaterial, as well as TDDFT for the molecular region was presented by Chulhai *et al.*[295] The authors were very impressively able to obtain an ensemble average SERS and SEHRS (surface enhances hyper-Raman spectroscopy) spectra using molecular dynamics to sample the conformational space of the nanoparticle-molecule system.[296]

Pipolo *et al.* presented a scheme using time-dependent configuration interaction (TDCI) instead of the usual DFT-based scheme for the modelization of the adsorbed molecule, combined with the usual classical representation of the nanoparticle.[275,297]

3.5 Concluding Remarks

In this chapter, we have encompassed the main theories and developments on the calculation of vibrational frequencies in molecular systems. As a result of decades of efforts and inspiration from many authors, the state of the art has achieved such a refinement and accuracy so as to transform the practice of vibrational spectroscopy, by providing invaluable tools for the interpretation of experimental data. For small sized species, the available techniques offer predictive power. One of the critical challenges remaining is the reduction of the computational demand associated with these methodologies, to make it feasible for the reliable description of anharmonicity and quantum effects in moderate size and complex systems. To conclude, it must be noted that the kind of developments reviewed in these pages are not only relevant in spectroscopy. The investigation of processes such as energy transfer and charge transport in polymers and biomolecules, which have started to be addressed from a microscopic standpoint both computationally and experimentally in recent years, requires a realistic treatment of molecular vibrations, often coupled to the electronic degrees of freedom. In this framework, the developments discussed herein gain a renewed significance.

References

1. A. Ali, E. C. Sittler, D. Chornay, B. Rowe and C. Puzzarini, Cyclopropenyl cation–the simplest Huckel's aromatic molecule–and its cyclic methyl derivatives in Titan's upper atmosphere. *Planetary and Space*, *Science*, 2013, **87**, 96–105.
2. T. Gautier, N. Carrasco, A. Mahjoub, S. Vinatier, A. Giuliani, C. Szopa, C. M. Anderson, J.-J. Correia, P. Dumas and G. Cernogora, Mid-and far-infrared absorption spectroscopy of Titan's aerosols analogues, *Icarus*, 2012, **221**, 320–327.
3. R. C. Fortenberry, X. Huang, A. Yachmenev, W. Thiel and T. J. Lee, On the use of quartic force fields in variational calculations, *Chem. Phys. Lett.*, 2013, **574**, 1–12.
4. D. J. Alonso de Armiño and C. M. Bustamante, A quartic force field coordinate substitution scheme using hyperbolic sine coordinates, *Int. J. Quantum Chem.*, 2017, **117**, e25390.
5. D. Strobusch and C. Scheurer, Adaptive sparse grid expansions of the vibrational Hamiltonian, *J. Chem. Phys.*, 2014, **140**, 074111.
6. M. R. Hermes and S. Hirata, Stochastic algorithm for size-extensive vibrational self-consistent field methods on fully anharmonic potential energy surfaces, *J. Chem. Phys.*, 2014, **141**, 244111.
7. M. Neff and G. Rauhut, Toward large scale vibrational configuration interaction calculations, *J. Chem. Phys.*, 2009, **131**, 124129.
8. M. R. Hermes and S. Hirata, Diagrammatic theories of anharmonic molecular vibrations, *Int. Rev. Phys. Chem.*, 2015, **34**, 71–97.

9. B. Thomsen, K. Yagi and O. Christiansen, Optimized coordinates in vibrational coupled cluster calculations, *J. Chem. Phys.*, 2014, **140**, 154102.
10. D. A. McQuarrie, *Statistical Mechanics*, Harper & Row, London, 1976, p. 641.
11. H. A. Kramers and W. Heisenberg, Über die Streuung von Strahlung durch Atome, *Z. Phys.*, 1925, **31**, 681–708.
12. G. Placzek, Intensität und polarisation der Ramanschen streustrahlung mehratomiger moleküle, *Z. Phys.*, 1931, **70**, 84–103.
13. D. A. Long, *The Raman Effect*, John Wiley & Sons, Ltd, Chichester, UK, 2002.
14. J. Behringer, Zur theorie des resonanz-Raman-effektes, *Z. Elektrochem. Ber. Bunsenges. Phys. Chem.*, 1958, **62**, 906–914.
15. A. Komornicki and J. W. McIver, An efficient ab initio method for computing infrared and Raman intensities: Application to ethylene, *J. Chem. Phys.*, 1979, **70**, 2014.
16. A. C. Albrecht, On the theory of Raman intensities, *J. Chem. Phys.*, 1961, **34**, 1476.
17. J. Guthmuller, Comparison of simplified sum-over-state expressions to calculate resonance Raman intensities including Franck-Condon and Herzberg-Teller effects, *J. Chem. Phys.*, 2016, **144**, 064106.
18. J. Neugebauer and B. A. Hess, Resonance Raman spectra of uracil based on Kramers--Kronig relations using time-dependent density functional calculations and multireference perturbation theory, *J. Chem. Phys.*, 2004, **120**, 11564–11577.
19. J. Neugebauer, E. J. Baerends, E. V. Efremov, F. Ariese and C. Gooijer, Combined theoretical and experimental deep-UV resonance Raman studies of substituted pyrenes, *J. Phys. Chem. A*, 2005, **109**, 2100–2106.
20. M. Thomas, F. Latorre and P. Marquetand, Resonance Raman spectra of ortho-nitrophenol calculated by real-time time-dependent density functional theory, *J. Chem. Phys.*, 2013, **138**, 044101.
21. L. Jensen, J. Autschbach and G. C. Schatz, Finite lifetime effects on the polarizability within time-dependent density-functional theory, *J. Chem. Phys.*, 2005, **122**, 224115.
22. L. Jensen, L. L. Zhao, J. Autschbach and G. C. Schatz, Theory and method for calculating resonance Raman scattering from resonance polarizability derivatives, *J. Chem. Phys.*, 2005, **123**, 174110.
23. H. Chen, J. M. J. McMahon, M. A. Ratner and G. C. Schatz, Classical electrodynamics coupled to quantum mechanics for calculation of molecular optical properties: A RT-TDDFT/FDTD approach, *J. Phys. Chem. C*, 2010, **114**, 14384–14392.
24. M. Thomas, M. Brehm, R. Fligg, P. Vöhringer and B. Kirchner, Computing vibrational spectra from ab initio molecular dynamics, *Phys. Chem. Chem. Phys.*, 2013, **15**, 6608–6622.
25. R. Futrelle and D. McGinty, Calculation of spectra and correlation functions from molecular dynamics data using the fast Fourier transform, *Chem. Phys. Lett.*, 1971, **12**, 285–287.

26. L.-H. Xu, Two decades of advances in high-resolution spectroscopy of large-amplitude motions in n-fold potential wells, as illustrated by methanol, *71st International Symposium on Molecular Spectroscopy*, 2016.
27. Y. I. Kurokawa, H. Nakashima and H. Nakatsuji, Solving the Schrödinger equation of hydrogen molecules with the free-complement variational theory: Essentially exact potential curves and vibrational levels of the ground and excited states of the Σ symmetry, *Phys. Chem. Chem. Phys.*, 2019.
28. Y. Guan, J. Gao, Y. Song, Y. Li, H. Ma and J. Song, Variational Effect and Anharmonic Torsion on Kinetic Modeling for Initiation Reaction of Dimethyl Ether Combustion, *J. Phys. Chem. A*, 2017, **121**, 1121–1132.
29. L. He, Q. Zhang, P. Lan, W. Cao, X. Zhu, C. Zhai, F. Wang, W. Shi, M. Li and X.-B. Bian, *et al.*, Monitoring ultrafast vibrational dynamics of isotopic molecules with frequency modulation of high-order harmonics, *Nat. Commun.*, 2018, **9**, 1108.
30. C. Fábri, R. Marquardt, A. G. Császár and M. Quack, Controlling tunneling in ammonia isotopomers, *J. Chem. Phys.*, 2019, **150**, 014102.
31. C. Léonard, N. C. Handy, S. Carter and J. M. Bowman, The vibrational levels of ammonia, *Spectrochim. Acta, Part A*, 2002, **58**, 825–838.
32. F. Gatti, C. Iung, C. Leforestier and X. Chapuisat, Fully coupled 6D calculations of the ammonia vibration-inversion-tunneling states with a split Hamiltonian pseudospectral approach, *J. Chem. Phys.*, 1999, **111**, 7236–7243.
33. J. M. Bowman, X. Huang and S. Carter, Full dimensional calculations of vibrational energies of H3O+ and D3O+, *Spectrochim. Acta, Part A*, 2002, **58**, 839–848.
34. M. W. Davies, M. Shipman, J. H. Tucker and T. R. Walsh, Control of pyramidal inversion rates by redox switching, *J. Am. Chem. Soc.*, 2006, **128**, 14260–14261.
35. T. Rajamäki, A. Miani and L. Halonen, Six-dimensional ab initio potential energy surfaces for H3O+ and NH3: Approaching the subwave number accuracy for the inversion splittings, *J. Chem. Phys.*, 2003, **118**, 10929–10938.
36. M. Neff and G. Rauhut, Towards black-box calculations of tunneling splittings obtained from vibrational structure methods based on normal coordinates, *Spectrochim. Acta, Part A*, 2014, **119**, 100–106.
37. T. Rajamäki, M. Kállay, J. Noga, P. Valiron and L. Halonen, High excitations in coupled-cluster series: vibrational energy levels of ammonia, *Mol. Phys.*, 2004, **102**, 2297–2310.
38. X. Huang, D. W. Schwenke and T. J. Lee, Rovibrational spectra of ammonia. I. Unprecedented accuracy of a potential energy surface used with nonadiabatic corrections, *J. Chem. Phys.*, 2011, **134**, 044320.

39. P. M. Hundt, B. Jiang, M. E. van Reijzen, H. Guo and R. D. Beck, Vibrationally promoted dissociation of water on Ni (111), *Science*, 2014, **344**, 504–507.
40. B. Jiang and H. Guo, Control of mode/bond selectivity and product energy disposal by the transition state: X + H2O (X = H, F, O (3P), and Cl) reactions, *J. Am. Chem. Soc.*, 2013, **135**, 15251–15256.
41. H. Guo and B. Jiang, The sudden vector projection model for reactivity: mode specificity and bond selectivity made simple, *Acc. Chem. Res.*, 2014, **47**, 3679–3685.
42. M. N. Siamwiza, R. C. Lord, M. C. Chen, T. Takamatsu, I. Harada, H. Matsuura and T. Shimanouchi, Interpretation of the doublet at 850 and 830 cm-1 in the Raman spectra of tyrosyl residues in proteins and certain model compounds, *Biochemistry*, 1975, **14**, 4870–4876.
43. J. L. McHale, Fermi resonance of tyrosine and related compounds. Analysis of the Raman doublet, *J. Raman Spectrosc.*, 1982, **13**, 21–24.
44. E. G. Buchanan, J. C. Dean, T. S. Zwier and E. L. Sibert III, Towards a first-principles model of Fermi resonance in the alkyl CH stretch region: Application to 1, 2-diphenylethane and 2, 2, 2-paracyclophane, *J. Chem. Phys.*, 2013, **138**, 064308.
45. E. L. Sibert III, D. P. Tabor, N. M. Kidwell, J. C. Dean and T. S. Zwier, Fermi resonance effects in the vibrational spectroscopy of methyl and methoxy groups, *J. Phys. Chem. A*, 2014, **118**, 11272–11281.
46. E. L. Sibert III, N. M. Kidwell and T. S. Zwier, A first-principles model of Fermi resonance in the alkyl CH stretch region: Application to hydronaphthalenes, indanes, and cyclohexane, *J. Phys. Chem. B*, 2014, **118**, 8236–8245.
47. M. Reppert and A. Tokmakoff, Computational amide I 2D IR spectroscopy as a probe of protein structure and dynamics, *Annu. Rev. Phys. Chem.*, 2016, **67**, 359–386.
48. J. K. Watson, Simplification of the molecular vibration-rotation Hamiltonian, *Mol. Phys.*, 1968, **15**, 479–490.
49. J. Tennyson and B. T. Sutcliffe, The abinitio calculation of the vibrational-rotational spectrum of triatomic systems in the close-coupling approach, with KCN and H2Ne as examples, *J. Chem. Phys.*, 1982, **77**, 4061–4072.
50. F. Gatti, C. Iung, M. Menou, Y. Justum, A. Nauts and X. Chapuisat, Vector parametrization of the n-atom problem in quantum mechanics. I. Jacobi vectors, *J. Chem. Phys.*, 1998, **108**, 8804–8820.
51. F. Gatti, C. Iung, M. Menou and X. Chapuisat, Vector parametrization of the N-atom problem in quantum mechanics. II. Coupled-angular-momentum spectral representations for four-atom systems, *J. Chem. Phys.*, 1998, **108**, 8821–8829.
52. M. Mladenović, Rovibrational Hamiltonians for general polyatomic molecules in spherical polar parametrization. II. Nonorthogonal descriptions of internal molecular geometry, *J. Chem. Phys.*, 2000, **112**, 1082–1095.

53. T. Carrington Jr, The advantage of writing kinetic energy operators in polyspherical curvilinear coordinates in terms of zi = cos φi, *J. Chem. Phys.*, 2000, **112**, 4413–4414.
54. F. Gatti and C. Iung, Exact and constrained kinetic energy operators for polyatomic molecules: The polyspherical approach, *Phys. Rep.*, 2009, **484**, 1–69.
55. J. Sielk, H. F. von Horsten, B. Hartke and G. Rauhut, Towards automated multi-dimensional quantum dynamical investigations of double-minimum potentials: Principles and example applications, *Chem. Phys.*, 2011, **380**, 1–8.
56. J. M. Bowman, Self-consistent field energies and wavefunctions for coupled oscillators, *J. Chem. Phys.*, 1978, **68**, 608.
57. J. M. Bowman, K. Christoffel and F. Tobin, Application of SCF-SI theory to vibrational motion in polyatomic molecules, *J. Phys. Chem.*, 1979, **83**, 905–912.
58. J. M. Bowman, The self-consistent-field approach to polyatomic vibrations, *Acc. Chem. Res.*, 1986, **19**, 202–208.
59. J. M. Bowman, T. Carrington and H.-D. Meyer, Variational quantum approaches for computing vibrational energies of polyatomic molecules, *Mol. Phys.*, 2008, **106**, 2145–2182.
60. K. Yagi, S. Hirata and K. Hirao, Vibrational quasi-degenerate perturbation theory: applications to fermi resonance in CO_2, H_2CO, and C_6H_6, *Phys. Chem. Chem. Phys.*, 2008, **10**, 1781–1788.
61. B. Thomsen, K. Yagi and O. Christiansen, Optimized coordinates in vibrational coupled cluster calculations, *J. Chem. Phys.*, 2014, **140**, 154102.
62. A. Roitberg, R. B. Gerber, R. Elber and M. A. Ratner, Anharmonic wave functions of proteins: quantum self-consistent field calculations of BPTI, *Science*, 1995, **268**, 1319–1322.
63. A. E. Roitberg, R. B. Gerber and M. A. Ratner, A vibrational eigenfunction of a protein: Anharmonic coupled-mode ground and fundamental excited states of bpti, *J. Phys. Chem. B*, 1997, **101**, 1700–1706.
64. K. Yagi, S. Hirata and K. Hirao, Vibrational quasi-degenerate perturbation theory: applications to fermi resonance in CO_2, H_2CO, and C_6H_6, *Phys. Chem. Chem. Phys.*, 2008, **10**, 1781–1788.
65. O. Christiansen, Møller-Plesset perturbation theory for vibrational wave functions, *J. Chem. Phys.*, 2003, **119**, 5773–5781.
66. I. Hamilton and J. Light, On distributed Gaussian bases for simple model multidimensional vibrational problems, *J. Chem. Phys.*, 1986, **84**, 306–317.
67. M. Neff and G. Rauhut, Toward large scale vibrational configuration interaction calculations, *J. Chem. Phys.*, 2009, **131**, 124129.
68. A. Szabo and N. S. Ostlund, *Modern Quantum Chemistry: Introduction to Advanced Electronic Structure Theory*, Courier Corporation, 2012.

69. J. M. Bowman, K. Christoffel and F. Tobin, Application of SCF-SI theory to vibrational motion in polyatomic molecules, *J. Phys. Chem.*, 1979, **83**, 905–912.
70. M. A. Ratner, V. Buch and R. Gerber, The semiclassical self-consistent-field (SC-SCF) approach to energy levels of coupled vibrational modes. II. The semiclassical state-interaction procedure, *Chem. Phys.*, 1980, **53**, 345–356.
71. T. C. Thompson and D. G. Truhlar, SCF CI calculations for vibrational eigenvalues and wavefunctions of systems exhibiting fermi resonance, *Chem. Phys. Lett.*, 1980, **75**, 87–90.
72. K. M. Christoffel and J. M. Bowman, Investigations of self-consistent field, scf ci and virtual stateconfiguration interaction vibrational energies for a model three-mode system, *Chem. Phys. Lett.*, 1982, **85**, 220–224.
73. S. Carter, J. M. Bowman and N. C. Handy, Extensions and tests of 'multimode': a code to obtain accurate vibration/rotation energies of many-mode molecules, *Theor. Chem. Acc.*, 1998, **100**, 191–198.
74. A. Samsonyuk and C. Scheurer, Configuration space partitioning and matrix buildup scaling for the vibrational configuration interaction method, *J. Comput. Chem.*, 2013, **34**, 27–37.
75. C. Qu and J. M. Bowman, Quantum approaches to vibrational dynamics and spectroscopy: is ease of interpretation sacrificed as rigor increases?, *Phys. Chem. Chem. Phys.*, 2019, **21**, 3397–3413.
76. D. A. Matthews, J. Vázquez and J. F. Stanton, Calculated stretching overtone levels and Darling–Dennison resonances in water: a triumph of simple theoretical approaches, *Mol. Phys.*, 2007, **105**, 2659–2666.
77. R. Burcl, S. Carter and N. C. Handy, On the representation of potential energy surfaces of polyatomic molecules in normal coordinates: II. Parameterisation of the force field, *Chem. Phys. Lett.*, 2003, **373**, 357–365.
78. R. Burcl, N. C. Handy and S. Carter, Vibrational spectra of furan, pyrrole, and thiophene from a density functional theory anharmonic force field, *Spectrochim. Acta, Part A*, 2003, **59**, 1881–1893.
79. V. Barone, Anharmonic vibrational properties by a fully automated second-order perturbative approach, *J. Chem. Phys.*, 2005, **122**, 014108.
80. E. L. Sibert III, Theoretical studies of vibrationally excited polyatomic molecules using canonical Van Vleck perturbation theory, *J. Chem. Phys.*, 1988, **88**, 4378–4390.
81. A. Messiah, *Quantum Mechanics: Two Volumes Bound as One*, 2014.
82. L. S. Norris, M. A. Ratner, A. E. Roitberg and R. Gerber, Moller–Plesset perturbation theory applied to vibrational problems, *J. Chem. Phys.*, 1996, **105**, 11261–11267.
83. J. O. Jung and R. B. Gerber, Vibrational wave functions and spectroscopy of (H2O) n, n= 2, 3, 4, 5: Vibrational self-consistent field with correlation corrections, *J. Chem. Phys.*, 1996, **105**, 10332–10348.

84. D. Z. Goodson and A. V. Sergeev, On the use of algebraic approximants to sum divergent series for Fermi resonances in vibrational spectroscopy, *J. Chem. Phys.*, 1999, **110**, 8205–8206.
85. S. Hirata, X. He, M. R. Hermes and S. Y. Willow, Second-Order Many-Body Perturbation Theory: An Eternal Frontier, *J. Phys. Chem. A*, 2013, **118**, 655–672.
86. J. Čížek and V. Špirko, Bludský, O. On the use of divergent series in vibrational spectroscopy. Two-and three-dimensional oscillators, *J. Chem. Phys.*, 1993, **99**, 7331–7336.
87. J. A. Faucheaux and S. Hirata, Higher-order diagrammatic vibrational coupled-cluster theory, *J. Chem. Phys.*, 2015, **143**, 134105.
88. M. R. Hermes and S. Hirata, Diagrammatic theories of anharmonic molecular vibrations, *Int. Rev. Phys. Chem.*, 2015, **34**, 71–97.
89. P. Knowles, K. Somasundram, N. Handy and K. Hirao, The calculation of higher-order energies in the many-body perturbation theory series, *Chem. Phys. Lett.*, 1985, **113**, 8–12.
90. O. Christiansen, Vibrational structure theory: new vibrational wave function methods for calculation of anharmonic vibrational energies and vibrational contributions to molecular properties, *Phys. Chem. Chem. Phys.*, 2007, **9**, 2942–2953.
91. O. Christiansen, A second quantization formulation of multimode dynamics, *J. Chem. Phys.*, 2004, **120**, 2140–2148.
92. O. Christiansen, *Frontiers of Quantum Chemistry*, Springer Singapore, Singapore, 2018, pp. 199–221.
93. C. Puzzarini and V. Barone, Diving for Accurate Structures in the Ocean of Molecular Systems with the Help of Spectroscopy and Quantum Chemistry, *Acc. Chem. Res.*, 2018, **51**, 548–556.
94. A. Baiardi, J. Bloino and V. Barone, Accurate Simulation of resonance-Raman spectra of flexible molecules: An internal coordinates approach, *J. Chem. Theory Comput.*, 2015, **11**, 3267–3280.
95. R. C. Fortenberry, Interstellar Anions: The Role of Quantum Chemistry, *J. Phys. Chem. A*, 2015, **119**, 9941–9953.
96. R. C. Fortenberry, Quantum astrochemical spectroscopy, *Int. J. Quantum Chem.*, 2017, **117**, 81–91.
97. J. Bloino, A. Baiardi and M. Biczysko, Aiming at an accurate prediction of vibrational and electronic spectra for medium-to-large molecules: An overview, *Int. J. Quantum Chem.*, 2017, **116**, 1543–1574.
98. T. K. Roy and R. B. Gerber, Vibrational self-consistent field calculations for spectroscopy of biological molecules: new algorithmic developments and applications, *Phys. Chem. Chem. Phys.*, 2013, **15**, 9468–9492.
99. K. Yagi, T. Taketsugu, K. Hirao and M. S. Gordon, Direct vibrational self-consistent field method: Applications to H 2 O and H 2 CO, *J. Chem. Phys.*, 2000, **113**, 1005–1017.
100. S. Carter, S. J. Culik and J. M. Bowman, Vibrational self-consistent field method for many-mode systems: A new approach and application to

the vibrations of CO adsorbed on Cu (100), *J. Chem. Phys.*, 1997, **107**, 10458–10469.
101. G. C. Schatz, The analytical representation of electronic potential-energy surfaces, *Rev. Mod. Phys.*, 1989, **61**, 669.
102. H. Rabitz and Ö. F. Aliş, General foundations of high-dimensional model representations, *J. Math. Chem.*, 1999, **25**, 197–233.
103. H.-D. Meyer, Studying molecular quantum dynamics with the multi-configuration time-dependent Hartree method. *Wiley Interdisciplinary Reviews: Computational Molecular, Science*, 2012, **2**, 351–374.
104. M. Griebel, *Sparse Grids and Related Approximation Schemes for Higher Dimensional Problems*, Citeseer, 2005.
105. D. Strobusch and C. Scheurer, Adaptive sparse grid expansions of the vibrational Hamiltonian, *J. Chem. Phys.*, 2014, **140**, 074111.
106. D. Strobusch, M. Nest and C. Scheurer, The adaptive hierarchical expansion of the kinetic energy operator, *J. Comput. Chem.*, 2013, **34**, 1210–1217.
107. S. Döpking, C. P. Plaisance, D. Strobusch, K. Reuter, C. Scheurer and S. Matera, Addressing global uncertainty and sensitivity in first-principles based microkinetic models by an adaptive sparse grid approach, *J. Chem. Phys.*, 2018, **148**, 034102.
108. B. Jiang, J. Li and H. Guo, Potential energy surfaces from high fidelity fitting of ab initio points: the permutation invariant polynomial-neural network approach, *Int. Rev. Phys. Chem.*, 2016, **35**, 479–506.
109. M. Majumder, S. A. Ndengue and R. Dawes, Automated construction of potential energy surfaces, *Mol. Phys.*, 2016, **114**, 1–18.
110. T.-S. Ho and H. Rabitz, Reproducing kernel Hilbert space interpolation methods as a paradigm of high dimensional model representations: Application to multidimensional potential energy surface construction, *J. Chem. Phys.*, 2003, **119**, 6433–6442.
111. R. Ramakrishnan and G. Rauhut, Semi-quartic force fields retrieved from multi-mode expansions: Accuracy, scaling behavior, and approximations, *J. Chem. Phys.*, 2015, **142**, 154118.
112. J. Ischtwan and M. A. Collins, Molecular potential energy surfaces by interpolation, *J. Chem. Phys.*, 1994, **100**, 8080–8088.
113. M. J. Jordan, K. C. Thompson and M. A. Collins, The utility of higher order derivatives in constructing molecular potential energy surfaces by interpolation, *J. Chem. Phys.*, 1995, **103**, 9669–9675.
114. K. C. Thompson, M. J. Jordan and M. A. Collins, Polyatomic molecular potential energy surfaces by interpolation in local internal coordinates, *J. Chem. Phys.*, 1998, **108**, 8302–8316.
115. K. Yagi, T. Taketsugu and K. Hirao, A new analytic form of ab initio potential energy function: An application to H_2O, *J. Chem. Phys.*, 2002, **116**, 3963–3966.
116. S. Y. Lin, P. Zhang and J. Z. Zhang, Hybrid many-body-expansion/Shepard-interpolation method for constructing ab initio potential

energy surfaces for quantum dynamics calculations, *Chem. Phys. Lett.*, 2013, **556**, 393–397.
117. T. J. Frankcombe, M. A. Collins and D. H. Zhang, Modified Shepard interpolation of gas-surface potential energy surfaces with strict plane group symmetry and translational periodicity, *J. Chem. Phys.*, 2012, **137**, 144701.
118. C. Crespos, M. Collins, E. Pijper and G. Kroes, Multi-dimensional potential energy surface determination by modified Shepard interpolation for a molecule–surface reaction: H2 + Pt (1 1 1), *Chem. Phys. Lett.*, 2003, **376**, 566–575.
119. O. Christiansen, Selected new developments in vibrational structure theory: potential construction and vibrational wave function calculations, *Phys. Chem. Chem. Phys.*, 2012, **14**, 6672–6687.
120. R. Dawes, D. L. Thompson, A. F. Wagner and M. Minkoff, Interpolating moving least-squares methods for fitting potential energy surfaces: A strategy for efficient automatic data point placement in high dimensions, *J. Chem. Phys.*, 2008, **128**, 084107.
121. C. E. Dateo, T. J. Lee and D. W. Schwenke, An accurate quartic force field and vibrational frequencies for HNO and DNO, *J. Chem. Phys.*, 1994, **101**, 5853–5859.
122. R. C. Fortenberry, X. Huang, T. D. Crawford and T. J. Lee, The 1 3A′ HCN and 1 3A′ HCO + Vibrational Frequencies and Spectroscopic Constants from Quartic Force Fields, *J. Phys. Chem. A*, 2012, **117**, 9324–9330.
123. R. C. Fortenberry, X. Huang, J. S. Francisco, T. D. Crawford and T. J. Lee, The trans-HOCO radical: Quartic force fields, vibrational frequencies, and spectroscopic constants, *J. Chem. Phys.*, 2011, **135**, 134301.
124. R. C. Fortenberry, X. Huang, T. D. Crawford and T. J. Lee, Quartic force field rovibrational analysis of protonated acetylene, C2H3 +, and its isotopologues, *J. Phys. Chem. A*, 2014, **118**, 7034–7043.
125. W. Meyer, P. Botschwina and P. Burton, A binitio calculation of near-equilibrium potential and multipole moment surfaces and vibrational frequencies of H + 3 and its isotopomers, *J. Chem. Phys.*, 1986, **84**, 891–900.
126. S. Carter and N. Handy, A theoretical determination of the rovibrational energy levels of the water molecule, *J. Chem. Phys.*, 1987, **87**, 4294–4301.
127. S. Carter and N. C. Handy, On the representation of potential energy surfaces of polyatomic molecules in normal coordinates, *Chem. Phys. Lett.*, 2002, **352**, 1–7.
128. G. D. Carney, L. L. Sprandel and C. W. Kern, Variational Approaches to Vibration-Rotation Spectroscopy for Polyatomic Molecules, *Adv. Chem. Phys.*, 1978, **37**, 305–379.
129. M. Cohen, S. Greita and R. McEarchran, Approximate and exact quantum mechanical energies and eigenfunctions for a system of coupled oscillators, *Chem. Phys. Lett.*, 1979, **60**, 445–450.

130. R. Gerber and M. Ratner, A semiclassical self-consistent field (SC SCF) approximation for eigenvalues of coupled-vibration systems, *Chem. Phys. Lett.*, 1979, **68**, 195–198.
131. B. C. Garrett and D. G. Truhlar, Semiclassical self-consistent-field method for reactive resonances, *Chem. Phys. Lett.*, 1982, **92**, 64–70.
132. A. Smith, W.-K. Liu and D. Noid, Vibrational levels of triatomic molecules-semiclassical self-consistent-field and classical spectral calculations, *Chem. Phys.*, 1984, **89**, 345–351.
133. G. Rauhut, Efficient calculation of potential energy surfaces for the generation of vibrational wave functions, *J. Chem. Phys.*, 2004, **121**, 9313–9322.
134. T. K. Roy and M. D. Prasad, A thermal self-consistent field theory for the calculation of molecular vibrational partition functions, *J. Chem. Phys.*, 2009, **131**, 114102.
135. T. Horn, R. Gerber and M. A. Ratner, Vibrational states of very floppy clusters: Approximate separability and the choice of good curvilinear coordinates for XeHe2, I2He, *J. Chem. Phys.*, 1989, **91**, 1813–1823.
136. L. L. Gibson, R. Roth, M. A. Ratner and R. Gerber, The semiclassical self-consistent field method for polyatomic vibrations: Use of hyperspherical coordinates for H_2O and CO_2, *J. Chem. Phys.*, 1986, **85**, 3425–3431.
137. T. Horn, R. Gerber, J. Valentini and M. A. Ratner, Vibrational states and structure of Ar3: The role of three-body forces, *J. Chem. Phys.*, 1991, **94**, 6728–6736.
138. Z. Bacic, R. Gerber and M. Ratner, Vibrational levels and tunneling dynamics by the optimal coordinates, self-consistent field method: a study of hydrocyanic acid. dblarw. hydroisocyanic acid, *J. Phys. Chem.*, 1986, **90**, 3606–3612.
139. S. Carter, N. C. Handy and J. M. Bowman, High torsional vibrational energies of H_2O_2 and CH_3OH studied by MULTIMODE with a large amplitude motion coupled to two effective contraction schemes, *Mol. Phys.*, 2009, **107**, 727–737.
140. J. M. Bowman, X. Huang, N. C. Handy and S. Carter, Vibrational Levels of Methanol Calculated by the Reaction Path Version of MULTIMODE, Using an ab initio, Full-Dimensional Potential, *J. Phys. Chem. A*, 2007, **111**, 7317–7321.
141. Y. Wang and J. M. Bowman, Towards an ab initio flexible potential for water, and post-harmonic quantum vibrational analysis of water clusters, *Chem. Phys. Lett.*, 2010, **491**, 1–10.
142. Y. Wang and J. M. Bowman, Ab initio potential and dipole moment surfaces for water. II. Local-monomer calculations of the infrared spectra of water clusters, *J. Chem. Phys.*, 2011, **134**, 154510.
143. Y. Wang and J. M. Bowman, Coupled-monomers in molecular assemblies: Theory and application to the water tetramer, pentamer, and ring hexamer, *J. Chem. Phys.*, 2012, **136**, 144113.

144. X. Cheng and R. P. Steele, Efficient anharmonic vibrational spectroscopy for large molecules using local-mode coordinates, *J. Chem. Phys.*, 2014, **141**, 104105.
145. T. C. Thompson and D. G. Truhlar, Optimization of vibrational coordinates, with an application to the water molecule, *J. Chem. Phys.*, 1982, **77**, 3031–3035.
146. K. Yagi, M. Keçeli and S. Hirata, Optimized coordinates for anharmonic vibrational structure theories, *J. Chem. Phys.*, 2012, **137**, 204118.
147. B. Podolsky, Quantum-mechanically correct form of Hamiltonian function for conservative systems, *Phys. Rev.*, 1928, **32**, 812.
148. R. Meyer and H. H. Günthard, General Internal Motion of Molecules, Classical and Quantum-Mechanical Hamiltonian, *J. Chem. Phys.*, 1968, **49**, 1510–1520.
149. H. M. Pickett, Vibration--Rotation Interactions and the Choice of Rotating Axes for Polyatomic Molecules, *J. Chem. Phys.*, 1972, **56**, 1715–1723.
150. A. Nauts and X. Chapuisat, Momentum, quasi-momentum and hamiltonian operators in terms of arbitrary curvilinear coordinates, with special emphasis on molecular hamiltonians, *Mol. Phys.*, 1985, **55**, 1287–1318.
151. X. Chapuisat, A. Nauts and J.-P. Brunet, Exact quantum molecular hamiltonians: Part I. Application to the dynamics of three particles, *Mol. Phys.*, 1991, **72**, 1–31.
152. J. Pesonen, Vibration–rotation kinetic energy operators: A geometric algebra approach, *J. Chem. Phys.*, 2001, **114**, 10598–10607.
153. J. Laane, M. A. Harthcock, P. Killough, L. Bauman and J. Cooke, Vector representation of large-amplitude vibrations for the determination of kinetic energy functions, *J. Mol. Spectrosc.*, 1982, **91**, 286–299.
154. Y. Scribano, D. M. Lauvergnat and D. M. Benoit, Fast vibrational configuration interaction using generalized curvilinear coordinates and self-consistent basis, *J. Chem. Phys.*, 2010, **133**, 094103.
155. D. Schäpers, B. Zhao and U. Manthe, Coordinate systems and kinetic energy operators for multi-configurational time-dependent Hartree calculations studying reactions of methane, *Chem. Phys.*, 2018, **509**, 37–44.
156. A. Nauts and D. Lauvergnat, Numerical on-the-fly implementation of the action of the kinetic energy operator on a vibrational wave function: application to methanol, *Mol. Phys.*, 2018, 1–9.
157. E. Castro, G. Avila, S. Manzhos, J. Agarwal, H. F. Schaefer and T. Carrington Jr, Applying a Smolyak collocation method to Cl2CO, *Mol. Phys.*, 2017, **115**, 1775–1785.
158. A. Perveaux, M. Lorphelin, B. Lasorne and D. Lauvergnat, Fast and slow excited-state intramolecular proton transfer in 3-hydroxychromone: a two-state story?, *Phys. Chem. Chem. Phys.*, 2017, **19**, 6579–6593.
159. P. J. Castro, A. Perveaux, D. Lauvergnat, M. Reguero and B. Lasorne, Ultrafast internal conversion in 4-aminobenzonitrile occurs sequentially along the seam, *Chem. Phys.*, 2018, **509**, 30–36.

160. F. Gatti, B. Lasorne, H.-D. Meyer and A. Nauts, *The Kinetic Energy Operator in Curvilinear Coordinates*, 2017, pp. 127–166.
161. G. M. Chaban, J. O. Jung and R. B. Gerber, Ab initio calculation of anharmonic vibrational states of polyatomic systems: Electronic structure combined with vibrational self-consistent field, *J. Chem. Phys.*, 1999, **111**, 1823–1829.
162. G. M. Chaban, J. O. Jung and R. B. Gerber, Anharmonic Vibrational Spectroscopy of Hydrogen-Bonded Systems Directly Computed from ab Initio Potential Surfaces:(H2O) n, n= 2, 3; Cl-(H2O) n, n= 1, 2; H+ (H2O) n, n= 1, 2; H2O- CH3OH, *J. Phys. Chem. A*, 2000, **104**, 2772–2779.
163. L. Pele, B. Brauer and R. B. Gerber, Acceleration of correlation-corrected vibrational self-consistent field calculation times for large polyatomic molecules, *Theor. Chem. Acc.*, 2007, **117**, 69–72.
164. L. Pele and R. B. Gerber, On the number of significant mode-mode anharmonic couplings in vibrational calculations: Correlation-corrected vibrational self-consistent field treatment of di-, tri-, and tetrapeptides, *J. Chem. Phys.*, 2008, **128**, 04B624.
165. N. Matsunaga, G. M. Chaban and R. B. Gerber, Degenerate perturbation theory corrections for the vibrational self-consistent field approximation: Method and applications, *J. Chem. Phys.*, 2002, **117**, 3541–3547.
166. K. Yagi and H. Otaki, Vibrational quasi-degenerate perturbation theory with optimized coordinates: Applications to ethylene and trans-1, 3-butadiene, *J. Chem. Phys.*, 2014, **140**, 084113.
167. K. Yagi and B. Thomsen, Infrared Spectra of Protonated Water Clusters, H+ (H2O) 4, in Eigen and Zundel Forms Studied by Vibrational Quasi-Degenerate Perturbation Theory, *J. Phys. Chem. A*, 2017, **121**, 2386–2398.
168. P. Daněček and P. Bouř, Comparison of the numerical stability of methods for anharmonic calculations of vibrational molecular energies, *J. Comput. Chem.*, 2007, **28**, 1617–1624.
169. E. Matito, J. M. Barroso, E. Besalú, O. Christiansen and J. M. Luis, The vibrational auto-adjusting perturbation theory, *Theor. Chem. Acc.*, 2009, **123**, 41–49.
170. M. Piccardo, J. Bloino and V. Barone, Generalized vibrational perturbation theory for rotovibrational energies of linear, symmetric and asymmetric tops: Theory, approximations, and automated approaches to deal with medium-to-large molecular systems, *Int. J. Quantum Chem.*, 2015, **115**, 948–982.
171. N. C. Handy and S. Carter, Large vibrational variational calculations using 'multimode'and an iterative diagonalization technique, *Mol. Phys.*, 2004, **102**, 2201–2205.
172. Y. Scribano and D. M. Benoit, Iterative active-space selection for vibrational configuration interaction calculations using a reduced-coupling VSCF basis, *Chem. Phys. Lett.*, 2008, **458**, 384–387.

173. M. Odunlami, V. Le Bris, D. Bégué, I. Baraille and O. Coulaud, A-VCI: A flexible method to efficiently compute vibrational spectra, *J. Chem. Phys.*, 2017, **146**, 214108.
174. R. Garnier, M. Odunlami, V. Le Bris, D. Bégué, I. Baraille and O. Coulaud, Adaptive vibrational configuration interaction (A-VCI): a posteriori error estimation to efficiently compute anharmonic IR spectra, *J. Chem. Phys.*, 2016, **144**, 204123.
175. R. Garnier, Dual vibration configuration interaction (DVCI). An efficient factorization of molecular Hamiltonian for high performance infrared spectrum computation, *Comput. Phys. Commun.*, 2019, **234**, 263–277.
176. M. Sibaev and D. L. Crittenden, Balancing accuracy and efficiency in selecting vibrational configuration interaction basis states using vibrational perturbation theory, *J. Chem. Phys.*, 2016, **145**, 064106.
177. P. Carbonnière, A. Dargelos and C. Pouchan, The VCI-P code: an iterative variation–perturbation scheme for efficient computations of anharmonic vibrational levels and IR intensities of polyatomic molecules, *Theor. Chem. Acc.*, 2010, **125**, 543–554.
178. G. Rauhut, Configuration selection as a route towards efficient vibrational configuration interaction calculations, *J. Chem. Phys.*, 2007, **127**, 184109.
179. C. König and O. Christiansen, Automatic determination of important mode–mode correlations in many-mode vibrational wave functions, *J. Chem. Phys.*, 2015, **142**, 144115.
180. Y. Scribano and D. M. Benoit, Iterative active-space selection for vibrational configuration interaction calculations using a reduced-coupling VSCF basis, *Chem. Phys. Lett.*, 2008, **458**, 384–387.
181. P. Seidler and O. Christiansen, Automatic derivation and evaluation of vibrational coupled cluster theory equations, *J. Chem. Phys.*, 2009, **131**, 234109.
182. P. Seidler, E. Matito and O. Christiansen, Vibrational coupled cluster theory with full two-mode and approximate three-mode couplings: The VCC model, *J. Chem. Phys.*, 2009, **131**, 034115.
183. A. Zoccante, P. Seidler, M. B. Hansen and O. Christiansen, Approximate inclusion of four-mode couplings in vibrational coupled-cluster theory, *J. Chem. Phys.*, 2012, **136**, 204118.
184. P. Seidler and O. Christiansen, Vibrational excitation energies from vibrational coupled cluster response theory, *J. Chem. Phys.*, 2007, **126**, 204101.
185. P. Seidler, M. Sparta and O. Christiansen, Vibrational coupled cluster response theory: A general implementation, *J. Chem. Phys.*, 2011, **134**, 054119.
186. B. Thomsen, M. B. Hansen, P. Seidler and O. Christiansen, Vibrational absorption spectra from vibrational coupled cluster damped linear response functions calculated using an asymmetric Lanczos algorithm, *J. Chem. Phys.*, 2012, **136**, 124101.

187. C. Puzzarini, A. Baiardi, J. Bloino, V. Barone, T. E. Murphy, H. D. Drew and A. Ali, Spectroscopic Characterization of Key Aromatic and Heterocyclic Molecules: A Route toward the Origin of Life, *Astron. J.*, 2017, **154**, 82.
188. A. Ali, E. Sittler, D. Chornay, B. Rowe and C. Puzzarini, Cyclopropenyl cation – the simplest Huckel's aromatic molecule – and its cyclic methyl derivatives in Titan's upper atmosphere. *Planetary and Space, Science*, 2013, **87**, 96–105.
189. A. Ali, E. Sittler, D. Chornay, B. Rowe and C. Puzzarini, Organic chemistry in Titan's upper atmosphere and its astrobiological consequences: I. Views towards Cassini plasma spectrometer (CAPS) and ion neutral mass spectrometer (INMS) experiments in space. *Planetary and Space, Science*, 2015, **109–110**, 46–63.
190. C. Puzzarini, A. Ali, M. Biczysko and V. Barone, Accurate spectroscopic characterization of protonated oxirane: A potential prebiotic species in Titanś atmosphere, *Astrophys. J.*, 2014, **792**, 118.
191. C. Puzzarini, M. Biczysko, J. Bloino and V. Barone, Accurate spectroscopic characterization of oxirane: A valuable route to its identification in Titan's atmosphere and the assignment of unidentified infrared bands, *Astrophys. J.*, 2014, **785**, 107.
192. C. Puzzarini, Rotational spectroscopy meets theory, *Phys. Chem. Chem. Phys.*, 2013, **15**, 6595–6607.
193. C. Puzzarini, M. L. Senent, R. Domnguez-Gómez, M. Carvajal, M. Hochlaf and M. M. Al-Mogren, Accurate spectroscopic characterization of ethyl mercaptan and dimethyl sulfide isotopologues: A route toward their astrophysical detection, *Astrophys. J.*, 2014, **796**, 50.
194. C. Puzzarini, A. Baiardi, J. Bloino, V. Barone, T. E. Murphy, H. D. Drew and A. Ali, Spectroscopic Characterization of Key Aromatic and Heterocyclic Molecules: A Route toward the Origin of Life, *Astron. J.*, 2017, **154**, 82.
195. A. F. Al-Refaie, R. I. Ovsyannikov, O. L. Polyansky, S. N. Yurchenko and J. Tennyson, A variationally calculated room temperature line-list for H2O2, *J. Mol. Spectrosc.*, 2015, **318**, 84–90.
196. A. F. Al-Refaie, A. Yachmenev, J. Tennyson and S. N. Yurchenko, ExoMol line lists – VIII. A variationally computed line list for hot formaldehyde, *Mon. Not. R. Astron. Soc.*, 2015, **448**, 1704–1714.
197. A. Owens and A. Yachmenev, RichMol: A general variational approach for rovibrational molecular dynamics in external electric fields, *J. Chem. Phys.*, 2018, **148**, 124102.
198. A. Owens, A. Yachmenev, S. N. Yurchenko and J. Küpper, Climbing the Rotational Ladder to Chirality, *Phys. Rev. Lett.*, 2018, **121**, 193201.
199. A. Owens, S. N. Yurchenko, A. Yachmenev, J. Tennyson and W. Thiel, A highly accurate ab initio potential energy surface for methane, *J. Chem. Phys.*, 2016, **145**, 104305.

200. A. Owens, S. N. Yurchenko, A. Yachmenev, J. Tennyson and W. Thiel, A global ab initio dipole moment surface for methyl chloride, *J. Quant. Spectrosc. Radiat. Transfer*, 2016, **184**, 100–110.
201. A. Owens, S. N. Yurchenko, W. Thiel and V. Špirko, Accurate prediction of the ammonia probes of a variable proton-to-electron mass ratio, *Mon. Not. R. Astron. Soc.*, 2015, **450**, 3191–3200.
202. A. Owens, S. N. Yurchenko, A. Yachmenev and W. Thiel, A global potential energy surface and dipole moment surface for silane, *J. Chem. Phys.*, 2015, **143**, 244317.
203. A. Owens, S. N. Yurchenko, A. Yachmenev, J. Tennyson and W. Thiel, Accurate ab initio vibrational energies of methyl chloride, *J. Chem. Phys.*, 2015, **142**, 244306.
204. A. Owens, A. Yachmenev, J. Küpper, S. N. Yurchenko and W. Thiel, The rotation–vibration spectrum of methyl fluoride from first principles, *Phys. Chem. Chem. Phys.*, 2019, **21**, 3496–3505.
205. S. N. Yurchenko, A. Yachmenev and R. I. Ovsyannikov, Symmetry-Adapted Ro-vibrational Basis Functions for Variational Nuclear Motion Calculations: TROVE Approach, *J. Chem. Theory Comput.*, 2017, **13**, 4368–4381.
206. A. Yachmenev and S. N. Yurchenko, Automatic differentiation method for numerical construction of the rotational-vibrational Hamiltonian as a power series in the curvilinear internal coordinates using the Eckart frame, *J. Chem. Phys.*, 2015, **143**, 014105.
207. A. Owens, A. Yachmenev, S. N. Yurchenko and J. Küpper, Climbing the Rotational Ladder to Chirality, *Phys. Rev. Lett.*, 2018, **121**, 193201.
208. M. J. Murray, H. M. Ogden, C. Toro, Q. Liu, D. A. Burns, M. H. Alexander and A. S. Mullin, State-Specific Collision Dynamics of Molecular Super Rotors with Oriented Angular Momentum, *J. Phys. Chem. A*, 2015, **119**, 12471–12479.
209. L. Yuan, C. Toro, M. Bell and A. S. Mullin, Spectroscopy of molecules in very high rotational states using an optical centrifuge, *Faraday Discuss.*, 2011, **150**, 101–111.
210. A. A. Milner, A. Korobenko, K. Rezaiezadeh and V. Milner, From Gyroscopic to Thermal Motion: A Crossover in the Dynamics of Molecular Superrotors, *Phys. Rev. X*, 2015, **5**, 031041.
211. A. A. Milner, A. Korobenko, J. W. Hepburn and V. Milner, Effects of ultrafast molecular rotation on collisional decoherence, *Phys. Rev. Lett.*, 2014, **113**, 043005.
212. H. Li and J. H. Jensen, Partial Hessian vibrational analysis: the localization of the molecular vibrational energy and entropy, *Theor. Chem. Acc.*, 2002, **107**, 211–219.
213. S. Jin and J. D. Head, Theoretical investigation of molecular water adsorption on the Al (111) surface, *Surf. Sci.*, 1994, **318**, 204–216.
214. M. D. Calvin, J. D. Head and S. Jin, Theoretically modelling the water bilayer on the Al (111) surface using cluster calculations, *Surf. Sci.*, 1996, **345**, 161–172.

215. J. D. Head, A vibrational analysis with Fermi resonances for methoxy adsorption on Cu (111) using ab initio cluster calculations, *Int. J. Quantum Chem.*, 2000, **77**, 350–357.
216. A. Ghysels, D. Van Neck, V. Van Speybroeck, T. Verstraelen and M. Waroquier, Vibrational modes in partially optimized molecular systems, *J. Chem. Phys.*, 2007, **126**, 224102.
217. A. Ghysels, V. Van Speybroeck, E. Pauwels, D. Van Neck, B. R. Brooks and M. Waroquier, Mobile Block Hessian approach with adjoined blocks: an efficient approach for the calculation of frequencies in macromolecules, *J. Chem. Theory Comput.*, 2009, **5**, 1203–1215.
218. A. Ghysels, D. Van Neck, B. R. Brooks, V. Van Speybroeck and M. Waroquier, Normal modes for large molecules with arbitrary link constraints in the mobile block Hessian approach, *J. Chem. Phys.*, 2009, **130**, 084107.
219. P. Durand, G. Trinquier and Y.-H. Sanejouand, A new approach for determining low-frequency normal modes in macromolecules, *Biopolym.: Orig. Res. Biomol.*, 1994, **34**, 759–771.
220. F. Tama, F. X. Gadea, O. Marques and Y.-H. Sanejouand, Building-block approach for determining low-frequency normal modes of macromolecules, *Proteins: Struct., Funct., Bioinf.*, 2000, **41**, 1–7.
221. M. C. González Lebrero, L. L. Perissinotti and D. A. Estrin, Solvent effects on peroxynitrite structure and properties from QM/MM simulations, *J. Phys. Chem. A*, 2005, **109**, 9598–9604.
222. M. Martínez, M.-P. Gaigeot, D. Borgis and R. Vuilleumier, Extracting effective normal modes from equilibrium dynamics at finite temperature, *J. Chem. Phys.*, 2006, **125**, 144106.
223. M. Eichinger, P. Tavan, J. Hutter and M. Parrinello, A hybrid method for solutes in complex solvents: Density functional theory combined with empirical force fields, *J. Chem. Phys.*, 1999, **110**, 10452–10467.
224. L. Walewski, P. Bala, M. Elstner, T. Frauenheim and B. Lesyng, Fast QM/MM method and its application to molecular systems, *Chem. Phys. Lett.*, 2004, **397**, 451–458.
225. M. Schwrer, C. Wichmann and P. Tavan, A polarizable QM/MM approach to the molecular dynamics of amide groups solvated in water, *J. Chem. Phys.*, 2016, **144**, 114504.
226. A. O. Tirler and T. S. Hofer, A Comparative study of $[CaEDTA]^{-2}$ and $[MgEDTA]^{-2}$: Structural and dynamical insights from quantum mechanical charge field molecular dynamics, *J. Phys. Chem. B*, 2015, **119**, 8613–8622.
227. K. Welke, H. C. Watanabe, T. Wolter, M. Gaus and M. Elstner, QM/MM simulations of vibrational spectra of bacteriorhodopsin and channelrhodopsin-2, *Phys. Chem. Chem. Phys.*, 2013, **15**, 6651–6659.

228. T. S. Hofer, A. B. Pribil, B. R. Randolf and B. M. Rode, Structure and dynamics of solvated Sn(II) in aqueous solution: An ab initio QM/MM MD approach, *J. Am. Chem. Soc.*, 2005, **127**, 14231–14238.
229. C. Kritayakornupong, K. Plankensteiner and B. M. Rode, Structure and dynamics of the Cr(III) ion in aqueous solution: Ab initio QM/MM molecular dynamics simulation, *J. Comput. Chem.*, 2004, **25**, 1576–1583.
230. C. Kritayakornupong, K. Plankensteiner and B. M. Rode, Structure and dynamics of the Cd^{2+} ion in aqueous solution: Ab initio QM/MM molecular dynamics simulation, *J. Phys. Chem. A*, 2003, **107**, 10330–10334.
231. A. Bochenkova, D. Firsov and A. Nemukhin, Hybrid DIM-based QM/MM approach applied to vibrational spectra and trapping site structures of HArF in solid argon, *Chem. Phys. Lett.*, 2005, **405**, 165–171.
232. R. Car and M. Parrinello, Unified approach for molecular dynamics and density-functional theory, *Phys. Rev. Lett.*, 1985, **55**, 2471–2474.
233. Q. Cui and M. Karplus, Molecular properties from combined QM/MM methods. I. Analytical second derivative and vibrational calculations, *J. Chem. Phys.*, 2000, **112**, 1133–1149.
234. C. Rovira, B. Schulze, M. Eichinger, J. D. Evanseck and M. Parrinello, Influence of the heme pocket conformation on the structure and vibrations of the Fe-CO bond in myoglobin: A QM/MM density functional study, *Biophys. J.*, 2001, **81**, 435–445.
235. M. C. González Lebrero, D. E. Bikiel, M. D. Elola, D. A. Estrin and A. E. Roitberg, Solvent-induced symmetry breaking of nitrate ion in aqueous clusters: A quantum-classical simulation study, *J. Chem. Phys.*, 2002, **117**, 2718–2725.
236. D. E. Bikiel, F. Di Salvo, M. C. González Lebrero, F. Doctorovich and D. A. Estrin, Solvation and structure of $LiAlH_4$ in ethereal solvents, *Inorg. Chem.*, 2005, **44**, 5286–5292.
237. C. M. Guardia, M. C. González Lebrero, S. E. Bari and D. A. Estrin, QM–MM investigation of the reaction products between nitroxyl and O_2 in aqueous solution, *Chem. Phys. Lett.*, 2008, **463**, 112–116.
238. C. Falvo, L. Daniault, T. Vieille, V. Kemlin, J.-C. Lambry, C. Meier, M. H. Vos, A. Bonvalet and M. Joffre, Ultrafast dynamics of carboxy-hemoglobin: Two-dimensional infrared spectroscopy experiments and simulations, *J. Phys. Chem. Lett.*, 2015, **6**, 2216–2222.
239. Y.-A. Yan and O. Kuhn, Geometric correlations and infrared spectrum of adenine-uracil hydrogen bonds in $CDCl_3$ solution, *Phys. Chem. Chem. Phys.*, 2010, **12**, 15695–15703.
240. J. Jeon, S. Yang, J.-H. Choi and M. Cho, Computational vibrational spectroscopy of peptides and proteins in one and two dimensions, *Acc. Chem. Res.*, 2009, **42**, 1280–1289.

241. H. C. Watanabe, M. Banno and M. Sakurai, An adaptive quantum mechanics/molecular mechanics method for the infrared spectrum of water: incorporation of the quantum effect between solute and solvent, *Phys. Chem. Chem. Phys.*, 2016, **18**, 7318–7333.
242. D. Hunt, V. M. Sanchez and D. A. Scherlis, A quantum-mechanics molecular-mechanics scheme for extended systems, *J. Phys.: Condens. Matter*, 2016, **28**, 335201.
243. M. A. Mroginski, F. Mark, W. Thiel and P. Hildebrandt, Quantum mechanics/molecular mechanics calculation of the Raman spectra of the phycocyanobilin chromophore in alpha-C-phycocyanin, *Biophys. J.*, 2007, **93**, 1885–1894.
244. M. A. Mroginski, D. H. Murgida and P. Hildebrandt, The chromophore structural changes during the photocycle of phytochrome: A combined resonance Raman and quantum chemical approach, *Acc. Chem. Res.*, 2007, **40**, 258–266.
245. M. A. Mroginski, D. Von Stetten, F. V. Escobar, H. M. Strauss, S. Kaminski, P. Scheerer, M. Günther, D. H. Murgida, P. Schmieder, C. Bongards and W. Gärtner, *et al.*, Chromophore structure of cyanobacterial phytochrome Cph1 in the Pr state: Reconciling structural and spectroscopic data by QM/MM calculations, *Biophys. J.*, 2009, **96**, 4153–4163.
246. M. A. Mroginski, S. Kaminski and P. Hildebrandt, Raman spectra of the phycoviolobilin cofactor in phycoerythrocyanin calculated by QM/MM methods, *ChemPhysChem*, 2010, **11**, 1265–1274.
247. M. Horch, J. Schoknecht, M. A. Mroginski, O. Lenz, I. Zebger and M. Feld, Resonance raman spectroscopy on [NiFe] Hydrogenase provides structural insights into catalytic intermediates and reactions, *J. Am. Chem. Soc.*, 2014, **136**, 9870–9873.
248. E. Siebert, Y. Rippers, S. Frielingsdorf, J. Fritsch, A. Schmidt, J. Kalms, S. Katz, O. Lenz, P. Scheerer and L. Paasche, *et al.*, Resonance Raman spectroscopic analysis of the active site and the proximal [4Fe-3S] cluster of an O_2-tolerant membrane-bound hydrogenase in the crystalline state, *J. Phys. Chem. B*, 2015, **119**, 13785–13796.
249. M. Fleischmann, P. J. Hendra and A. J. McQuillan, Raman spectra of pyridine adsorbed at a silver electrode, *Chem. Phys. Lett.*, 1974, **26**, 163–166.
250. D. L. Jeanmaire and R. P. Van Duyne, Surface raman spectroelectrochemistry Part I. Heterocyclic, aromatic, and aliphatic amines adsorbed on the anodized silver electrode, *J. Electroanal. Chem.*, 1977, **84**, 1–20.
251. S. Nie and S. R. Emory, Probing single molecules and single nanoparticles by surface enhanced raman scattering, *Science*, 1997, **275**, 1102–1106.
252. K. Kneipp, Y. Wang, H. Kneipp, I. Itzkan, R. Dasari and M. Feld, Population pumping of excited vibrational states by spontaneous

surface-enhanced Raman scattering, *Phys. Rev. Lett.*, 1996, **76**, 2444–2447.

253. K. Kneipp, Y. Wang, H. Kneipp, L. T. Perelman, I. Itzkan, R. R. Dasari and M. S. Feld, Single molecule detection using surface-enhanced Raman scattering (SERS), *Phys. Rev. Lett.*, 1997, **78**, 1667–1670.
254. D. Cialla, A. März, R. Böhme, F. Theil, K. Weber, M. Schmitt and J. Popp, Surface-enhanced Raman spectroscopy (SERS): Progress and trends, *Anal. Bioanal. Chem.*, 2012, **403**, 27–54.
255. A. B. Zrimsek, N. Chiang, M. Mattei, S. Zaleski, M. O. McAnally, C. T. Chapman, A.-I. Henry, G. C. Schatz and R. P. Van Duyne, Single-molecule chemistry with surface- and tip-enhanced Raman spectroscopy, *Chem. Rev.*, 2016, 7583–7613.
256. F. Pozzi, S. Zaleski, F. Casadio, M. Leona, J. Lombardi and R. Van Duyne, *Nanoscience and Cultural Heritage*, Atlantis Press, Paris, 2016, pp. 161–204.
257. A. I. Henry, B. Sharma, M. F. Cardinal, D. Kurouski and R. P. Van Duyne, Surface-enhanced Raman spectroscopy biosensing: In vivo diagnostics and multimodal imaging, *Anal. Chem.*, 2016, **88**, 6638–6647.
258. P. Ball, Martin Fleischmann (1927–2012), *Nature*, 2012, **489**, 34.
259. M. Moskovits, Persistent misconceptions regarding SERS, *Phys. Chem. Chem. Phys.*, 2013, **15**, 5301.
260. M. Moskovits, Surface roughness and the enhanced intensity of Raman scattering by molecules adsorbed on metals, *J. Chem. Phys.*, 1978, **69**, 4159.
261. M. Moskovits, Enhanced Raman scattering by molecules adsorbed on electrodes–a theoretical model, *Solid State Commun.*, 1979, **32**, 59–62.
262. J. Gersten and A. Nitzan, Electromagnetic theory of enhanced Raman scattering by molecules adsorbed on rough surfaces, *J. Chem. Phys.*, 1980, **73**, 3023–3037.
263. P. Aravind and H. Metiu, The enhancement of raman and fluorescent intensity by small surface roughness. changes in dipole emission, *Chem. Phys. Lett.*, 1980, **74**, 301–305.
264. M. Kerker, D.-S. Wang and H. Chew, Surface enhanced Raman scattering (SERS) by molecules adsorbed at spherical particles, *Appl. Opt.*, 1980, **19**, 3373.
265. M. Moskovits, Surface-enhanced spectroscopy, *Rev. Mod. Phys.*, 1985, **57**, 783–826.
266. V. M. Shalaev and M. I. Stockman, Fractals: optical susceptibility and giant raman scattering, *Z. Phys. D Atoms, Mol. Clusters*, 1988, **10**, 71–79.
267. J. Kottmann and O. Martin, Plasmon resonant coupling in metallic nanowires, *Opt. Express*, 2001, **8**, 655–663.

268. K. L. Kelly, E. Coronado, L. L. Zhao and G. C. Schatz, The optical properties of metal nanoparticles: The influence of size, shape, and dielectric environment, *J. Phys. Chem. B*, 2003, **107**, 668–677.
269. E. Prodan, C. J. Radloff, N. J. Halas and P. Norlander, A hybridization model for the plasmon response of complex nanostructures, *Science*, 2003, **302**, 419–422.
270. A. Otto, J. Billmann, J. Eickmans, U. Ertürk and C. Pettenkofer, The "adatom model" of SERS (Surface Enhanced Raman Scattering): The present status, *Surf. Sci.*, 1984, **138**, 319–338.
271. F. J. Adrian, Charge transfer effects in surface-enhanced Raman scattering, *J. Chem. Phys.*, 1982, **77**, 5302.
272. A. M. Michaels, M. Nirmal and L. E. Brus, Surface enhanced Raman spectroscopy of individual rhodamine 6G molecules on large Ag nanocrystals, *J. Am. Chem. Soc.*, 1999, **121**, 9932–9939.
273. S. Corni and J. Tomasi, Enhanced response properties of a chromophore physisorbed on a metal particle, *J. Chem. Phys.*, 2001, **114**, 3739–3751.
274. S. Corni and J. Tomasi, Theoretical evaluation of Raman spectra and enhancement factors for a molecule adsorbed on a complex-shaped metal particle, *Chem. Phys. Lett.*, 2001, **342**, 135–140.
275. S. Corni and J. Tomasi, Surface enhanced Raman scattering from a single molecule adsorbed on a metal particle aggregate: A theoretical study, *J. Chem. Phys.*, 2002, **116**, 1156–1164.
276. D. J. Masiello and G. C. Schatz, On the linear response and scattering of an interacting molecule-metal system, *J. Chem. Phys.*, 2010, **132**, 064102.
277. D. J. Masiello and G. C. Schatz, Many-body theory of surface-enhanced Raman scattering, *Phys. Rev. A: At., Mol., Opt. Phys.*, 2008, **78**, 042505.
278. V. Arcisauskaite, J. Kongsted, T. Hansen and K. V. Mikkelsen, Charge transfer excitation energies in pyridine-silver complexes studied by a QM/MM method, *Chem. Phys. Lett.*, 2009, **470**, 285–288.
279. S. M. Morton and L. Jensen, A discrete interaction model/quantum mechanical method for describing response properties of molecules adsorbed on metal nanoparticles, *J. Chem. Phys.*, 2010, **133**, 074103.
280. J. L. Payton, S. M. Morton, J. E. Moore and L. Jensen, A discrete interaction model/quantum mechanical method for simulating surface-enhanced Raman spectroscopy, *J. Chem. Phys.*, 2012, **136**, 214103.
281. J. L. Payton, S. M. Morton, J. E. Moore and L. Jensen, A hybrid atomistic electrodynamics-quantum mechanical approach for simulating surface-enhanced Raman scattering, *Acc. Chem. Res.*, 2014, **47**, 88–99.
282. J. E. Moore, S. M. Morton and L. Jensen, Importance of correctly describing charge-transfer excitations for understanding the chemical effect in SERS, *J. Phys. Chem. Lett.*, 2012, **3**, 2470–2475.

283. J. M. Rinaldi, S. M. Morton and L. Jensen, A discrete interaction model/quantum mechanical method for simulating nonlinear optical properties of molecules near metal surfaces, *Mol. Phys.*, 2013, **111**, 1322–1331.
284. D. V. Chulhai and L. Jensen, Simulating surface-enhanced Raman optical activity using atomistic electrodynamics-quantum mechanical models, *J. Phys. Chem. A*, 2014, **118**, 9069–9079.
285. Z. Rinkevicius, X. Li, J. A. R. Sandberg, K. V. Mikkelsen and H. Ågren, A hybrid density functional theory/molecular mechanics approach for linear response properties in heterogeneous environments, *J. Chem. Theory Comput.*, 2014, **10**, 989–1003.
286. Z. Rinkevicius, X. Li, J. A. Sandberg and H. Ågren, Non-linear optical properties of molecules in heterogeneous environments: a quadratic density functional/molecular mechanics response theory, *Phys. Chem. Chem. Phys.*, 2014, **16**, 8981–8989.
287. Z. Rinkevicius, J. A. R. Sandberg, X. Li, M. Linares, P. Norman and H. Ågren, Hybrid complex polarization propagator/molecular mechanics method for heterogeneous environments, *J. Chem. Theory Comput.*, 2016, **12**, 2661–2667.
288. X. Li, V. Carravetta, C. Li, S. Monti, Z. Rinkevicius and H. Ågren, Optical properties of gold nanoclusters functionalized with a small organic compound: Modeling by an integrated quantum-classical approach, *J. Chem. Theory Comput.*, 2016, **12**, 3325–3339.
289. K. S. Yee, Numerical solution of initial boundary value problems involving Maxwell's equations in isotropic media, *IEEE Trans. Antennas Propag.*, 1966, **14**, 302–307.
290. K. Lopata and D. Neuhauser, Multiscale Maxwell-Schrödinger modeling: A split field finite-difference time-domain approach to molecular nanopolaritonics, *J. Chem. Phys.*, 2009, **130**, 104707.
291. Z. Zeng, Y. Liu and J. Wei, Recent advances in surface-enhanced raman spectroscopy (SERS): Finite-difference time-domain (FDTD) method for SERS and sensing applications, *TrAC, Trends Anal. Chem.*, 2016, **75**, 162–173.
292. H. Chen, M. G. Blaber, S. D. Standridge, E. J. Demarco, J. T. Hupp, M. A. Ratner and G. C. Schatz, Computational modeling of plasmon-enhanced light absorption in a multicomponent dye sensitized solar cell, *J. Phys. Chem. C*, 2012, **116**, 10215–10221.
293. J. Mullin and G. C. Schatz, Combined linear response quantum mechanics and classical electrodynamics (QM/ED) method for the calculation of surface-enhanced Raman spectra, *J. Phys. Chem. A*, 2012, **116**, 1931–1938.
294. J. Mullin, N. Valley, M. G. Blaber and G. C. Schatz, Combined quantum mechanics (TDDFT) and classical electrodynamics (Mie Theory) methods for calculating surface enhanced raman and hyper-raman spectra, *J. Phys. Chem. A*, 2012, **116**, 9574–9581.

295. D. V. Chulhai, X. Chen and L. Jensen, Simulating ensemble-averaged surface-enhanced Raman scattering, *J. Phys. Chem. C*, 2016, **120**, 20833–20842.
296. Z. Hu, D. V. Chulhai and L. Jensen, Simulating surface-enhanced hyper-Raman scattering using atomistic electrodynamics-quantum mechanical models, *J. Chem. Theory Comput.*, 2016, **12**, 5968–5978.
297. S. Pipolo and S. Corni, Real-time description of the electronic dynamics for a molecule close to a plasmonic nanoparticle, *J. Phys. Chem. C*, 2016, **120**, 28774–28781.

CHAPTER 4

Isotope Effects as Analytical Probes: Applications of Computational Theory

PIOTR PANETH AND AGNIESZKA DYBALA-DEFRATYKA*

Institute of Applied Radiation Chemistry, Lodz University of Technology, Zeromskiego 116, 90-924 Lodz, Poland
*Email: agnieszka.dybala-defratyka@p.lodz.pl

4.1 Introduction

4.1.1 Isotope Effects: What Are They?

Formally speaking, every change of a property induced by a change of an isotope can be considered as an isotope effect.[1] This general definition makes the term "isotope effect" very broad. For example, the fact that the protium atom is stable while the tritium atom is radioactive can be classified as an isotope effect. Similarly, the shift of the resonance signal for OD *versus* OH in an infrared spectroscopy (IR) spectrum (see other chapters in this volume) is also an isotope effect. In this chapter, we will focus on the thermodynamics of chemical, biochemical, and physicochemical processes. However, even this narrowing of the subject is not sufficient. For example, magnetic nuclei can significantly change the dynamics of these processes[2] Furthermore, in some cases, dynamics may differ owing to the symmetry of the system rather than the nuclear properties. An example of such a system is the decomposition of ozone, in which the change of the isotopic

Theoretical and Computational Chemistry Series No. 20
Computational Techniques for Analytical Chemistry and Bioanalysis
Edited by Philippe B. Wilson and Martin Grootveld

Published by the Royal Society of Chemistry, www.rsc.org

composition of ^{17}O and ^{18}O (*vs.* ^{16}O) is practically identical.[3] These isotope effects are called mass-independent.

The above discussion clarifies the points we will not discuss in this chapter. In this way, we narrowed the area of phenomena connected with a change of an isotope to only these cases in which thermodynamic changes are induced by the mass difference of competing isotopes. These are sometimes referred to as mass-dependent isotope effects. There is a number of classifications of these isotope effects. The major one depends on the process in question. Consider a single-step (elemental) reaction:

$$R \rightleftarrows [TS]^{\neq} \rightarrow P \tag{4.1}$$

in which R denotes reactants, TS a transition state (the properties of which are usually labeled with $\neq$), and P denotes products. Such a reaction can be characterized by a rate constant k. For isotopic variants of the reactants the rate constants will be different. Usually, this is indicated by subscripts L and H corresponding to light and heavy species, respectively. The ratio k_L/k_H then defines the **kinetic isotope effect (KIE)** on a reaction. These isotope effects are of special interest in chemistry and biochemistry as their magnitudes report on the properties of the transition state, a species otherwise not amenable for direct experimental scrutiny. Knowledge of the structure of the transition state is desired, for example in rational drug design of transition state analogs.[4]

Similarly, when the process studied is an equilibrium:

$$R \rightleftarrows P \tag{4.2}$$

the associated isotope effect is the ratio of the isotopic equilibrium constants K_L/K_H called the equilibrium isotope effect (EIE). As the equilibrium can be considered as a net result of two reactions in a forward and reverse direction, as one might expect the equilibrium isotope effects are usually smaller than the kinetic isotope effects (KIEs). Another important difference, that has a significant bearing on the theoretical calculations of isotope effects is the fact that KIEs relate the properties of the reactants to those of the transition state, while EIEs relate the properties of the reactants to the properties of the products.

As illustrated above, we have considered a first-order reaction. The order of the reaction may affect the experimental determination of a KIE, but is not a problem in theoretical predictions of isotope effects. Thus, we note here, for the record only, that in the majority of cases first order treatment is an adequate approximation for reactions of higher order when heavy isotopologue is present in trace amounts (usually a typical case as for example the natural abundance of ^{13}C is only 1.1% and the abundance of heavy isotopes of other elements is usually even smaller). This is because the probability of the reaction occurring between two heavy species is negligibly small.

Another important classification is connected with the position of the isotopic atom in the studied system. Isotope effects, in which bonds are formed or broken to the isotopic atom are called **primary** isotope effects.

Otherwise, they are called **secondary** isotope effects. In this classification, we define one other type; the **solvent** isotope effect in which the isotopic atom belongs to the solvent. Thus far, almost exclusively solvent deuterium isotope effects have been studied (reactions in D_2O *versus* H_2O) while some attempts to capture rate differences in $H_2{}^{18}O$ *versus* $H_2{}^{16}O$ have been made.[5]

In general, primary isotope effects (IEs) are larger than secondary ones. There are, however, exceptions. Consider a simple S_N2 reaction:

$$Cl^- + H_3C\text{–}Br \rightleftarrows [Cl^{\delta-}\cdots(H_3C)\cdots Br^{\delta-}]^{\neq} \rightarrow Cl\text{–}CH_3 + Br^- \tag{4.3}$$

There are two factors controlling KIEs (see Section 4.1.2 for details). One is the ratio of the isotopic imaginary frequencies, called the temperature independent factor. It always favors the light species (owing to the ratio of reduced masses and the equality of the force constants). However, the magnitude of the so-called temperature dependent factor depends on the number of bonds to the isotopic atom and their strength. In the above example the C–Br bond is being broken (therefore the strength of the bonds to bromine diminishes) which also favors the light species (owing to differences in the zero point energy difference between isotopologues). Thus, both factors favor light species for the leaving group atom and the corresponding kinetic isotope effect (KIE) will be significantly larger than unity (we call such isotope effects **normal**). On the contrary, a new bond is formed to the incoming nucleophile and thus the temperature dependent factor will be smaller than unity. As the temperature independent factor is still larger than unity three results are possible, the KIE will be slightly normal, there will be no isotope effect (when the factors cancel each other) or the KIE will be slightly lower than unity (we refer to such cases as **inverse** isotope effects).

Thus far, we have considered a single-step reaction. However, many chemical and all enzymatic reactions are complex. At first glance the computational consequence of this situation is only the fact that one should individually model all elemental reactions involved in the complex kinetic scheme. However, the question arises of how to compare the results of calculations with the experimental data. We will discuss it using an example of the simplest enzymatic reaction in which the reactants, upon binding to an enzyme, form a (Michaelis–Menten) complex with the enzyme, ER, which then reacts to form products and a free enzyme which can reenter the catalytic cycle:

$$R + E \underset{k_{off}}{\overset{k_{on}}{\leftrightarrow}} ER \overset{k_{cat}}{\rightarrow} E + P \tag{4.4}$$

The overall rate constant (also called the "net" or apparent rate constant), k_{app}, is related to three individual rate constants by the so-called commitment to catalysis factor, C, given by the ratio k_{cat}/k_{off}. Within the steady-state approximation, the isotope effect on k_{app} (denoted as AKIE) is given by:

$$\mathrm{AKIE} = \mathrm{KIE_{on}}\frac{\mathrm{KIE_{cat}} / \mathrm{KIE_{off}} + C}{1 + C} \tag{4.5}$$

In extreme cases, when C is close to zero (k_{off} is much larger than k_{cat}) AKIE is equal to KIE_{cat} times the equilibrium isotope effect on binding ($BIE = KIE_{cat}/KIE_{off}$), and when C is large AKIE approaches KIE_{on}. KIE_{cat} is called the intrinsic isotope effect and corresponds to the chemical reaction occurring in the active site of the enzyme. IEs upon binding are frequently neglected (see examples below showing that this is not the general case) and thus the equation for the apparent KIE can be simplified to:

$$\mathrm{AKIE} = \frac{\mathrm{KIE_{cat}} + C}{1 + C} \tag{4.6}$$

There is another important point regarding enzymatic reactions. As they follow saturation kinetics there are two kinetic parameters describing their rate. The first one is the V_{max}/K_M ratio which corresponds to the rate constant at a small concentration of the reactant, and V_{max}, which corresponds to the rate constant with saturating concentrations of the reactant. These two rates report on different events of the catalytic cycle; the first one "sees" events through the first irreversible step while the second one reports on steps up to the rate-limiting step of the whole process. Most experiments are carried out using a competitive method (*i.e.*, both isotopes are present simultaneously in the reaction mixture) which allows for measuring only the isotope effects on V_{max}/K_M.

Finally, before going into the theory behind calculations and examples of practical use of isotope effects in mechanistic studies, a comment is needed on their magnitudes. Generally speaking, isotope effects are small, thus they are frequently reported as a percentage deviation from unity. For example, 5% KIE represents an isotope effect k_L/k_H equal to 1.05. With such small values care has to be taken, not only in experimental determinations (this is usually achieved by using isotope ratio mass spectrometry),[6] but also in calculations (which will be illustrated by examples). On the contrary, hydrogen isotope effects can be quite large as the mass difference between isotopologues matches the atomic mass. Although primary deuterium KIEs are typically in the range 5–8, much larger values, even over 50 have been reported.[7] For processes with isotope effects of this magnitude, a change in the mechanisms has been implied. In fact, the true mechanism remains the same, but experimentally this situation may look like a massive change of the process; imagine a reaction with a deuterium isotope effect of 100 and a half-life of say 7 h. In three half-lives, in which practically all the light reactant has gone (87.5% to be exact) only about 2% of the deuterated species will react, a change that might be undetectable. Thus, from the experimental standpoint, one may draw a conclusion that the protiated species undergoes a reaction, while the deuterated one does not react.

Although the above presentation should suffice for following the remaining material in this chapter, below we introduce a few other terms used in studies of isotope effects that the reader may find helpful while reading references and other publications in the field of isotope effects.

In typical studies, one might like to learn the value of a particular isotope at a particular position. This is quite a simple task theoretically, but frequently is not included in the experimental work, which for example may lead to isotopic substitution of two equivalent positions. For isotope effects which are small compared to unity (practically all isotope effects except primary hydrogen isotope effects), one invokes a rule of geometric mean that describes the additivity of isotope effects. It states that the overall isotope effect on an *n* equivalent position equals *n* times the isotope effect "per position". One should, however, remember that in the case of reactions the equivalency of isotopic positions can be lost. Consider for example dehalogenation of dichloroacetylene $ClC{\equiv}CCl \rightarrow ClC{\equiv}C + Cl$. The isotope effect for the displaced chlorine atom will formally be a primary KIE, while for the other there will be a secondary KIE. An example of chlorine is not coincidental as this is one of the very few elements which has a heavy isotopologue present in a large abundance in nature (^{35}Cl:^{37}Cl is about 3 : 1). As mentioned above, this complicates the experimental analysis. Furthermore, this composition induces two different competitions; intermolecular associated with $^{35}ClC{\equiv}C^{35}Cl$ *versus* $^{37}ClC{\equiv}C^{37}Cl$, and intramolecular as $^{35}ClC{\equiv}C^{37}Cl$ can expel chlorine in two isotopically different ways.

The idea of using more than one isotopic species has been used in many variants in enzymology.[8] They are collectively called isotope effects on isotope effects and they were used to elucidate the value of the intrinsic KIE – the holy grail of such studies. The main idea behind this approach is to change the value of the commitment factor upon a second isotopic substitution. It is not possible to directly elucidate the value of the intrinsic KIE from the apparent one. However, if the second substitution in the molecule, or in the solvent, significantly perturbs *C* (usually such a change would be caused by deuteration) then we can write an additional equation:

$$\mathrm{AKIE_D} = \frac{\mathrm{KIE_{cat}} + \mathrm{KIE_D}C}{1 + \mathrm{KIE_D}C} \tag{4.7}$$

The measurements of AKIE, $\mathrm{AKIE_D}$, and $\mathrm{KIE_D}$ allow for solving simultaneously eqn (4.6) and (4.7) and recalculating the intrinsic isotope effect.

For another reason, more than one isotope is frequently used in environmental studies. The problem here is connected with low availability of the reactant and even lower availability of the product(s). Therefore, recovered reactant samples are subjected to isotopic analysis of as many elements as possible. This is most readily possible nowadays for measurements of the isotopic composition of carbon and nitrogen (and sulfur) with the aid of modern isotope-ratio mass spectrometers. As isotopic information obtained in this way has the same time stamp for each isotope so-called dual slope plots[9] are created in which the slope change indicates a change in the mechanism. To conclude this introduction, let us point out that the more isotope effects are studied for a given process the more reliable the results obtained. This is best exemplified by the use of IEs (referred to as isotopic fractionation) in product authentication. For example, in wine analysis, two position-specific hydrogen compositions (in methyl and methylene groups), carbon and oxygen isotopic compositions are used.[10]

Theoretically predicted KIEs provide detailed information on transition state structures and thus allow either the confirmation or discarding of the mechanistic scenario underlying the studied reaction. For a multistep reaction, each single step isotopic 'response', of each specific position within a molecular species taking part in a reaction, can be revealed allowing the assessment of the contribution of each single atom of a molecule to the overall calculated isotope effect to be determined using this method. This helps to avoid erroneous mechanistic interpretations and fills the gaps formed owing to the inaccessibility of the experimental isotopic data for either certain molecular positions or entire reaction steps. Equilibrium isotope effects, on the other hand, can provide valuable information about the interactions between reactants and the surrounding media (*i.e.* solvent or protein active site) and contribute to knowledge of, for instance, the non-catalytic steps of an overall enzymatic reaction.

4.1.2 Theory of Isotope Effects: How Do We Compute Them?

Early attempts to predict isotope effects were based on using bond orders and tabulated force constants.[11] Later on, this approach was refined by using molecular mechanics methods that allowed the generation of force fields and include bond energies in the model. This resulted in the BEBOVIB software and a method to calculate the isotope effect implemented therein. KIEs were predicted based on the structure of the reactant and the transition state of a reaction. Different protocols were used for proposing/guessing an appropriate transition state model.[12] However, as force constants and bond orders were defined by the user the entire procedure often led to the failure of the method.

Taking into account that isotope effects result from the quantum mechanical nature of nuclear motion, even the early developments in electronic structure computational methods constituted a promising tool for calculating IEs. In calculations of IEs, the most important quantities are the energies of the ground electronic state of the molecular system in the function of nuclear coordinates and the vibrational frequencies of this system. For the configurations that represent either a minimum or a saddle point on the energy surface using the same level of theory (semi-empirical, Hartree–Fock, density functional theory (DFT), post Hartree–Fock) the second derivatives of energy with respect to Cartesian displacements from the molecular stationary points are found. In the case of maxima, more sophisticated search algorithms to locate the transition state are used. The sets of second derivatives are then used to obtain the harmonic vibrational frequencies. The resulting $3N \times 3N$ (in which N is the number of atoms in a molecular system) Cartesian force constant matrix (called Hessian) yields six zero frequencies (five for a linear molecule) and $3N - 6$ ($3N - 5$ for a linear molecule) nonzero frequencies which correspond to the normal vibrations of the molecule.

Subsequently, using the Bigeleisen eqn (4.8), along with the rigid-rotor-harmonic-oscillator approximation and by neglecting tunneling effects, IEs can be calculated from the vibrational frequencies for stationary points (reactants and transition states in the case of KIE, and reactants and products in the case of EIE). Assuming that the Teller–Redlich product rule[13] that separates the translational and rotational motions from vibrational motions within the harmonic approximation applies, one can calculate the KIE from the following equation:

$$\frac{k_L}{k_H} = \frac{\nu_L^{\neq}}{\nu_H^{\neq}} \prod_i^{3n^R-6} \frac{u_{iH}^R \cdot \sinh\left(\frac{u_{iL}^R}{2}\right)}{u_{iL}^R \cdot \sinh\left(\frac{u_{iH}^R}{2}\right)} \prod_i^{3n^{\neq}-7} \frac{u_{iL}^{\neq} \cdot \sinh\left(\frac{u_{iH}^{\neq}}{2}\right)}{u_{iH}^{\neq} \cdot \sinh\left(\frac{u_{iL}^{\neq}}{2}\right)} \tag{4.8}$$

in which ν denotes the isotopic frequencies for the reactants and transition state, $u = h\nu/kT$; $\nu^{\neq}$ represents the imaginary frequencies of crossing the barrier; T is the temperature; and h and k are the Planck's and Boltzmann's constants, respectively. A similar expression, in which there is no ratio of imaginary frequencies, and the second multiplication is carried out over $3n - 6$ vibrational degrees of freedom of the product, holds for the EIEs.

For systems of up to 100 atoms, modern computational chemists are equipped with a whole palette of electronic structure methodologies to optimize the coordinates of a system and calculate its energy and Hessian at the same, selected quantum level of theory. If reactions in the condensed phase are of interest, however, then most likely a different computational treatment will be required as the systems involving an explicitly-defined environment are larger than the same system in the gas phase. Naturally this very often leads to a reduction in the level of theory and finding the best approximation to include the effects of the surrounding media. Another frequently applied approach is treating the environment implicitly as an electrostatic field that has an effect on the behavior of the solute and neglects its possible fluctuations and their influence. This approach substantially lowers the computational cost and it is simple to use. However, despite numerous successful examples of improving or enabling interpretation of observable IEs it sometimes does not allow for reproducing the measured magnitudes of KIEs.[14] There could be several reasons for the failure of such an approach. One of them is that KIEs calculations rely only on individual transition states by excluding the explicit treatment of the first solvation shell and appropriate averaging over multiple configurations. However, it is neither always absolutely necessary nor feasible to use full explicit solvation models whenever environmental effects are of interest.[15] It will of course depend on a system, but there are species (ionic or polar) for which it turns out to be essential as the geometry of the reacting species may largely differ from those obtained using a gas-phase approximation. Although many methods have been developed during the recent decades,[16] the current status of the solvation models shows that the prediction of the solvation energy of neutral species is quite straightforward and can be easily achieved by applying a solvation model in the form of the structureless field (so-called continuum or implicit solvation model). These models, while less computationally demanding and easier to apply than models comprising the explicit presence of solvent molecules, do not account for interactions between a solute and solvent molecules in the first solvation shell, which are expected to be particularly important for ionic species.[17] In order to overcome this and other shortcomings, such as the effect of charge transfer for example, so-called mixed (explicit solvent molecules added in the continuum model calculations) models have been proposed.[16a,17a,18] One, however, should be careful when deciding on a model as the indirect choice of other parameters related to building up cavities, atomic radii, and so forth may influence the direction and magnitude of the modeled isotope effects.[19] Force constants matrix (Hessian) calculations constitute another challenge. Although using a continuum solvation model does not increase the size of a system, any approach involving explicit treatment of surrounding media surely does, and if a process of interest requires including many solvent molecules, then a choice of method to perform normal modes analysis to obtain Hessian and partition functions, and thence IEs, is limited to protocols allowing computations only for a subset of atoms (isotopically sensitive, so to speak).[20] In the case of processes taking place in condensed phases, additional intermolecular interactions (solute-environment) of low frequencies will also play an important role and need appropriate theoretical treatment. This may influence both the choice of computational method and the reaction dynamics methodology. The quality of the calculated vibrational frequencies and the subsequent IEs will, thus, depend largely on a selected approximation. In larger systems, when only a subset of atoms can be used for the Hessian calculation, the method of treating low-frequency modes is also crucial. Very recently, Williams highlighted the importance of including six external degrees of freedom in the computation of a Hessian as they may constitute the response of a system to its environment (when represented explicitly), and in larger systems, these eigenvalues are usually nonzero.[19] An example of the kinetic isotope effects on the dechlorination of β-HCH by LinB and the use of an ensemble of transition states protocols will be demonstrated (Section 4.2.3).

4.1.3 Beyond Transition State Theory

Despite many approximations and generalizations, transition state theory (TST) turned out to be very useful and is sometimes the only available tool that can provide useful information regarding the pathways by which a chemical reaction occurs. TST as an approximate theory was advanced for bimolecular reactions taking place in the gas-phase. It is based on two assumptions, namely an equilibrium assumption and a dynamical bottleneck assumption. Within the first one, transition state species, defined as a hypersurface in the phase space that separates reactants from products that originate from reactants that are in a local equilibrium with them. The other assumption is that any system passing through the transition state does so only once before the next collision, or before it is stabilized or thermalized as a product. It is called a dynamic bottleneck assumption or no-recrossing assumption. If the local equilibrium assumption and classical mechanics are valid, TST always overestimates the rate constant because it over-counts the reactive trajectories; it counts all forward-crossing trajectories so it does not miss any contributions to the equilibrium rate constant, but if recrossing occurs it counts some trajectories that do not contribute and/or it over-counts some trajectories that do. Thus, classical mechanical transition state theory provides an upper bound to the classical mechanical equilibrium rate constant.[21] Therefore, in other words, in classical mechanics $k(T)$ is always greater than or equal to the local equilibrium rate constant, therefore a procedure that would allow optimization of the transition state dividing surface in a way that would maximize the free energy and minimize the calculated rate constant is needed. This approach is called the variational transition state theory (VTST).[22] It is designed in a way that the reference point along the minimum energy path (MEP) is being moved variationally either backwards or forwards from the transition state (TS) structure until the rate constant is minimized:

$$k^{\mathrm{VTST}} = \frac{k_{\mathrm{B}}T}{h\Phi^{R}(T)} \min_{z_0,\Omega} Q^{\mathrm{GT}}(T, z_0, \Omega) e^{-V_{\mathrm{RP}}/(z_0)/\mathrm{RT}} \tag{4.9}$$

In which GT is the generalized transition state, R denotes reactants, $\Phi^R(T)$ is the partition function of the reactants per unit volume, z_0 is the value of the reaction coordinate at which the dividing surface crossed the reaction path, is the shape and orientation of the TS, Q^{GT} is the partition function of the generalized transition state, and V_{RP} is the molar potential energy on the reaction path. In eqn (4.9), it is assumed that all vibrational partition functions are quantized.

Apart from incorporating variational effects, this theory also enables the inclusion of the multidimensional corner-cutting tunneling (MT) contribution to deal with many important and common hydrogen species transfers. Transmission coefficients calculated this way can be expressed as a product:

$$\gamma = \kappa(T)\Gamma(T) \tag{4.10}$$

In which $\kappa(T)$ accounts for tunneling, and Γ for classical or quasiclassical reflection, which is due to the fact that not all systems that reach transition state configuration give rise to products.

Multiplying eqn (4.9) by the transmission coefficient (eqn (4.10)) will result in the rate constant including multidimensional tunneling, $k^{\mathrm{VTST/MT}}$. Here, we highlight a very recent article reviewing the fundamentals of VTST and its recent development along with some modern applications.[23] Although tunneling and non-classical reflection have opposite effects on the rate constant (the former increases it by allowing lower-energy states to become reactive and the latter decreases it by reducing the reactivity of higher-energy systems) and they might be assumed to cancel, tunneling effects tend to predominate over the non-classical reflection because of the higher Boltzmann population of the lower-energy systems. This fact makes the inclusion of quantum mechanical tunneling a critical issue for accurately predicting the rate constants. In the context of a many-atom system, tunneling through the barrier may occur along one or more coordinates, and the mass for each case may be considered as the reduced mass of the normal mode. Thus, tunneling effects can be present even when the reaction coordinate itself is dominated only by a heavy-atom motion. Highly accurate prediction of transmission coefficients including many degrees of freedom is a very difficult quantum mechanical problem. A simplifying approximation is to consider tunneling only in the degree of freedom corresponding to the reaction coordinate. This assumption allows the reaction surface to be treated as a one-dimensional problem. Although such an approach is an oversimplification it enabled handling of the mathematical treatment of the problem and the avoidance of too many complications and unknown parameters. Various levels of approximation are available. The simplest and most commonly used methods are those of Wigner[24] and Bell.[25] In both these methods the profile of the barrier is modeled by a truncated parabola shape.

$$\kappa(T) = 1 + \frac{1}{24}\left[\frac{hIm(\nu^{\neq})}{k_{\mathrm{B}}T}\right]^2 \tag{4.11}$$

In which $\nu\neq$ is the imaginary frequency associated with the reaction coordinate. Im(x) denotes that only the imaginary part of the frequency is taken, therefore it is treated as a real number rather than a complex one. It was found that the Wigner correction works well, provided that $h\mathrm{Im}(\nu\neq) \ll k_{\mathrm{B}}T$. It turned out to be valid only when the correction is small, typically less than a factor of 2. The work by Bell was found to have a larger region of validity, but the expressions used turned out to be discontinuous and tended to oscillate. Skodje and Truhlar[26] provided a more robust approximation to κ by generalizing Bell's method:

$$\kappa(T) = \frac{\beta\pi/\alpha}{\sin(\beta\pi/\alpha)} - \frac{\beta}{\beta-\alpha}\mathrm{e}^{(\beta-\alpha)(\Delta V^{\neq}-\mathrm{V})}\ \text{for}\ \beta \le \alpha \tag{4.12}$$

$$\kappa(T) = \frac{\beta}{\beta-\alpha}\mathrm{e}^{[(\beta-\alpha)(\Delta V^{\neq}-\mathrm{V})-1]}\ \text{for}\ \beta \ge \alpha \tag{4.13}$$

In which $\alpha = \frac{2\pi}{hIm(\nu^{\neq})}$ and $\beta = \frac{1}{k_{\mathrm{B}}T}$. V is 0 for an exoergic reaction and the positive zero-point-including the potential energy difference between reactants and products for an endoergic reaction. The method by Skodje and Truhlar was found to be applicable to unsymmetrical barriers and useful for barriers with shapes other than parabolic. Using the parabolic barrier description can be justified and appropriate in cases in which only the potential energy surface near the top of the barrier is of interest. Otherwise, other possible barrier shapes should be explored. One of the approaches involves fitting the reaction coordinate to the so-called Eckart potential.[27] The Eckart potential permits an exact, analytical solution to the probability of tunneling through the barrier (and of nonclassical reflection) from the time-independent Schrödinger equation for systems of fixed energy E. A very good estimate of κ in the limit of tunneling along a single dimension can be obtained by numerical integration over all energies weighted by the Boltzmann probability of the reacting systems having a particular energy at a given temperature T. However, in many cases even the best one-dimensional simplified treatment underestimates the full tunneling contribution as tunneling may occur through dimensions of the potential energy surface other than the reaction coordinate. Fortunately, practical multidimensional tunneling approximations are available.

Another method that is worth mentioning for treating the quantum mechanical effects was developed by Warshel and co-workers[28] on the basis of the centroid path integral approach introduced by Gillan[29] and extended by Voth and co-workers.[30] Warshel and co-workers in their development of the method focused on the evaluation of the quantum mechanical free energy of activation for an adiabatic reaction in the condensed phase and the introduction of the free energy perturbation and umbrella sampling approaches to the centroid calculations. This led to the development of the quantized classical path (QCP) method. The QCP approach starts with the evaluation of the quantum mechanical free energy of activation, $\Delta G_q^{\neq}$. Some quantum mechanical properties can be calculated using Feynman's path integral formulation[31] in which each quantized particle (this can be a particle transferred along with its donor and acceptor) is represented by a closed ring (or necklace) of quasiparticles, which are sequentially connected by harmonic springs and each experiences a fraction of the external potential acting on the real particle. The effective quantum mechanical potential of such a quantized particle is given by the equation:

$$V_q = \frac{mP}{2}\frac{(k_B T)^2}{h^2}\sum_{J=1}^{P}(x_j - x_{j+1})^2 + \frac{1}{P}\sum_{j=1}^{P}V_{cl}(x_j) \tag{4.14}$$

In which P is the number of quasiparticles (with coordinate x_j) in the ring (or necklace), m is the mass of the real atom, and V_{cl} is the classical potential on the atom. For the interaction between the quantized atoms, quasiparticle j in one necklace interacts with the corresponding jth bead in the other. Therefore, the total quantum mechanical partition function can be obtained by running classical trajectories of the quasiparticles with the potential V_q. The probability of it being along a reaction coordinate can be found by evaluating the probability distribution for the center of mass of the quasiparticles (centroid)[29,30] which can be determined using the QCP approach in which classical trajectories (of the classical particles only) are propagated on the classical potential surface and the positions of the classical atoms are then used for the quantum mechanical partition function. Therefore, in other words, the quantum mechanical free energy profile is evaluated by a centroid approach which is constrained to move on the classical potential. The discrete Feynman path integral method has already been applied many times in nuclear quantum effects treatment. For instance, path-integral molecular dynamics (PIMD) has become a method for calculating the equilibrium properties of a quantum many-body system. Different implementations of PIMD do exist in modern simulation packages. More information will be given along with illustrative examples.

4.2 Examples

Once the theory of reactivity has been decided upon (usually it is TST or VTST although Marcus theory has also been used for the prediction of IEs,[32] calculations of IEs require a number of decisions to be made. The first and most obvious one is the theory level to be used. Following recent trends, the majority of calculations related to reactivity are carried out using the DFT level of theory expressed in at least a double-zeta basis set. The particular combination, as in the case of other quantum-chemical calculations, depends strongly on the system studied and the computational resources (and to some extent on the availability of the desired theory level in the available programs). Benchmarks of particular theory levels should be inspected for their performance for the system of interest or closely related ones. If possible, results should be confirmed with experimental data.

As implicated by eqn (4.8) the minimal information points necessary to calculate the value of a kinetic isotope effect are the isotopic frequencies for reactant(s) and the transition state. For calculations of EIEs the vibrational data for the product(s) rather than the transition state are necessary. This restricts practical calculations of IEs to systems for which Hessian can be effectively computed. This poses problems for models of processes in the condensed phases when solvent molecules are explicitly included in the model. The problem is even more severe in the case of enzymatic reactions. In recent years the approach of choice is the use of hybrid calculations that combine quantum mechanics (QM) calculations of the most important fragment in the system, while treating the remaining part at a lower level of theory, usually using molecular mechanics (MM).[33] The problem of the size of the whole model, as well as the QM part, is still under debate.[34] Obviously, also this problem is influenced by the hardware available at the moment of study. For example, in our studies of methylmalonyl-CoA mutase at the turn of the century we were able to address the problem of the hydrogen atom transfer using QM cluster models comprising 36 to 119 atoms[35] while only a few years later it was possible to construct the model including the whole enzyme[36] described using the QM/MM scheme with appropriate treatment of open-shell species.

Below we present several examples that illustrate the different methodologies and problems associated with calculations of the IEs. All these examples are taken from our own studies. We start with the largest, hydrogen KIEs on a chemical reaction, and then move successively to smaller and smaller ones including enzyme-catalyzed reactions, and finally arriving at IEs associated with physical processes of phase change and adsorption.

4.2.1 Kinetic Isotope Effects in Multipath VTST: Application to a Hydrogen Abstraction Reaction

Within the transition state theory, the description of the behavior of the systems can be affected by the fact that a number of approximations are used, namely the Born–Oppenheimer, rigid-rotor, and harmonic oscillator approaches. Recent papers by Truhlar and coworkers, and others[37] highlighted the importance of the proper description of the systems containing several conformations (at the reactants and transition state structures). Usually, all the conformers belonging to a given channel interconvert between each other by torsional internal rotations with modest energetic barriers. The most common and straightforward approach is to use the rigid-rotor harmonic-oscillator description to treat these internal modes, so each of the stationary point configurations is treated independently. However, this is not always the best choice as it may lead to substantial errors in the evaluation of thermal rate constants and KIEs. Therefore, various approximations have been proposed to include the torsional anharmonicity in thermodynamics calculations, in particular, to improve the theoretical description of combustion reactions.[38] This makes the calculation of the vibrational partition function more challenging. A combustion reaction (gas-phase), although very often studied over a wide range of temperatures, occurs at high temperatures and it is at those temperatures that significant deviations from the harmonic behavior of modes are expected. In contrast, torsional anharmonic effects may have a minor effect at lower (around room) temperatures, but there are issues that still may involve substantial deviations from the separable harmonic oscillator approximation, such as the coupling between hindered rotors. One of the recent extensions of VTST called multi-path VTST (MP-VTST) allows treating flexible molecules with multiple transition states and the interconversion between them.[39,40]

An example of the hydrogen abstraction reaction from ethanol by atomic hydrogen in an aqueous solution at room temperature (eqn (4.15)) and the application of MP-VTST to predict the deuterium isotope effects is demonstrated.

$$CH_3CH_2OH + H^\bullet \rightarrow CH_3CHOH + H_2 \tag{4.15}$$

Briefly, within this methodology a kinetic isotope effect can be calculated as a product of the following factors:

$$\mathrm{KIE} = \mathrm{KIE}_{trans} \times \mathrm{KIE}_{rv}^{\neq} \times \mathrm{KIE}_{vtun} \tag{4.16}$$

In which

$$\mathrm{KIE}_{trans} = \frac{\Phi_{\mathrm{rel},D}}{\Phi_{\mathrm{rel},H}} \tag{4.17}$$

and Φ is a partition function including the relative translational motion of reactants;

$$\mathrm{KIE}_{rv}^{\neq} = \frac{Q_{R,D}}{Q_{R,H}} \frac{Q_{\mathrm{rot},H}^{\neq} Q_{\mathrm{vib},H}^{\neq}}{Q_{\mathrm{rot},D}^{\neq} Q_{\mathrm{vib},D}^{\neq}} \tag{4.18}$$

Q_{rot} and Q_{vib} are the rotational and vibrational partition functions of the reactants (R) and transition states (≠), respectively. *H* and *D* denote the hydrogen and deuterium species.

Finally,

$$\mathrm{KIE}_{\mathrm{vtun}} = \frac{\gamma_H^{\mathrm{CVT/MT}}}{\gamma_D^{\mathrm{CVT/MT}}} \tag{4.19}$$

The multipath CVT/MT rate constant is calculated based on the following equation:

$$k^{\mathrm{MP-CVT/MT}} = \sum_{i=1}^{k} \alpha_{\mathrm{tor},k} k_{\mathrm{har},k}^{\mathrm{CVT/MT}}(T) \tag{4.20}$$

In which α_{tor} is a ratio of the anharmonic corrections at the transition state of the *k*th conformational reaction channel and of the reactants.

The molecule of ethanol can adopt three different conformations, namely two *gauche* and one *trans* with respect to the methyl group (Figure 4.1). Each of them can lead to a distinguishable transition state which can then interconvert between each other (Figure 4.2). Although in the gas phase at high temperatures ethanol can undergo hydrogen abstraction from other positions than the one shown as eqn (4.15), it was, however, found that below 500 K this one mostly contributes to the overall rate constant. Three different isotopic substitutions scenarios were considered (eqn (4.21)–(4.23)) and the KIEs were measured.[41,42]

$$CH_3CD_2OH + H^{\bullet} \rightarrow CH_3CDOH + HD \tag{4.21}$$

$$CH_3CH_2OD + D^{\bullet} \rightarrow CH_3CHOD + HD \tag{4.22}$$

$$CD_3CD_2OD + D^{\bullet} \rightarrow CD_3CDOD + D_2 \tag{4.23}$$

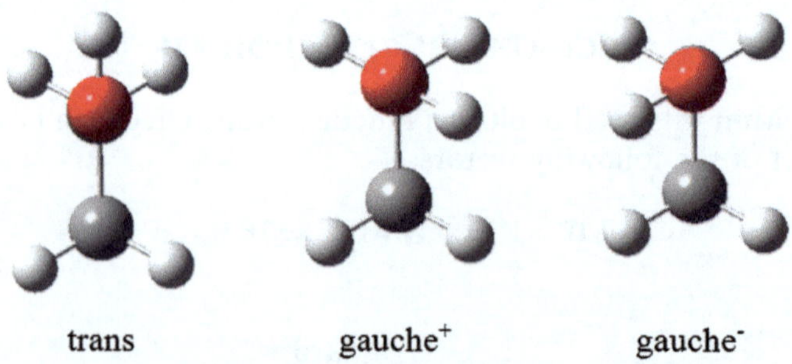

Figure 4.1 Possible conformations of ethanol.

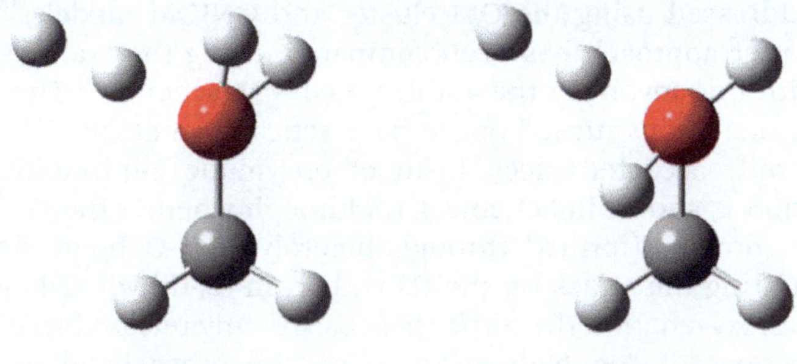

Figure 4.2 Optimized transition state conformations at the PCM/MPWB1K/6-31+G(d,p)[44] level of theory.

Table 4.1 Various factors contributing to the overall KIEs (KIE_{calc}) and individual contributions of each of the transition state conformers (KIE_i).

Reaction	TS conformer	KIE_{trans}	$KIE^{\neq}_{rv}$	KIE_{vtun}	KIE_i	KIE_{calc}	KIE_{exp}
5	*trans*	1.003	6.958	1.120	4.360	7.66	7.38
	gauche	1.003	6.845	1.095	3.304		
6	*trans*	2.740	0.197	1.832	0.570	1.00	0.73
	gauche	2.740	0.199	1.861	0.432		
7	*trans*	2.757	1.394	1.369	2.934	5.16	6.80
	gauche	2.757	1.383	1.327	2.223		

The reactions in eqn (4.21) and (4.23) resulted in normal KIEs of 7.4 and 6.8, respectively, whereas the reaction shown in eqn (4.22) was a source of the inverse isotope effect of 0.7.

In order to gain more knowledge on the contribution of each of the transition state conformers to the overall KIE, the MP-VTST methodology along with the continuum solvation model was applied.[43] Within the performed analysis it was shown that the substitution resulting in the reaction shown in eqn (4.21) is characterized by a large rovibrational effect, small-tunneling, and negligible translational contributions (Table 4.1). The reaction shown in eqn (4.23) is accompanied by substantial rovibrational and translational factors. In contrast, the reaction shown in eqn (4.22) has a very significant rovibrational effect, but in the inverse sense (0.197 ÷ 0.199, Table 4.1) and the largest tunneling contribution amongst all three considered substitutions. Furthermore, it was found that the percentage contribution of each transition state to the total KIE is independent of the isotopic substitution.

4.2.2 Comparison of Different QM/MM Approaches: Nitrobenzene Dioxygenase

Studies of the enzymatic decomposition of nitroaromatic pollutants allowed for a few comparisons of computational methodologies. In these studies competition between dioxygenation and monooxygenation of nitrotoluene has been addressed using the QM cluster and ONIOM models.[45] Furthermore, the latter approach has been compared, using the example of nitrobenzene, with one involving the additive QM/MM scheme.[46] The results of the hybrid calculations turned out to be practically identical.

The generally accepted mechanism of enzymatic 2-nitrotoluene (2NT) decomposition is shown in Scheme 4.1, although whether the Fe^{III}–OOH or HO–Fe^{V}=O complex (formed through heterolytic O–O bond cleavage) is actually initiating the attack on the C3 carbon in *cis*-dihydroxylation is still debated. In dioxygenation the nitrite is subsequently released leading to the catechol product.[47] Monohydroxylation, on the other hand, follows the oxygen rebound mechanism, wherein methyl hydrogen is abstracted to form a radical intermediate that then recombines with the resulting hydroxyl group to form nitrobenzyl alcohol. These two reactions are competitive, and calculated intrinsic KIEs can serve as reference values that would be useful in the identification of the predominant degradation pathway.

^{13}C and ^{2}H KIEs associated with 2NT *cis*-dihydroxylation and monohydroxylation catalyzed by nitrobenzene dioxygenase (NBDO) were computed using cluster and QM/MM models of reactants and transition states at the sextet and quartet spin state surfaces. As these results are not readily available in the literature we will expand on the methodology of calculations. Enzymatic models were obtained[48] using the electronic embedding scheme of the ONIOM method,[49] with the QM region treated at the B3LYP/LACVP* level[50] and the MM region described using the AMBER ff99SB force field.[51] Calculations using mechanical embedding yielded unrealistic spin densities. Final energies were calculated using the larger LAC3VP + * basis set and corrected with LACVP* zero-point energies. The Skodie–Truhlar tunneling correction[26] was used in calculations of the hydrogen KIEs for the monooxidation pathway which involves hydrogen transfer. The QM part included Fe, hydroperoxo, 2NT, and the sidechains of His-206, His-211, and Asp-360. Furthermore, the sidechain of Asn-258, which positions the substrate for oxidation through the H bond interaction,[52] (Figure 4.3) was also included in the models. ONIOM calculations were repeated with the B97-D functional,[53] to evaluate the effect of dispersion interactions. The cluster model was equivalent to the QM region of the enzymatic model. The protein environment in this model was

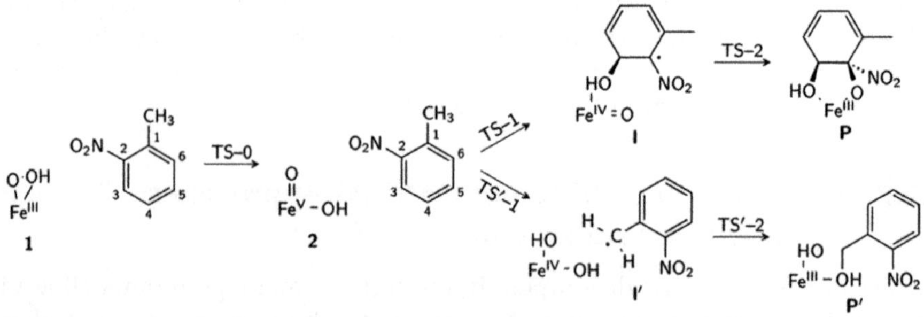

Scheme 4.1 *Cis*-dihydroxylation and monohydroxylation mechanisms.

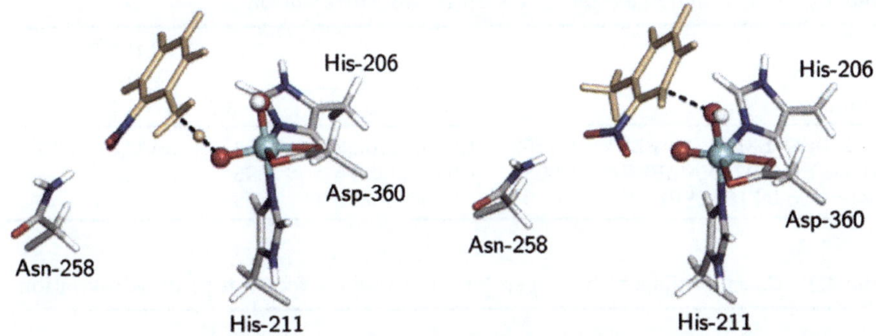

Figure 4.3 Transition state structures for H abstraction (TS′–1) in monohydroxylation (left) and for C–O bond formation (TS–1) in *cis*-dihydroxylation (right).

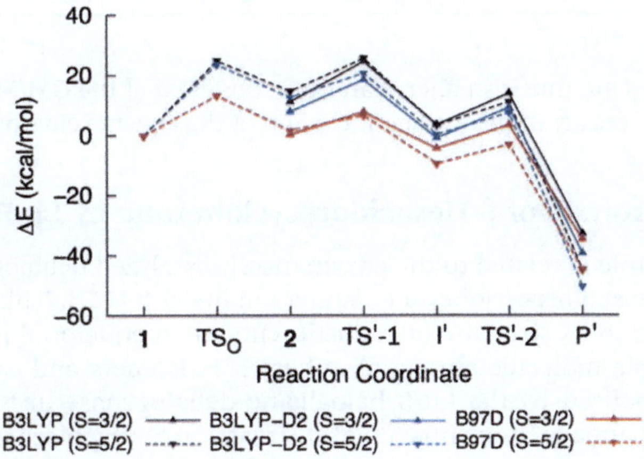

Figure 4.4 Energy profile of monohydroxylation calculated with ONIOM-EE. Adapted from ref. 45. Published by the PCCP Owner Societies.

approximated as a homogeneous continuum medium using the polarizable continuum model[54] with a dielectric constant of $\varepsilon = 5.62$.

The energy profile of monohydroxylation indicates that the reaction can proceed in either the sextet or quartet state. As illustrated by Figure 4.4 (in which additional results obtained with B3LYP-D2 are provided for comparison) the dispersion reduces the activation barriers. The optimized ONIOM(B97-D:AMBER) structures are characterized by closer contact between the substrate and active site, which generally led to later (more product-like) transition states. However, the results collected in Table 4.2 indicate that KIE values are insensitive to the model and theory level used.

In contrast to the case of monohydroxylation, the cluster approach is insufficient for determining the theoretical KIEs associated with *cis*-dihydroxylation. As it can be seen from the results collected in Table 4.3 for the quartet state, for which the cluster calculation yielded the wrong spin distribution, the

Table 4.2 CH_3-kinetic isotope effects on monohydroxylation.

Method	^{13}C		^{2}H		^{2}H (with tunneling)	
	$S=5/2$	$S=3/2$	$S=5/2$	$S=3/2$	$S=5/2$	$S=3/2$
ONIOM-EE (B3LYP:AMBER)	1.0103	1.0106	6.58	6.64	30.1	78.7
ONIOM-EE (B97-D:AMBER)	1.0087	1.0114	5.76	5.95		
Cluster model (B3LYP)	1.0094	1.0106	5.98	6.35		

Table 4.3 C3 and adjacent hydrogen kinetic isotope effects on *cis*-dihydroxylation.

Method	^{13}C		^{2}H	
	$S=5/2$	$S=3/2$	$S=5/2$	$S=3/2$
ONIOM-EE (B3LYP:AMBER)	1.0321	1.0290	0.9752	0.9721
ONIOM-EE (B97-D:AMBER)	1.0379	1.3710	0.9814	0.9895
Cluster model(B3LYP)	1.0322	1.0125	0.9963	0.9962

calculated KIEs are much smaller than those obtained at the QM/MM level, indicating the necessity of the explicit inclusion of the enzyme environment.

4.2.3 Hydrolysis of β-Hexachlorocyclohexane by LinB

Another example is related to the enzymatically catalyzed dehalogenation of one of the hexachlorocyclohexane isomers, namely β-HCH. This particular isomer, owing to its structure (all substituents in an equatorial position), is the most stable molecule among all other HCH isomers and can be quite readily metabolized by the LinB haloalkane dehalogenase in the environment.[55] It has already been shown that the reaction taking place in the active site of LinB is a nucleophilic substitution within which Asp108 plays the role of the nucleophile (Figure 4.5).[56] As a result, one of the C–Cl bonds is cleaved and the eliminated chloride is stabilized by two neighboring residues, Trp109 and Asn38 ***via*** the hydrogen bond network.

With a use of the QM/MM MD methodology 1D and 2D free energy surfaces were modeled at the PM3[57]/MM (MM = ff99SB,[51b] TIP3P[58] level of theory). Subsequently, energetic barriers of 30 and 31 kcal mol^{-1}, respectively, were corrected at the DFT level representing the region treated quantum mechanically. The final free energies of activation of 16 and 19 kcal mol^{-1} were in very good agreement with the value derived from the rate constant and was equal to 17 kcal mol^{-1}.[59] Both the free energy profiles led to the conclusion that the formation of the O–C bond between Asp108 and the substrate and the cleavage of the C–Cl bond occur concertedly. Confirming this, the accuracy of the active site models used in this study ^{13}C and ^{37}Cl-KIEs were predicted for the dechlorination reaction. Using ten enzyme-transition state complexes already optimized at the DFT level of theory for the QM region representation, the intrinsic reaction coordinate calculations were performed in order to locate the corresponding enzyme-reactant and enzyme-product complexes. Then, the

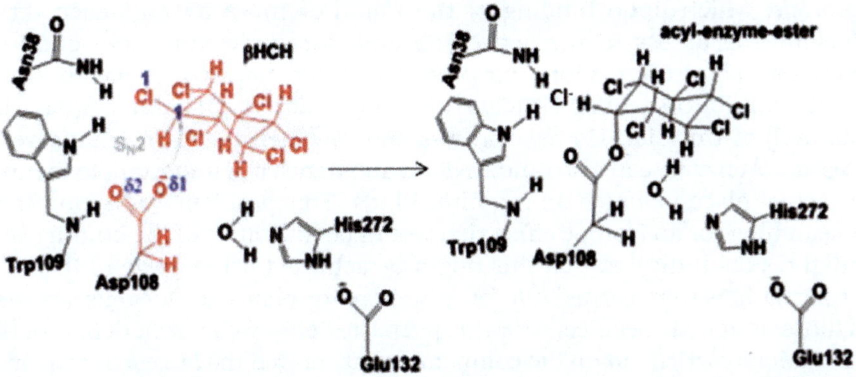

Figure 4.5 Dechlorination step of the β-HCH isomer transformation in the active site of LinB. Atoms shown in red comprise the region of the overall system treated quantum mechanically.
Reproduced from ref. 56 with permission from Elsevier, Copyright 2014.

Table 4.4 Primary ^{13}C and ^{37}Cl-KIEs for the dechlorination of β-HCH in the active site of LinB predicted at the M06-2X/6-31+G(d,p):[60] CHARGE/MM level at 300 K.

^{13}C	^{37}Cl
1.0491 ± 0.0017	1.0079 ± 0.0004

normal modes analysis was carried out for those ten sets of reactants-transition states pairs. The resulting KIEs are shown in Table 4.4.

A more recent computational study on the metabolism of 1-chloropropane by LinB showed that the dechlorination and the following hydrolysis of the substrate is a rate-determining step of the overall reaction catalyzed by this dehalogenase,[61] the predicted magnitudes of the intrinsic KIEs might be treated as the apparent effects assuming that no masking by other non-chemical steps (transport, binding, product release) occurs. Nevertheless, the computed average ^{37}Cl-KIEs turned out to have very similar magnitudes to the values obtained previously for other enzymatic S_N2 dehalogenations.[62]

4.2.4 Binding Isotope Effects

Isotope effects for the association of a reactant with an enzyme to form a Michaelis–Menten complex are frequently ignored in biochemical studies. This is frequently (although not always)[63] a reasonable assumption in studies of large KIEs. Values of binding isotope effects (BIEs) can, however, be a useful tool in studies of enzymatic reaction mechanisms.[64] Below, we discuss one of their applications for the example of HIV-1 reverse transcriptase, HIV-1 RT. This enzyme is one of three key enzymes in the proliferation cycle of the virus and thus a frequent target of the design of drugs that can block its activity.

HIV-1 RT can be inhibited by blocking its active site or the allosteric pocket. In the latter case, a small molecule acts as a wedge on the hinges of two large domains, which upon binding of the inhibitor move towards each other blocking the access to the active site and thus preventing transcription. Therefore, significant efforts have been directed towards the synthesis of suitable inhibitors (called NNRTIs (non-nucleoside reverse transcriptase inhibitors)) of this site. However, it was shown experimentally[65] that several sites at the enzyme can also bind NNRTIs and hence it is important to identify the actual place in which an inhibitor binds. This problem was approached by searching for an isotope effect that would be indicative of the binding site. Initial results indicated that this might be achieved using oxygen BIEs.[66]

Calculations were carried out for inhibitors for pharmaceutical use, as well as those designed specifically for this purpose. Below we present details of the calculations carried out on the compound that showed the largest dependence of the ^{18}O-BIE on the binding site[67] and at the same time exhibited improved solubility and the best inhibitory properties with an IC_{50} of 0.3 μM.[68]

The initial coordinates of HIV-1 RT were taken from the Protein Data Bank (entries with PDB ID: 2RKI[69] with a ligand bound in the allosteric cavity, 4IFY[65] with a ligand bound to the knuckles, 4ICL[65] with the ligand bound using incoming nucleotide binding, and 4KFB[65] with the ligand bound to an NNRTI adjacent binding site). A toolbox of programs was used in refining this model.[51,70–72] Native ligands were replaced by 4-[[2-[[5-(2-chlorophenyl)-4-(2,4-dimethylphenyl)-1,2,4-triazol-3-yl]sulfanyl]acetyl]amino]benzoic acid (L-3 illustrated in Figure 4.7). Chlorine atoms were added to neutralize the enzyme-ligand complexes, which were soaked in a 140×100×100 Å^3 box of explicit water molecules using a three-center TIP3P (Param 9) model[73] and optimized using the fDynamo.[74] The enzyme and inhibitor heavy atoms were restrained by means of a Cartesian harmonic umbrella. Subsequently, 2 ns of the QM/MM molecular dynamics (MD) NVT simulation were performed at 300 K. Atoms of ligands selected for the QM part were treated at the AM1[75] level of theory, while the remaining part of the system was described with the AMBER[51] force field. For non-bonding interactions, the cut-off rule was applied using a smooth switching function with a radius range from 14.5 to 16 Å. As reference systems, inhibitors in a 140×100×100 Å^3 box of TIP3P water molecules have been prepared and equilibrated during 200 ps of QM(AM1)/MM(TIP3P) MD simulations using the fDynamo.

Binding isotope effects were computed using the equation

$$\mathrm{BIE} = \mathrm{e}^{-(\Delta G_L - \Delta G_H)/\mathrm{RT}} \tag{4.24}$$

In which R is a gas constant, T is the absolute temperature, and

$$\Delta G_L = G_L^{E\cdot L} - G_L^I \tag{4.25}$$

$$\Delta G_H = G_H^{E\cdot L} - G_H^I \tag{4.26}$$

are Gibbs free energies (G) for isotopologues of the inhibitor, in the aqueous solution (superscript I) and are bound in the receptor (superscript $E{\cdot}I$).

Only atoms which were used to define the Hessian in accordance with the cut-off rule that takes advantage of the local nature of isotope effects[76] were included in the calculations. BIEs have been computed for eleven stationary structures (from 0 to 10) of the inhibitor in water, and eleven structures of the inhibitor bound in the protein. The enzyme-ligand structures were extracted from the last 100 ps of 2 ns QM/MM MD, and reference water-ligand structures were taken from the last 20 ps of 200 ps QM/MM MD. They were optimized at AM1/AMBER:TIP3P and used for BIEs calculations without scaling of the vibrational frequencies. Optimization in the condensed media (protein with a ligand in water) makes translational and rotational terms of partition functions negligible in this case. Full (3N×3N) Hessians have been used in the projection procedure to eliminate translational and rotational components, which give rise to small nonzero frequencies. Individual BIE values were obtained using a combination of vibrational data of all 11 structures of the ligand in aqueous solution with those for 11 structures of the ligand bound to the enzyme. This procedure yielded 121 (11×11) individual BIE values, which were averaged and the corresponding standard deviations were calculated.

Figure 4.6 illustrates L-3 bound in the hydrophilic RNase active site (top panel), containing two magnesium cations and a number of water molecules.

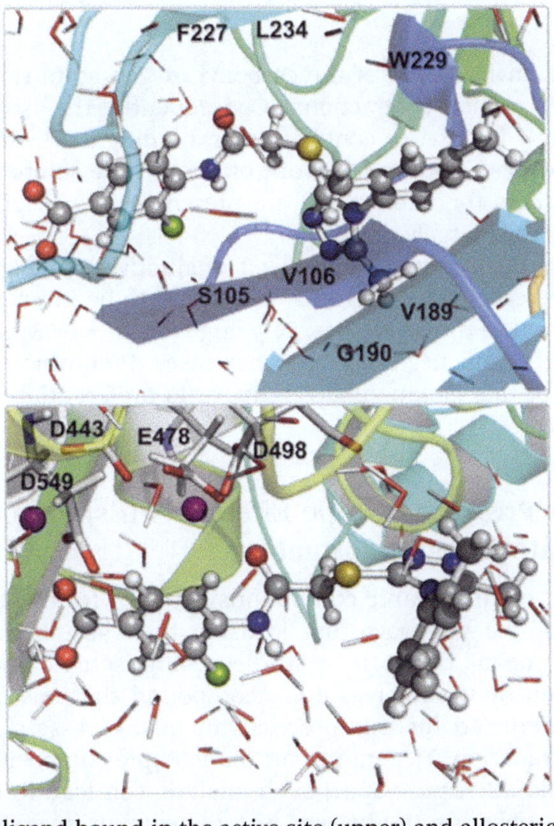

Figure 4.6 L-3 ligand bound in the active site (upper) and allosteric site (bottom) of HIV-1 RT.

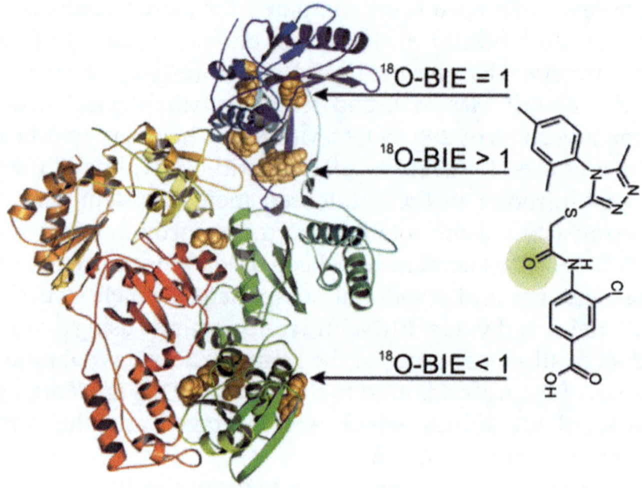

Figure 4.7 Carbonyl oxygen binding isotope effects on the association of L-3 in different sites of HIV-1 RT.
Reproduced from ref. 78 with permission.

This makes the environment of the carbonyl oxygen atom rich in hydrogen bonds and electrostatic interactions. Consequently, the isotope effect on binding is slightly inverse. In contrast, upon binding the hydrophobic allosteric pocket hydrogen bonding network present in the solution is lost as no immediate partners for interactions are present in the neighborhood. The isotope effect is thus largely positive.

The obtained results are summarized qualitatively in Figure 4.7. Most importantly, the carbonyl oxygen BIE was found to be inverse when the inhibitor was bound to the active site, very large (1.010 ± 0.03) when bound to the allosteric site, and negligible in other cases. Preliminary experimental results (1.015) of this isotope effect support its preferential binding in the allosteric pocket.[77]

4.2.5 Vapor Pressure Isotope Effects (VPIEs) Predicted Using the Path Integral Formalism

Evaporation of volatile organic compounds (VOCs)[79] from the pure organic phase is one of the pathways that leads to their attenuation in the environment and means they can be studied using isotopic analysis. As this process is driven by vapor pressure, a compound that has isotope effects measured or predicted for this process can help to assess the degree of compound degradation. Depending on the isotopic composition of a compound phase (liquid *versus* vapor) such analysis can lead to either normal (liquid phase enriched in the heavier isotopologue) or inverse (vapor phase enriched in the heavier isotopologue) vapor pressure isotope effects.

As an example of selected brominated solvents, bromobenzene (BB) and dibromomethane (DBM), the evidence of carbon and bromine isotopic fractionation was demonstrated using both measurements and theoretical predictions.[80] Knowledge of the magnitude and direction of VPIEs is essential to characterize the structure and nature of phases in more detail, in particular, a liquid one, and provide information on the intermolecular interactions governing each system, as well as changes that any studied compound undergoes at the transition from one phase to the other. Furthermore, one of the aims of the study was also to test the performance of available theoretical approaches to compute VPIEs. Three methods were applied, among them were the path integral formalism implemented in the AMBER simulation package.[51]

For each solvent, a proper model of a liquid phase was constructed by building periodic cubic solvent boxes of 40 Å. Then, the boxes were subjected to energy minimization, heating to 300 K and subsequent equilibration before they were used for production runs, which in this particular type of simulation means path integral molecular dynamics simulations using the normal mode PIMD protocol[81] implemented in AMBER. A vapor phase model of each compound comprised an individual molecule of the kind handled in the gas phase. For the purpose of predicting VPIEs, path integral quantum transition state theory[82] was used. Within this theory, a thermodynamic integration with respect to the mass was computed during which interpolation between the masses (λ) of light $(0)^{12}C$ and ^{79}Br and heavy $(1)^{13}C$ and ^{81}Br isotopologues using in-between values were used. ^{13}C and ^{81}Br isotope effects were calculated for each position within the respective molecule, and subsequently, the isotope effect was computed by averaging in the case of the carbon VPIE for BB and bromine VPIE for DBM. Different imaginary time slices were tested in order to find the best set of parameters for modeling the organobromine species under study. Within all the tested possibilities, a kind of agreement between the measured and predicted values was obtained in only very limited settings, and the best match is presented in Table 4.5. The resulting low-level consistency should only be looked at with respect to the direction of the isotope effect, and as such probably should not be treated as a complete failure. Instead, it may be taken as a challenging case calling for further analysis and calculations. In particular, within the same study other computational approaches such as the quantum mechanical cluster model and the two-layer ONIOM scheme[49] were also used to predict the same VPIEs and the final outcome from both of them were by far the more satisfying.

Table 4.5 Average (bulk) ^{13}C and ^{81}Br-VPIEs for evaporation of bromobenzene and dibromomethane measured and predicted using various computational approaches at 300 K.

	Theory							
	PIMD		QM cluster		ONIOM		Experiment	
Compound	^{13}C	^{81}Br	^{13}C	^{81}Br	^{13}C	^{81}Br	^{13}C	^{81}Br
Bromobenzene	1.0037	1.0000	1.0004	1.0001	1.0008	1.0002	1.0004	1.0009
Dibromomethane	0.9972	1.0001	0.9995	1.0001	0.9996	1.0001	0.9994	1.0008

There are several possible sources of these discrepancies. First of all, within the PIMD simulation the potential energy surface for entire systems was generated using a classical/empirical force field which is not completely free of nuclear quantum effects, as those can be implicitly embedded in the parameters used to construct a force field.[83] This fact influences the potential of the liquid phase, but simulations using it are much more efficient than the ones performed with the use of *ab initio* MD and can be applied to much larger systems nowadays. Secondly, an important issue is also that both systems, as they consist of only one type of molecule (either BB or DBM), were entirely quantized during the PIMD simulations in order to better predict the bulk properties of a solvent, which is in contrast to studies in which the QM/MM partitioning is used in order to describe a solute and a condensed phase at a different level of theory (see for instance Liu *et al.*).[84] Not making this choice for the systems under study was kind of a compromise between the accuracy of the theory level used for describing the potential energy surface and the system size required for the proper description of solvent properties. The study rather focused on the better description of the quantum mechanical nature of the nuclei in isotopic substitution in the systems without restricting the quantization of atoms to only one (solute) molecule, which in the case of pure-phase solvents would probably mean one central molecule within a box. However, this approach may turn out to be more appropriate when evaporation from impure organic phases is of interest, for instance, evaporation of organic substances dissolved in water. Thirdly, in two other computational approaches either the partitioning scheme was used, in which one central molecule in a model was treated quantum mechanically at various levels of theory (DFT, MP2, and CCSD) and the rest of the molecules were represented by the AMBER force field parameters, or clusters of different sizes (number of molecules ranging from 3 to 18) were fully treated quantum mechanically by the PM7 Hamiltonian.[85] In these two cases, neither the nuclear quantum effects nor solvent molecule sampling were incorporated. Therefore, basically only the quasiclassical nature of isotope effects was taken into account, whereas in the case of PIMD treatment of the systems some of the effects that hugely contribute to isotope effects, such as the zero-point energy (ZPE) difference caused by molecular vibrations altered by the isotope exchange in the system, were not properly, if at all, included. A previous study by one of the authors of this chapter on the Finkelstein reaction in a condensed phase, in which bromine was substituted by iodine in 2-bromoethylbenzene, clearly showed that ^{81}Br-KIE and ^{13}C-KIE can be 100% and 85%, respectively dominated by the effect on ZPE.[86] Therefore, this should be explicitly taken into account in any future studies using the path integral formalism engaged to predict isotope effects for the processes taking place in a condensed phase.

4.2.6 Isotope Effects Associated with Adsorption on Graphene

We conclude this Chapter with the case of benzene adsorption on graphene. This is the hardest and the most demanding case as only van der Waals

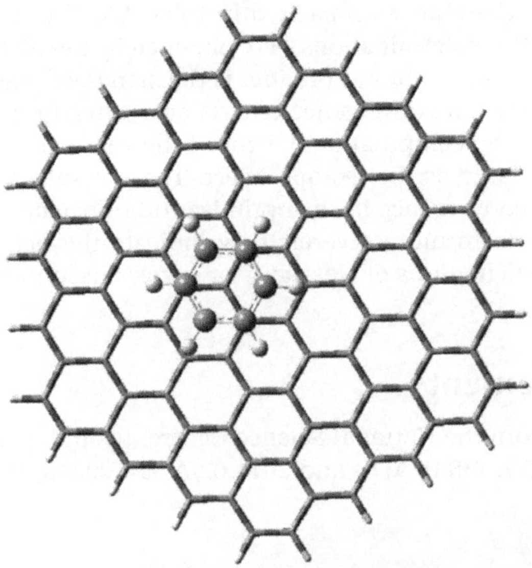

Figure 4.8 Most stable orientation of benzene over the graphene.

Table 4.6 Influence of calculation conditions on the precision of deuterium EIEs.

"Reactant"			"Product"			
Optimization convergence[a]	CPHF convergence	Grid[b]	Optimization convergence	CPHF convergence	Grid	EIE error
0.000300	exp-10	fine	0.000010	exp-10	Fine	0.0004
0.000010	exp-10	fine	0.000001	exp-10	Fine	0.0000_7
0.000300	exp-12	ultra fine	0.000010	exp-12	ultra fine	0.0006
0.000010	exp-12	ultra fine	0.000001	exp-12	ultra fine	0.0000_7
0.000300	exp-10	fine	0.000300	exp-12	ultra fine	0.0019
0.000010	exp-10	fine	0.000010	exp-12	ultra fine	0.0021
0.000001	exp-10	fine	0.000001	exp-12	ultra fine	0.0021

[a]Atomic units.
[b]Grid definition after Gaussian.[89]

interactions are at play and the isotope effects are very small. However, this system allows for a critical detailed analysis of computational protocols which should be used in order to obtain reliable values of IEs.

Although the energetic landscape of benzene orientation over the graphene surface is very flat, calculations on large sheets[87] favor the position of benzene, in which one hydrogen atom overlaps the center of the ring as illustrated in Figure 4.8.

Following the results reported by Wang[87] the ωB97-XD/def2-TZVPP level of theory was used.[88] It was shown that the deuterium and carbon equilibrium isotope effects on the adsorption of benzene on the graphene surface are indeed very small; 1.0028 and 0.9997 (averaged per position), respectively. However, the most important results for the present discussion were obtained in the analysis of the influence of the calculation conditions on the resulting value of the EIEs. These are summarized in Table 4.6. The reported values of inaccuracy in the EIE determinations were obtained by calculating an "isotope effect" in which the reactant and product is the same benzene molecule, but optimized with different convergence criteria and different grids.

As can be seen, the computational error of the value of the isotope effect can be almost as large as the isotope effect. These results indicate the importance of tight convergence in the optimization of reactants and products (transition state). Even more severely they indicate the need for using the same criteria in calculations of Hessians for species used in the calculations of isotope effects.

Acknowledgements

Partial support from the National Science Centre, Poland (grants 2014/14/E/ST4/00041, SONATA BIS to ADD and 2011/02/A/ST4/00246, MAESTRO to PP) is acknowledged.

References

1. M. Wolfsberg, A. Van Hook and P. Paneth, *Isotope Effects in the Chemical, Geological and Bio Sciences*, Springer, London, 2010.
2. A. L. Buchachenko, *Chem. Rev*, 1995, **95**, 2507.
3. (a) R. N. Clayton, L. Grossman and T. K. Mayeda, *Science*, 1973, **182**, 485; (b) M. H. Thiemens and J. E. Heidenreich III, *Science*, 1983, **219**, 1073.
4. V. L. Schramm, *Annu. Rev. Biochem.*, 2011, **80**, 703.
5. A. G. Cassano, V. E. Anderson and M. E. Harris, *Biochemistry*, 2004, **43**, 10547.
6. Z. Muccio and G. P. Jackson, *Analyst*, 2009, **134**, 213.
7. M. H. Glickman, J. S. Wiseman and J. P. Klinman, *J. Am. Chem. Soc.*, 1994, **116**, 793.
8. J. D. Hermes, C. A. Roeske, M. H. O'Leary and W. W. Cleland, *Biochemistry*, 1982, **21**, 5106.
9. P. Paneth, *Environ. Chem.*, 2012, **9**, 67.
10. G. Ciepielowski, B. Pacholczyk-Sienicka, T. Frączek, K. Klajman, P. Paneth and Ł. Albrecht, *J. Sci. Food Agric.*, 2018, **99**, 263.
11. (a) M. Wolfsberg and M. J. Stern, *Pure Appl. Chem.*, 1964, **8**, 225; (b) L. B. Sims and D. E. Lewis, in *Isotopes in Organic Chemistry*, ed. E. Buncel and C. C. Lee, Elsevier, Amsterdam, vol. 6, 1984, pp. 162–259; (c) M. J. Stern and M. Wolfsberg, *J. Chem. Phys.*, 1966, **45**, 4105.
12. (a) J. Rodgers, D. A. Femac and R. L. Schowen, *J. Am. Chem. Soc.*, 1982, **104**, 3263; (b) T. E. Casamassina and W. P. Huskey, *J. Am. Chem. Soc*, 1993, **115**, 14; (c) P. J. Berti, *Methods Enzymol.*, 1999, **308**, 355.
13. (a) E. Teller, as attributed by, W. R. Angus, C. R. Bailey, J. B. Hale, C. K. Ingold, A. H. Leckie, C. G. Raisin, J. W. Thompson and C. L. Wilson, *J. Chem. Soc.*, 1936, 971; (b) O. Redlich, *Z. Physik. Chem. B*, 1935, **28**, 371.

14. (a) F. Jensen, *J. Phys. Chem.*, 1996, **100**, 16892; (b) Fang, *et al.*, *Chem Eur. J.*, 2003, **9**, 2696; (c) P. Adamczyk, A. Dybala-Defratyka and P. Paneth, *Environ. Sci. Technol.*, 2011, **45**, 3006.
15. A. Grzybkowska, R. Kaminski and A. Dybala-Defratyka, *Phys. Chem. Chem. Phys.*, 2014, **16**, 15164.
16. (a) C. J. Cramer and D. G. Truhlar, *Chem. Rev.*, 1999, **99**, 2161; (b) M. Orozcoand and F. J. Luque, *Chem. Rev.*, 2000, **100**, 4187; (c) J. Tomasi, B. Mennucci and R. Cammi, *Chem. Rev*, 2005, **105**, 2999.
17. (a) C. P. Kelly, C. J. Cramer and D. G. Truhlar, *J. Phys. Chem. A*, 2006, **110**, 2493; (b) E. F. da Silva, H. F. Svendsen and K. M. Merz, *J. Phys. Chem. A*, 2009, **113**, 6404, and references therein; (c) V. S. Bryantsev, M. S. Diallo and W. A. Goddard III, *J. Phys. Chem. B*, 2008, **112**, 9709.
18. J. R. Pliego Jr. and J. M. Riveros, *J. Phys. Chem. A*, 2001, **105**, 7241.
19. P. B. Wilson, P. J. Weaver, I. R. Greig and I. H. Williams, *J. Phys. Chem. B*, 2015, **119**, 802.
20. (a) A. Ghysels, V. Van Speybroeck, E. Pauwels, S. Catak, B. R. Brooks, D. Van Neck and M. Waroquier, *J. Comp. Chem.*, 2010, **31**, 994, and references cited therein; (b) I. H. Williams, *J. Chem. Theory Comput.*, 2012, **8**, 542.
21. M. Kreevoy and D. G. Truhlar, in *Investigation of Rates and Mechanisms of Reactions*, ed. C. F. Bernasconi, Wiley and Sons, New York 4th edn, part I, and reference therein, 1986.
22. (a) D. G. Truhlar and B. C. Garrett, *Acc. Chem. Res.*, 1980, **13**, 440; (b) D. G. Truhlar, A. D. Isaacson and B. C. Garrett, in*Theory of Chemical Reaction Dynamics*, ed. M. Baer, CRC, Boca Raton, FL, vol. 4, 1985, pp. 65–137; (c) A. Fernandez-Ramos, A. Ellingson, B. C. Garrett and D. G. Truhlar, *Rev. Comput. Chem.*, 2007, **23**, 125.
23. J. L. Bao and D. G. Truhlar, *Chem. Soc. Rev.*, 2017, **46**, 7548.
24. E. Wigner, *Z. Phys. Chem. B*, 1932, **19**, 203.
25. R. P. Bell, *Trans. Faraday Soc.*, 1959, **55**, 1.
26. R. T. Skodje and D. G. Truhlar, *J. Phys. Chem.*, 1981, **85**, 624.
27. C. Eckart, *Phys. Rev.*, 1930, **35**, 1303.
28. J.-K. Hwang, Z. T. Chu, A. Yadav and A. Warshel, *J. Phys. Chem.*, 1991, **95**, 8445.
29. M. J. Gillan, *J. Phys. Chem. Solid State*, 1987, **20**, 3621.
30. G. Voth, D. Chandler and W. Miller, *J. Chem. Phys.*, 1989, **91**, 7749.
31. R. Feynman, *Statistical Mechanics*, Benjamin, New York, 1972.
32. J. P. Klinman and A. R. Offenbacher, *Acc. Chem. Res.*, 2018, **51**, 1966.
33. A. Warshel and M. Levitt, *J. Mol. Biol.*, 1976, **103**, 227.
34. M. Lundberg, T. Kawatsu, T. Vreven, M. J. Frisch and K. Morokuma, *J. Chem. Comput. Chem.*, 2008, **5**, 220.
35. A. Dybala-Defratyka and P. Paneth, *J. Inorg. Biochem.*, 2001, **86**, 681.
36. A. Dybala-Defratyka, P. Paneth, R. Banerjee and D. G. Truhlar, *Proc. Natl. Acad. Sci. U. S. A.*, 2007, **104**, 10779.
37. (a) R. Meana-Pañeda and A. Fernández-Ramos, *J. Am. Chem. Soc.*, 2012, **134**, 346; (b) T. Yu, J. Zheng and D. G. Truhlar, *J. Phys. Chem. A*, 2012,

116, 297; (c) J. Zheng and D. G. Truhlar, *J. Chem. Theory Comp.*, 2013, **9**, 2875.

38. (a) W. Wang and Y. Zhao, *J. Chem. Phys.*, 2012, **137**, 214306; (b) J. Zheng, Y. Tao, E. Papajak, I. M. Alecu, S. L. Mielke and D. G. Truhlar, *Phys. Chem. Chem. Phys.*, 2011, **13**, 10885; (c) T. Yu, J. Zheng and D. G. Truhlar, *Chem. Sci.*, 2011, **2**, 2199; (d) X. Xu, E. Papajak, J. Zheng and D. G. Truhlar, *Phys. Chem. Chem. Phys.*, 2012, **14**, 4204; (e) P. Seal, E. Papajak and D. G. Truhlar, *J. Phys. Chem. Lett.*, 2012, **3**, 264; (f) A. F. Ramos, *J. Chem. Phys.*, 2013, **138**, 134112.
39. R. Meana-Pañeda and A. Fernandez-Ramos, *J. Am. Chem. Soc.*, 2012, **134**, 346, correction: 2012, 134, 7193.
40. T. Yu, J. Zheng and D. G. Truhlar, *J. Phys. Chem. A*, 2012, **116**, 297.
41. E. Roduner and D. M. Bartels, *Phys. Chem.*, 1992, **96**, 1037.
42. A. M. Lossack, E. Roduner and D. M. Bartels, *J. Phys. Chem. A*, 1998, **102**, 7462.
43. L. Simon-Carballido, T. Vinicius Alves, A. Dybala-Defratyka and A. Fernandez-Ramos, *J. Phys. Chem. B*, 1911, **2016**, 120.
44. (a) V. Barone, M. Cossi and J. Tomasi, *J. Comput. Chem.*, 1998, **19**, 404; (b) Y. Zhao and D. G. Truhlar, *J. Phys. Chem. A*, 2004, **108**, 6908; (c) W. J. Hehre, R. Ditchfield and J. A. Pople, *J. Chem. Phys.*, 1972, **56**, 2257.
45. (a) I. Geronimo, PhD thesis, Lodz University of Technology, 2014; (b) I. Geronimo and P. Paneth, *Phys. Chem. Chem. Phys.*, 2014, **16**, 13889–13899.
46. S. Pati, Sarah, H.-P. Kohler, A. Pabis, P. Paneth, R. Parales and T. Hofstetter, *Env. Sci. Technol.*, 2016, **50**, 6708.
47. A. Pabis, I. Geronimo and P. Paneth, *J. Phys. Chem. B*, 2014, **118**, 3245.
48. A. Pabis, I. Geronimo, D. M. York and Piotr Paneth, *J. Chem. Theory Comput.*, 2014, **10**, 2246.
49. T. Vreven, K. S. Byun, I. Komáromi, S. Dapprich, J. A. Montgomery Jr., K. Morokuma and M. J. Frisch, *J. Chem. Theory Comput.*, 2006, **2**, 815.
50. (a) A. D. Becke, *J. Chem. Phys.*, 1992, **97**, 9173; (b) A. D. Becke, *J. Chem. Phys.*, 1993, **98**, 5648; (c) C. Lee, W. Yang and R. G. Parr, *Phys. Rev. B*, 1988, **37**, 785; (d) P. J. Hay and W. R. Wadt, *J. Chem. Phys.*, 1985, **82**, 270.
51. (a) D. A. Case, T. A. Darden, T. E. Cheatham III, C. L. Simmerling, J. Wang, R. E. Duke, R. Luo, R. C. Walker, W. Zhang, K. M. Merz, B. Roberts, S. Hayik, A. Roitberg, G. Seabra, J. Swails, A. W. Goetz, I. Kolossvary, K. F. Wong, F. Paesani, J. Vanicek, R. M. Wolf, J. Liu, X. Wu, S. R. Brozell, T. Steinbrecher, H. Gohlke, Q. Cai, X. Ye, J. Wang, M. J. Hsieh, G. Cui, D. R. Roe, D. H. Mathews, M. G. Seetin, R. SalomonFerrer, C. Sagui, V. Babin, T. Luchko, S. Gusarov, A. Kovalenko and P. A. Kollman, *AMBER 12*, University of California, San Francisco, 2012; (b) V. Hornak, R. Abel, A. Okur, B. Strockbine, A. Roitberg and A. Simmerling, *Proteins*, 2006, **65**, 712.
52. R. Friemann, M. M. Ivkovic-Jensen, D. U. Lessner, C.-L. Yu, D. T. Gibson, R. E. Parales, H. Eklund and S. Ramaswamy, *J. Mol. Biol.*, 2005, **348**, 1139.

53. S. Grimme, *J. Comput. Chem.*, 2006, **27**, 1787.
54. (a) E. Cancès, B. Mennucci and J. Tomasi, *J. Chem. Phys.*, 1997, **107**, 3032; (b) B. Mennucci, E. Cancès and J. Tomasi, *J. Phys. Chem. B*, 1997, **101**, 10506.
55. Y. Nagata, Z. Prokop, Y. Sato, P. Jerabek, A. Kumar, Y. Ohtsubo, M. Tsuda and J. Damborsky, *Appl. Environ. Microbiol.*, 2005, **71**, 2183.
56. R. N. Manna and A. Dybala-Defratyka, *Arch. Biochem. Biophys.*, 2014, **562**, 43.
57. J. J. P. Stewart, *J. Comp. Chem.*, 1989, **10**(209), 221.
58. W. L. Jorgensen, J. Chandrasekhar, J. D. Madura, R. W. Impey and M. L. Klein, *J. Chem. Phys.*, 1983, **79**, 926.
59. D. R. B. Brittain, R. Pandey, K. Kumari, P. Sharma, G. Pandey, R. Lal, M. L. Coote, J. G. Oakeshott and C. J. Jackson, *Chem. Commun.*, 2011, **47**, 976.
60. (a) Y. Zhao and D. G. Truhlar, *Theor. Chem. Acc.*, 2008, **120**, 215; (b) M. J. Frisch, J. A. Pople and J. S. Binkley, *J. Chem. Phys*, 1984, **80**, 3265.
61. X. Tang, R. Zhang, Y. Li, Q. Zhang and W. Wang, *Bioorg. Chem.*, 2017, **73**, 16.
62. P. Paneth, in *Isotopes in Chemistry and Biology*, ed. A. Kohen and H. H. Limbach, CRC Press, Inc, Baton Rouge, ch. 35, 2006, pp. 875–891.
63. A. Siwek, R. Omi, K. Hirotsu, K. Jitsumori, N. Esaki, T. Kurihara and P. Paneth, *Arch. Biochem. Biophys.*, 2013, **540**, 26.
64. K. Świderek and P. Paneth, *Chem. Rev.*, 2013, **113**, 7851.
65. J. D. Bauman, D. Patel, C. Dharia, M. W. Fromer, S. Ahmed, Y. Frenkel, R. S. Vijayan, J. T. Eck, W. C. Ho, K. Das, A. J. Shatkin and E. Arnold, *J. Med. Chem.*, 2013, **56**, 2738.
66. A. Krzemińska, P. Paneth, V. Moliner and K. Świderek, *J. Phys. Chem. B*, 2015, **119**, 917.
67. A. Krzemińska, T. Frączek and Paneth, *Arch. Biochem. Biophys.*, 2017, **635**, 87.
68. T. Frączek, R. Kamiński, A. Krakowiak, E. Naessens, B. Verhasselt and P. Paneth, *J. Enzyme Inhib. Med. Chem.*, 2017, **33**, 9.
69. T. A. Kirschberg, M. Balakrishnan, W. Huang, R. Hluhanich, N. Kutty, A. C. Liclican, D. J. McColl, N. H. Squires and E. B. Lansdon, *Bioorg. Med. Chem. Lett.*, 2008, **18**, 1131.
70. A. C. Camproux, R. Gautier and P. Tufféry, *J. Mol. Biol.*, 2004, **339**, 591.
71. (a) H. Li, A. D. Robertsonand and J. H. Jensen, *Proteins*, 2008, **73**, 765–783; (b) M. H. Olsson, C. R. Sondergaard, M. Rostkowski and J. H. Jensen, *J. Chem. Theory Comput.*, 2011, **7**, 525.
72. R. A. Friesner, R. B. Murphy, M. P. Repasky, L. L. Frye, J. R. Greenwood, T. A. Halgren, P. C. Sanschagrin and D. T. Mainz, *J. Med. Chem.*, 2006, **49**, 6177.
73. M. W. Mahoney and W. L. Jorgensen, *J. Chem. Phys.*, 2000, **112**, 8910.
74. M. J. Field, M. Albe, C. Bret, F. Proust-De Martin and A. Thomas, *J. Comput. Chem.*, 2000, **21**, 1088.
75. M. J. S. Dewar, E. G. Zoebisch and E. F. Healy, *J. Am. Chem. Soc.*, 1985, **107**, 3902.

76. G. D. Ruggiero, S. J. Guy, S. Martí, V. Moliner and I. H. Williams, *J. Phys. Org. Chem.*, 2004, **17**, 592.
77. R. Kaminski, T. Fraczek, A. Paneth and P. Paneth, unpublished work.
78. A. Krzemińska-Kowalska, PhD thesis, Lodz University of Technology, 2018.
79. (a) Directive 2004/42/CE of the European Parliament and of the Council of 21 April 2004 on the limitation of emissions of volatile organic compounds due to the use of organic solvents in certain paints and varnishes and vehicle refinishing products and amending Directive 1999/13/EC (https://eur-lex.europa.eu/eli/dir/2004/42/oj access data 7/12/2018); (b) L. Huang, N. C. Sturchio, T. Abrajano, L. J. Heraty and B. Holt, *Org. Geochem.*, 1999, **30**, 777; (c) A. Kornilova, L. Huang, M. Saccon and J. Rudolph, *Atmos. Chem. Phys.*, 2016, **16**, 11755.
80. L. Vasquez, M. Rostkowski, F. Gelman and A. Dybala-Defratyka, *J. Phys. Chem. B*, 2018, **122**, 7353.
81. B. J. Berne and D. Thirumalai, *Annu. Rev. Phys. Chem.*, 1986, **37**, 401.
82. G. A. Voth, D. Chandler and W. H. Miller, *J. Chem. Phys.*, 1989, **91**, 7749.
83. A. D. MacKerell Jr., D. Bashford, M. Bellott, R. L. Dunbrack, J. D. Evanseck, M. J. Field, S. Fischer, J. Gao, H. Guo, S. Ha, D. Joseph-McCarthy, L. Kuchnir, K. Kuczera, F. T. Lau, C. Mattos, S. Michnick, T. Ngo, D. T. Nguyen, B. Prodhom, W. E. Reiher, B. Roux, M. Schlenkrich, J. C. Smith, R. Stote, J. Straub, M. Watanabe, J. Wiórkiewicz-Kuczera, D. Yin and M. Karplus, *J. Phys. Chem. B*, 1998, **102**, 3586.
84. M. Liu, K. N. Youmans and J. Gao, *Molecules*, 2018, **23**, 2644.
85. J. J. P. Stewart, *J. Molec. Modeling*, 2013, **19**, 1.
86. S. Żaczek, F. Gelman and A. Dybala-Defratyka, *J. Phys. Chem. A*, 2017, **121**, 2311.
87. W. Wang, T. Sun, Y. Zhang and Y.-B. Wang, *J. Comput. Chem.*, 1763, **2015**, 36.
88. M. Pokora and P. Paneth, *Molecules*, 2018, **23**, 2981.
89. M. J. Frisch, G. W. Trucks, H. B. Schlegel, G. E. Scuseria, M. A. Robb, J. R. Cheeseman, G. Scalmani, V. Barone, G. A. Petersson, H. Nakatsuji, X. Li, M. Caricato, A. V. Marenich, J. Bloino, B. G. Janesko, R. Gomperts, B. Mennucci, H. P. Hratchian, J. V. Ortiz, A. F. Izmaylov, J. L. Sonnenberg, D. Williams-Young, F. Ding, F. Lipparini, F. Egidi, J. Goings, B. Peng, A. Petrone, T. Henderson, D. Ranasinghe, V. G. Zakrzewski, J. Gao, N. Rega, G. Zheng, W. Liang, M. Hada, M. Ehara, K. Toyota, R. Fukuda, J. Hasegawa, M. Ishida, T. Nakajima, Y. Honda, O. Kitao, H. Nakai, T. Vreven, K. Throssell, J. A. Montgomery Jr., J. E. Peralta, F. Ogliaro, M. J. Bearpark, J. J. Heyd, E. N. Brothers, K. N. Kudin, V. N. Staroverov, T. A. Keith, R. Kobayashi, J. Normand, K. Raghavachari, A. P. Rendell, J. C. Burant, S. S. Iyengar, J. Tomasi, M. Cossi, J. M. Millam, M. Klene, C. Adamo, R. Cammi, J. W. Ochterski, R. L. Martin, K. Morokuma, O. Farkas, J. B. Foresman and D. J. Fox, *Gaussian 16, Revision A.03*, Gaussian, Inc., Wallingford CT, 2016.

CHAPTER 5

Applications of Computational Intelligence Techniques in Chemical and Biochemical Analysis

MILES GIBSON,[a] BENITA PERCIVAL,[a] MARTIN GROOTVELD,[a] KATY WOODASON,[a] JUSTINE LEENDERS,[a] KINGSLEY NWOSU,[a] SHINA CAROLINE LYNN KAMERLIN[b] AND PHILIPPE B. WILSON*[c]

[a] Leicester School of Pharmacy, Faculty of Health and Life Sciences, De Montfort University, The Gateway, Leicester LE8 9BH, UK; [b] Department of Chemistry – BMC, Uppsala University, BMC Box 576, S-751 23 Uppsala, Sweden; [c] School of Animal, Rural and Environmental Sciences, Nottingham Trent University, Brackenhurst Campus, Brackenhurst Lane, Southwell, Nottinghamshire NG25 0QF, UK

*Email: philippe.wilson@ntu.ac.uk

5.1 Historical Use of Artificial Intelligence

The idea of a machine or mechanism exhibiting intelligent behaviour to test the hypothesis of logical thought was historically seen as merely a theoretical possibility. Events post-World War II led to the invention of what we now call modern computing. The influence on modern computing by visionaries such as Alan Turing, Howard Aiken, the Moore school and their associated laboratories, as well as IBM and Bell laboratories, allowed for a platform

Theoretical and Computational Chemistry Series No. 20
Computational Techniques for Analytical Chemistry and Bioanalysis
Edited by Philippe B. Wilson and Martin Grootveld

Published by the Royal Society of Chemistry, www.rsc.org

which enabled the growth of the computational power and programming language necessary to handle artificial intelligence (AI) applications. However, AI is not simply a measure of the computational power and problem-solving ability, but also the capacity of a 'machine' to understand the fundamentals of intelligent thoughts and actions.

Herb Simon formulated the basis of information processing theory, whereby the behaviour of a rational individual can be calculated, based on the premise that the decision made is a conclusion of the values of certain variables under consideration.[1] Therefore, if the variables of decision are specified and a value given, it is possible for a factual decision to be obtained through logic. Simon highlighted three components required to establish the framework for problem-solving behaviour: the information processing system, the task environment and the problem space, as shown and further defined in Figure 5.1.[1] This section will provide an overview of the history and development of AI, machine learning (ML) and neural networks (NN), as well as further discussions regarding the application and impact that these

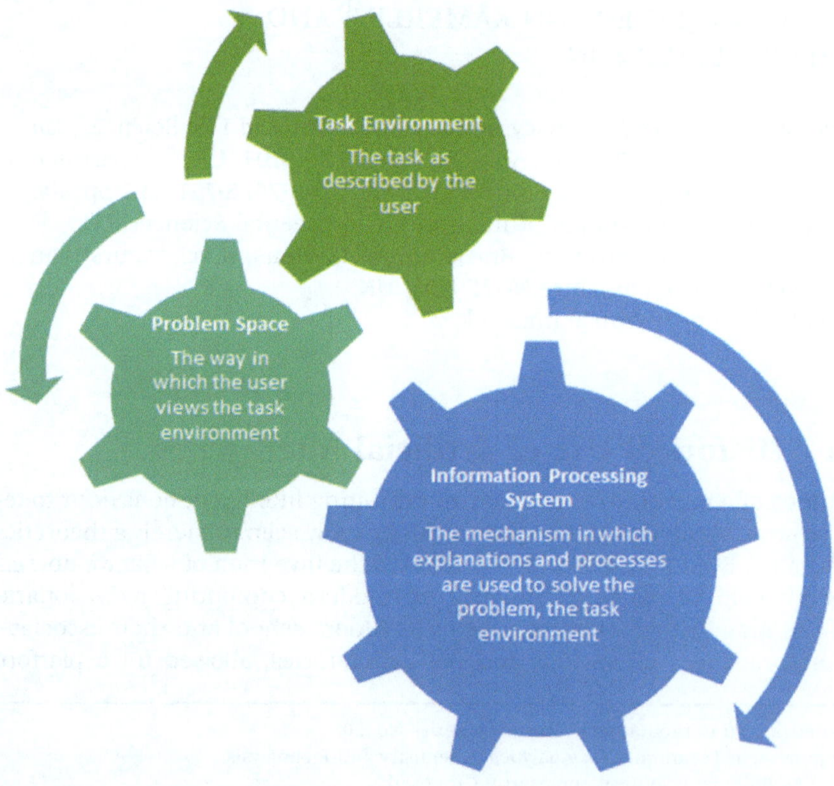

Figure 5.1 The conceptual framework for problem solving developed by Simon in 1978.

techniques have had on very large datasets comprising hyperdimensional patterns of molecules and their concentrations arising from modern-day, 'state-of-the-art' analytical/bioanalytical chemistry platforms.

Artificial intelligence as a concept has been theorised by philosophers for centuries, with literary works dating as far back as Homer's *Illiad* and the autonomous golden tripods of Hephaestus,[2] as well as Plato's *Meno* and the creations of Daedalus known to exhibit wanderlust if not kept securely fastened.[3] The influence of a multitude of disciplines has led to the evolution of modern-day AI, with the early works of Thomas Hobbes, Gottfried Leibniz and Rene Descartes laying foundations in philosophy, mathematics and the mechanisms of the human mind.[4–6] Indeed, Hobbes is viewed as one of the key influencers of AI with his rationale "*By ratiocination, I mean computation*". Moreover, Hobbes theorised that the human brain processes thoughts through a means of symbolic operations or 'parcels'. These parcels are not numbers or figures, but a language of the brain formulating the conceived input by means of association and connections. Therefore, the most optimal thought patterns are ones carried out under clear, rational and methodical directives. Hence, the processing of thoughts within a specified ruleset, such as the set of rules that dictate human reasoning, can be deemed mechanical in nature.[4]

In the 19th century, Charles Babbage developed the difference and analytical engines, the first devices capable of automatic calculation of mathematical functions.[7] The analytical engine, although not fully developed during Babbage's lifetime, possessed fundamental design characteristics that are still used in modern computers. Furthering the work of Babbage, Ada Lovelace theorised the creation of codes to be used in conjunction with the analytical engine, to manage numbers, letters and symbols within mathematical functions.[8] In 1854, George Boole published his work on '*The Laws of Thought*' which outlined the concept that in language, the operations of reasoning can be conducted using a system of 'signs'. Moreover, these signs can be applied as algebraic and logical functions in mathematical calculations. These principles, known as Boolean algebra or logic, form the foundations upon which computer programming and electrical engineering are based, amongst others.[9]

Claude Shannon described how Boolean algebra could be utilised as electronic relays and switches, and later published his seminal work '*The Mathematical Theory of Communication*', more commonly referred to as information theory, in 1948. Therein, Shannon defined information in terms of mathematics, outlined the schematic for a communication system, and also how this information could be transmitted from one destination to another.[10] In 1950, Alan Turing published an influential paper in the philosophy journal *Mind*, in which he posed the question "can machines think?", and outlined his concept of the imitation game, a means of testing whether an interrogator would be able to distinguish between human and machine through a series of questions posed.[11] This work demonstrated the potential, through programming, for a machine to behave intelligently.

The creation of the Stochastic Neural Analog Reinforcement Calculator (SNARC) by Marvin Minsky in 1951 was the first iteration of an NN learning machine, in which the device would increase the weighting or favourability of a synapse based on its success in performing a specific task.[12] Previous pioneering work in the field of neurophysiology[13,14] supported the formulation and development of this NN system.

The term 'AI' was first coined by John McCarthy during the collaborative Dartmouth research project.[15] This included many notable attendees of the AI, ML and NN community, such as Claude Shannon, Marvin Minksy, Warren McCulloch, Herb Simon and Allen Newell. The focus of the collaborative project was to precisely define each aspect of intelligence, wherein features of human intellect such as learning, creativity and self-improvement can be simulated by machines.[15] Frank Rosenblatt published his model of the *perceptron* in 1958, a single layer NN capable of binary classification, a system in which data or patterns that exist either side of a linear separating hyperplane can be identified.[16] The limitations of the perceptron were later highlighted by Minsky and Papert in *Perceptrons*, wherein the functionality of Rosenblatt's perceptron and its subsequent multilayer NN variants were heavily criticised for their functionality.[17]

In the 1970s, public interest in and perception of AI decreased significantly, causing an 'AI Winter'. Indeed, during this time funding for AI projects dwindled, and innovation in the field began to stagnate from the subsequent effects caused by a lack of public interest which led to a reduction of financing from sources such as the Defence Advanced Research Projects Agency (DARPA), which was previously a large supporter of this field. However, during this period, Simon and Newell published their work *Human Problem Solving*,[18] which developed upon their 1958 paper *Elements of a Theory of Human Problem Solving*,[19] wherein they tested the potential for programming a digital computer to perform problem-solving tasks deemed too difficult for humans. Indeed, *Human Problem Solving* laid out the basis for an information processing system, which was subsequently updated and published in *Information Processing Theory of Human Problem Solving*.[1]

The fields of ML and NN saw a number of substantial discoveries in the 1980s, including neocognitron, a self-organising learning NN with the ability to recognise stimulus patterns by geometric similarity,[20] Hopfield networks, a single layer NN system that featured content addressable memory capabilities, which allowed the system to recall whole patterns based on only a portion or noisy version of the data,[21] and Q learning, a reinforcement learning method that updates the weighting of a decision path based on whether the decision arising therefrom leads to a reward or not, the current weighting of the path and the weighting of further paths now being made available to it.[22] Further developments in this area arose in 1986 with Rumelhart's application of backpropagation, an expansion upon NN systems, wherein 'hidden' units continually adjust the weighting or bias of connections within the network, thus reducing the difference between the theoretical and measured outputs.[23]

With the continual advancement of computer hardware, including processing power and storage, AI products began to actively engage with the consumer market in the 21st century. Indeed, AI has seen a surge of popularity and funding in recent years, with giant corporations such as Amazon, Microsoft, Google, Apple and so forth all featuring continuous development of innovative products containing intelligent systems. This growth in the field of AI led to a monumental breakthrough with the development of deep learning in 2015.[24] Figure 5.2 shows the timeline of these progressions throughout the years.

5.2 Early Adoption in the Chemistry and Bioanalysis Areas

The fields of chemistry and bioanalysis are an ideal medium for the application of AI systems. Chemical disciplines often require collation, interpretation and statistical analysis of large amounts of data, providing both the required sample size in which an intelligent machine may learn, and a method in which the chemist may solve complicated problems, such as classification of differing groups or sub-groups. One of the earliest applications of AI in the field of chemistry is the DENDRAL project, derived from the term dendritic algorithm. This was an AI project undertaken in the early 1960s. The basis of this was to apply heuristic programming for experimental analysis in empirical science,[25] but was predominantly applied to molecular structural elucidation problems encountered in organic chemistry.[26] In the publication series *Applications of Artificial Intelligence for Chemical Inference I-XL* (1969–1982), the development and application of AI to fields such as mass spectrometry (MS), nuclear magnetic resonance spectroscopy (NMR), stereochemistry and structure elucidation were discussed. However, an in-depth analysis of these early works is beyond the scope of this chapter, and has therefore been included as a means of reference only. During the early 1970s, ML techniques were applied in a wide range of fields including polarography,[27] low-resolution MS[28–30] and pharmacology.[31]

Further developments arose in the application of statistical analysis techniques. Indeed, the *k*-nearest neighbours (KNN) pattern recognition technique was explored for analysis of the NMR pattern vectors in order to determine the molecular structures of compounds from unknown spectra.[32] Cluster analysis was utilized for the interpretation of MS data acquired on alkyl thiolesters using the shortest spanning path (SSP) method for linear ordering of clustered data. However, the application was found to be limited to characterisation of only straight chain alkyl thiolesters.[33] Furthermore, pattern recognition and cluster analysis were simultaneously featured in the analysis of the pharmacological activity of tranquilizers and sedatives,[34] as well as X-ray fluorescence analysis of trace elements found in archaeological obsidian glass samples.[35] Additional early developments were made in the field of infectious disease therapy, and the MYCIN system, coined from the suffix of antibiotics, was formulated in 1977 as a clinical consultant program

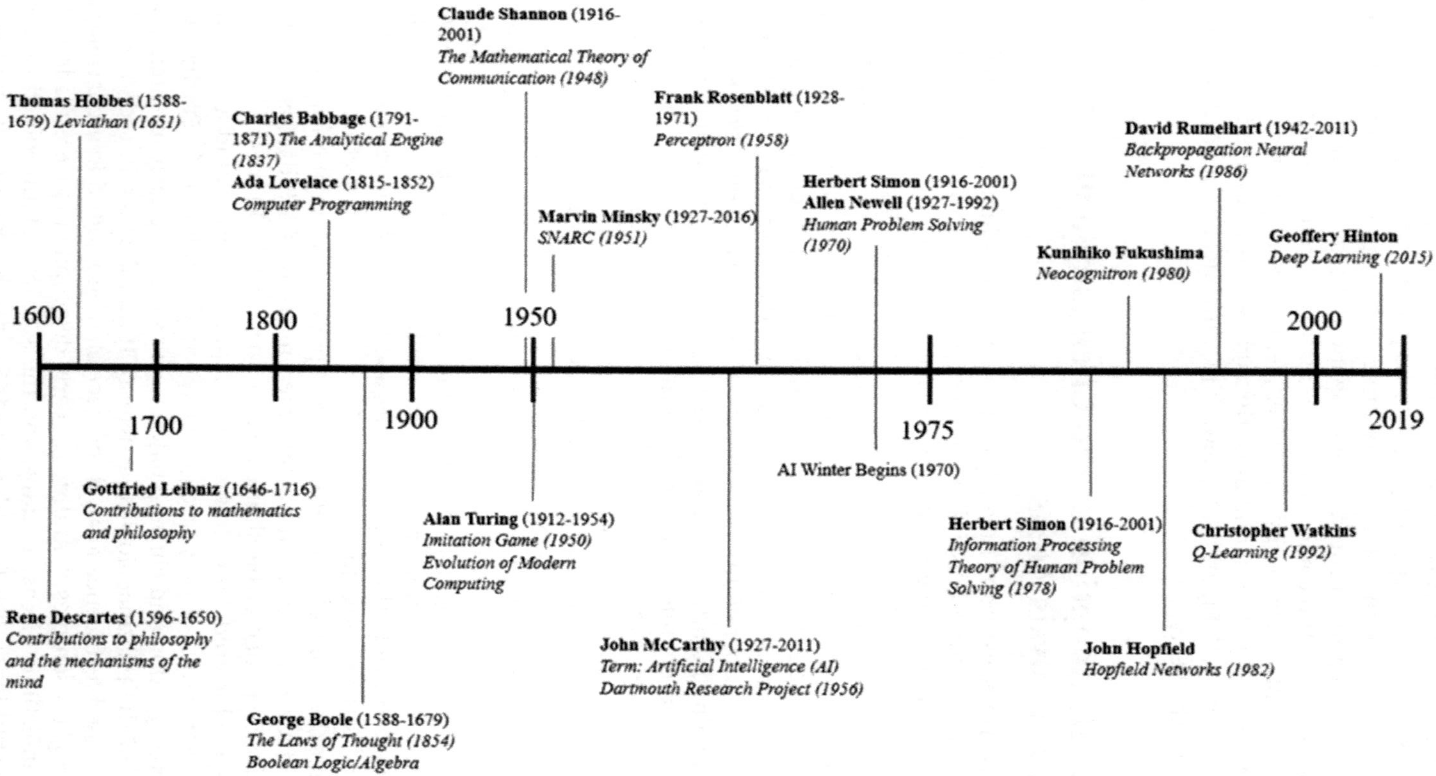

Figure 5.2 An overview of the progression of AI throughout the years, including some of the key influencers and their respective developments and contributions.

of therapy selection for patients with a range of defined infections. Indeed, the concept of the MYCIN project was a learning program that would receive general medical information from a medical professional, and thereafter formulate its own patient-specific diagnoses, and also provide advice and explanations for its decision.[36]

5.3 Recent Applications of Machine Learning in Chemistry

Artificial intelligence has evident applications for use in analytical methods and techniques across a multitude of chemical and biochemical disciplines, including forensics, organic chemistry, chemometrics, and its generally more sophisticated cousin, metabolomics. This section will describe the implementation of AI in this area, firstly by defining the learning method used, and exploring some of the breadth of algorithms utilised within these learning techniques. This section will further discuss a broad range of applications within chemistry, including an in-depth analysis of a case study with considerations as to which approach is suitable and why. Applications in metabolomics have been previously explored in chapter 2, and will not be discussed in depth herein, but some consideration of the applications available will be explored.

Machine learning can be split into four distinct groups of learning algorithms: supervised, unsupervised, semi-supervised and deep learning. A supervised learning technique utilises known input and output variables and an algorithm in order to learn the function required to map the input variables to the output ones.[37] The theory behind a supervised learning technique is that by using a training set, the mapping function can be refined, so that upon addition of new input data, an accurate prediction of output data can be obtained. Supervised learning algorithms are applied to datasets that exhibit classification or regression problems. Classification involves types of problems in which the output variables are required to be categorised, for example disease or non-disease states in clinical research. A regression problem involves prediction of a numerical value based on previous data, for example concentration determination through a calibration curve, or perhaps the severity of a disease process from clinical and/or molecular biomarker variables. Commonly used algorithms include support vector machine (SVM), multilayer neural networks, KNN, linear regression (LR), decision tree (DT) and random forest (RF) strategies.

Alternatively, unsupervised learning techniques only consist of input variables, and the presence of any underlying distributions or structures within the dataset are determined by the applied algorithm.[38] Unsupervised learning algorithms are applied to problems involving clustering or association within datasets. Clustering problems involve their analysis for underlying groupings, with the goal of separating groups based on common attributes. Association, on the other hand, refers to the determination of rules that are linked or describe large portions of input data. Examples of unsupervised learning algorithms include *k*-means clustering (KMC) and independent component analysis (ICA).

A semi-supervised approach is often more appropriate for datasets containing a large amount of variable input data with only partially-available output data. In this example, unsupervised techniques can be used to analyse a large dataset for patterns and distributions to which supervised techniques can then be applied in order to determine the relative output variable associated. The term semi-supervised refers to a combination of supervised and unsupervised algorithms applied to the specified problem. The supervised component trains the model on labelled data, which includes a known outcome variable, and the unsupervised component learns from the unlabelled data present that does not contain an outcome variable in order to more accurately understand the data structure. Semi-supervised learning is primarily used to increase the size of the training data when an insufficient amount of labelled data is present to create an accurate model. Certain assumptions are made within the unlabelled data in order to determine any underlying distribution present. The three major assumptions that form the basis for semi-supervised techniques are smoothness, cluster and manifold assumptions. Firstly, the smoothness assumption, also referred to as the continuity assumption, indicates that two data points in close proximity to one another would likely share the same label or grouping. Cluster assumption postulates that data tends to form groups, and by extension, data within the same group will likely share the same label. Finally, manifold assumptions presume that multidimensional data points are likely to form on a low dimensional Euclidean space or manifold.[39] Euclidean space refers to a set of data or data points represented in a three-dimensional topological space.

Finally, deep learning is a quite recently developed technique that utilises a NN approach consisting of multiple layers containing individual algorithms or components at each layer. Deep learning will be discussed in more depth below.

Further algorithms can be used for processing, such as feature learning, sparse dictionary learning, anomaly detection, decision tree learning, and association rule learning. Once the algorithms are defined, data can be modelled suitably using NN, SVM, RF and so forth. The use of such algorithms often requires validation and testing of the sensitivity and specificity, such as the area underneath the receiver operating characteristic curve (AUROC) value. Access to modelling programs is available online, and these are predominantly free; examples include the SVM^{light} package and MATLAB's free SVM toolbox.[40]

5.3.1 Support Vector Machines

Support vector machines (SVMs) are a form of supervised machine learning algorithm predominantly used for facilitating compound classification, ranking and regression-based predictions.[41] SVMs employ kernel functions for pattern recognition of datasets, and the most widely used kernel function within this strategy is the radial basis function (RBF), although other kernel-based functions include linear, polynomial and sigmoidal ones. An example of a linear kernel-based SVM approach was used by Granda *et al.*[42] for the prediction of new reactions using an organic synthesis robot. This robot was equipped with real-time sensors, a flow benchtop NMR analysis system, an MS facility, and an attenuated total-reflection infrared spectroscopy system. This system successfully recorded a reactivity classification accuracy of 86%.[42]

Input data are projected into a high-dimensional feature space using the applied kernel function in an attempt to represent them in a linearly separable manner. Once linearly separable, a hyperplane is introduced to separate the feature space which splits the dataset into two classification sets. The hyperplane position is determined by the greatest margin of distance between these classification sets; the data points used in this margin determination are known as support vectors, and lie upon a support hyperplane. In some instances, two classes will not be fully linearly separable. However, in these cases SVM applies a hyperplane using a soft margin which maximises the margin distance, whilst also minimising the sample misclassification rate.

Support vector machines have been utilised as tools for determining the accessibility of synthetic chemical compounds in drug discovery. These SVM methods are trained on diverse libraries of chemical compounds (known to be synthetically accessible) for the identification of those with and without precursor knowledge of a specific set of reactions and reactants.[43] In this particular study, two SVM approaches, namely RS_{SVM} and DR_{SVM}, were employed. The former considered defined reactions and starting materials, whilst the latter assessed synthetic accessibility. Both RS_{SVM} and DR_{SVM} approaches performed well, achieving a score of 0.888 and 0.952 respectively using receiver operation characteristics (ROC) as the cross-validation methodology, and were able to rank the difficulty of molecular synthesis correctly. The DR_{SVM} model was more effective than the RS_{SVM} one as there was no bias towards the reactions. The libraries varied in size from 830 to 22 4278 compounds, which demonstrated the ability of SVM to handle large datasets.

Moreover, SVMs have also been applied using the quantitative structure-activity relationship (QSAR) approach, a technique for the prediction of molecular behaviour within biological, environmental or physiochemical systems. QSAR involves the correlation of molecular feature changes to predefined molecular descriptors. In view of these features, QSAR finds application within fields such as medicinal chemistry, for example drug discovery and design,[44] environmental chemistry[45] and toxicology,[46] as a means of statistical optimisation through its ability to monitor the biological activities of compounds. QSAR has seen use for the prediction of anti-HIV activity in thiobenzimidazolone (TIBO) derivatives, the success or functionality of which is determined by each compound's ability to protect MT-4 cells against cytopathic effects exerted by the HIV virus.[47] Applicability within anti-HIV methodologies has been further explored for prediction of the bioactivity in HIV1 integrase strand transfer inhibitors,[48] and the screening of HIV-associated neurocognitive disorders.[49] Prediction of apoptosis protein sequences,[50] as well as prediction of "hot spot" residues in protein-protein binding interactions.[51]

More recently, SVM approaches have been applied for the classification of enzymes such as β-lactamases.[52] The *one-versus-rest* approach was used to formulate models utilising a range of parameters and kernels. The leave-one-out cross validation (LOOCV) or jackknife testing systems were applied in order to assess the performance, as it is a rigorous form of evaluation. Furthermore, sensitivity, specificity, accuracy and Matthews correlation coefficient (MCC) analyses were also performed, using eqn (5.1) to (5.4) respectively. These are suitable for single label systems, in which the TP and TN descriptors represent true positive and true negative values respectively, that is the number of proteins that were correctly classified. FP and FN represent the number of false positive and false negative numbers of proteins classified correctly and incorrectly classified respectively.

$$\text{Sensitivity} = \frac{\text{TP}}{\text{TP} + \text{FN}} \times 100\,\% \tag{5.1}$$

$$\text{Specificity} = \frac{\text{TN}}{\text{TN} + \text{FP}} \times 100\,\% \tag{5.2}$$

$$\text{Accuracy} = \frac{\text{TP} + \text{TN}}{\text{TP} + \text{TN} + \text{FP} + \text{FN}} \times 100\,\% \tag{5.3}$$

$$\text{MCC} = \frac{(\text{TP} \times \text{TN}) - (\text{FP} \times \text{FN})}{\sqrt{(\text{TP} + \text{FP})(\text{TP} + \text{FN})(\text{TN} + \text{FP})(\text{TN} + \text{FN})}} \times 100\,\% \tag{5.4}$$

This testing was performed at two levels, the first distinguishing between β-lactamase and non-β-lactamase systems. The second level determined the class of β-lactamase, that is classes A, B, C or D. The results acquired from the first level of testing showed an overall accuracy of 89.91%, with an MCC value of 0.60 and an AUC value of 0.91. The second level of testing showed a high accuracy (≥60%) and an AUC ≥ 0.63 for each group, with MCC levels ranging from 0.18 to 0.65.

Indeed, the application of SVMs to the field of forensic science has been used as a means for age prediction of forensic samples through analysis of 11CpG loci for DNA methylation; Figure 5.3 shows a comparison between the regression models used in this study. It can be observed that, in this case, the SVM model was proven to be more robust than its counterpart multivariate linear and non-linear regression approaches and back propagation NN.[53] Determination of anthropological craniometrics have also previously been shown as a robust application of SVMs in comparison to other regression models.[54] Further applications within the field of forensic science

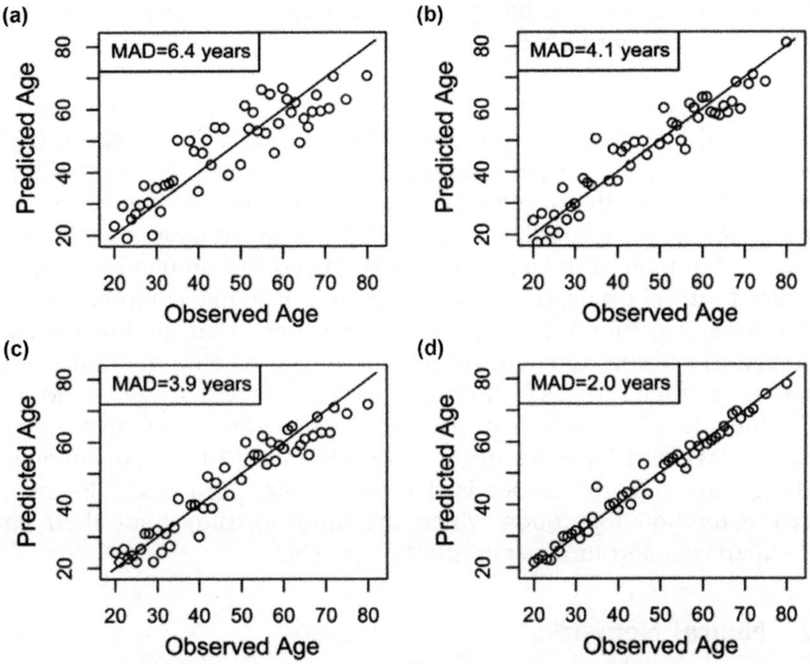

Figure 5.3 A comparative learning algorithm approach for the analysis of DNA methylation sites in forensic samples: (a) multilinear regression; (b) multivariate non-linear regression; (c) backpropagation neural network model; and (d) support vector machine regression model.

include firearms evidence for association of primer shear indentation topography present on expended 9-mm bullet cases with their respective firearm counterpart using confocal microscopy.[55] SVM has been employed in tandem with other machine learning techniques, specifically KNN, linear discrimination analysis (LDA) and partial least-squares discriminant analysis (PLS-DA) in a combinational approach for the determination of the geographical origin of medicinal herbs, using inductively coupled plasma atomic emission spectroscopy (ICP-AES)/MS and ^{1}H NMR spectroscopy.[56]

SVMs have been utilised in conjunction with a number of instrumental techniques such as dynamic surface-enhanced Raman spectroscopy (D-SERS) for the determination of drug concentrations of methamphetamine in human urine.[57] RBF kernels were applied in view of the fast training phase of this algorithm, as well as a grid search method for optimisation. Urine with and without 'spiked' methamphetamine and MDMA in various concentrations was used as a calibration model. The performance was evaluated using a 5-fold cross-validation methodology. The SVM divided the data into five equal subsets, one of which was for validation purposes and the remaining subsets were used for training. Validation of the model identified three extra samples that were indeed abusers of methamphetamine. The accuracy was shown to be more than 90% using this model.

Metabolomics investigations have used SVMs for discrimination of steroid metabolites in urine,[58] in which the application of SVMs with a longitudinal approach to the profiling of steroid samples was capable of discerning between abnormal and normal samples of these agents. The combination approach allowed for the screening of the most sensitive differences between abnormal and normal samples. Zsila *et al.*[59] developed an approach for the prediction of human serum albumin (HSA) IIA and IIIA protein binding-sites properties to novel drugs, drug candidates and drug-like compounds.[59]

SVMs have also found application in semi-supervised methods such as fault diagnosis studies of chemical processes, considering the faults of the Tennessee Eastman Process (TEP) benchmark as a model. A study by Monroy *et al.*[60] included ICA as a feature extraction method, Gaussian mixture models (GMM) with Bayesian information criterion (BIC) for unsupervised clustering, and SVM for supervised classification purposes.[60] Figure 5.4 shows the methodology upon which the Monroy study built their combined supervised and unsupervised SVM approach.

5.3.2 Neural Networks

Artificial neural networks (ANN) are a branch of supervised and unsupervised ML methods that function by the use of neurons similar to the neurological network system within the human brain. A neuron, by the most basic of definitions, is a binary decision maker whereby, the value for it can exist as either 1 or 0, indicating a result such as true/false or on/off, for example. The decision variable of the neuron is assigned a weighting based upon its greater or lesser importance within the network. To determine an output variable within a supervised ANN technique, an activation function is required, which when calculated with the sum of input values and their corresponding weights, returns a value for the desired output. In simpler NN systems, the activation function assigns negative input values to 0 and positive ones to 1, and may include a bias value. Some of the more common activation functions include sigmoidal, hyperbolic tangent, rectified linear and radial-based functions.

Neurons are typically grouped together into a simultaneously operating array known as a NN layer. A multilayer NN will consist of an input layer, an output layer and a number of hidden layers between the two. Neurons connect with all those present in the preceding and following layers, but do not interconnect within the same layer. The input values and weightings of a NN are optimised by training the network on a given dataset. The network is allowed to calculate the theoretical output value, which is then compared to the actual known figure, and adjustments are made to the NN in order to self-enhance optimisation. These adjustments are calculated based on a cost function. The function analyses the entire network, and modifies the weights and biases in order to achieve a greater similarity between the theoretical and

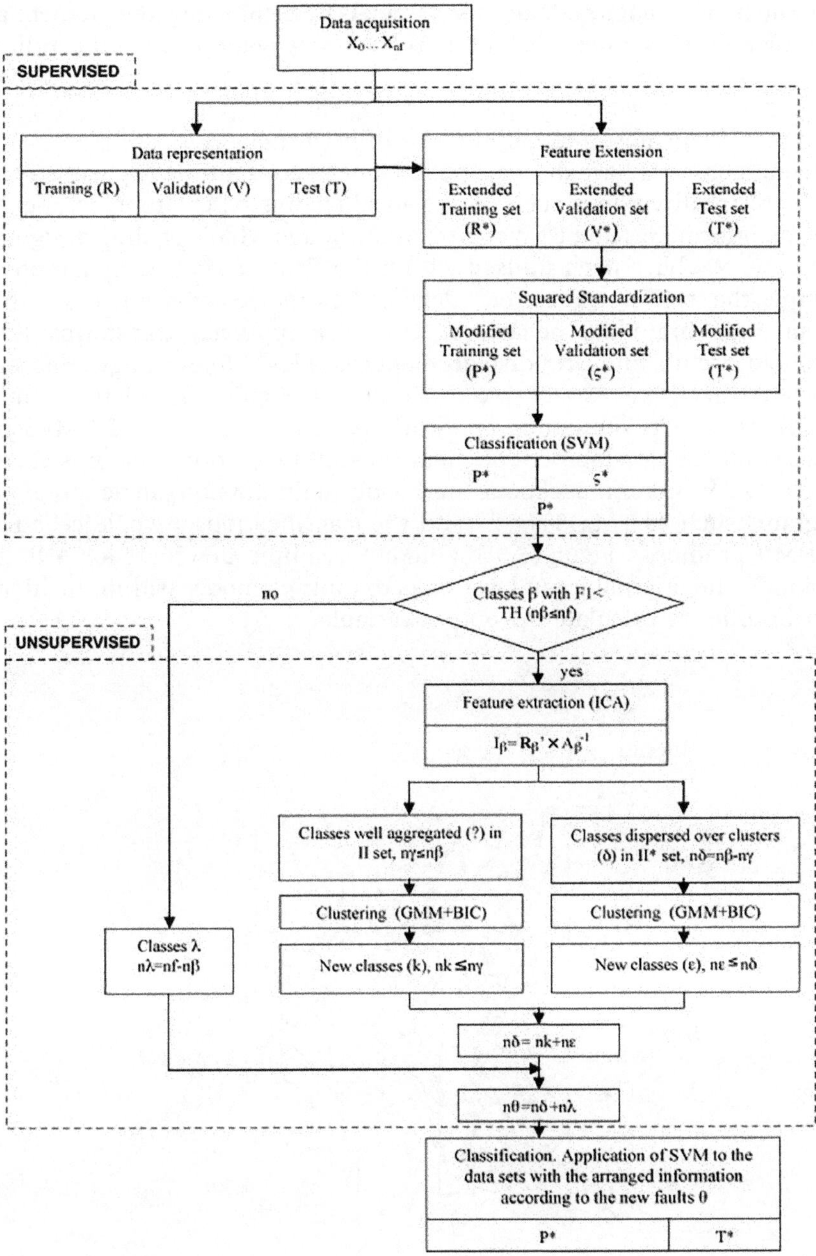

Figure 5.4 Combined supervised and unsupervised methodology used for the determination of faults within chemical processes.
Reproduced from ref. 60 with permission from Elsevier, Copyright 2010.

true values. The magnitude of this change is determined by the gradient and learning rate. This process is known as back-propagation and as it continues, the system becomes more accurate until no further increase in weighting or bias can be achieved, an output indicating that the network weighting and bias are at their most optimal in terms of performance.

Neural networks have diverse applications in a multitude of chemistry and biochemistry disciplines such as analytical chemistry, environmental chemistry, bioinformatics and DNA computation, systems biology, drug design and discovery. NN have been utilised within the field of DNA computation: for example, the study by Qian *et. al.*[61] detailed how molecular systems can exhibit autonomous brain-like behaviour. This NN approach transforms linear threshold circuits into DNA displacement cascades.[61] Interestingly, this study also noted that this NN system is compatible with Hopfield associative memory.

ANN have seen numerous applications within the field of NMR spectroscopy for the development of chemical shift prediction programs such as SPARTA+,[62] protein backbone and side-chain torsion angle prediction programs such as TALOS-N,[63,64] and the classification of whole cells from ^{1}H NMR profiles.[65] Figure 5.5 highlights a simple flowchart for ANN formation.[62] The original input layer links to multiple nodes within the hidden layer in order to calculate the output variable.

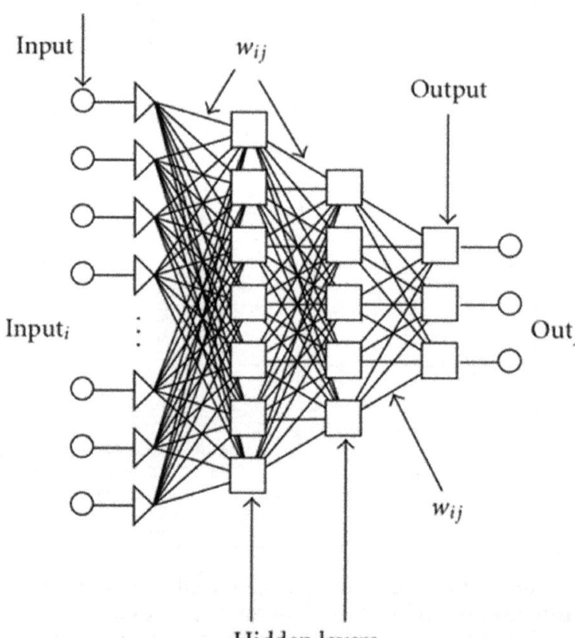

Figure 5.5 Schematic of ANN architecture.
Reproduced from ref. 65, https://doi.org/10.1155/2011/158094, under the terms of a CC BY 3.0 license, https://creativecommons.org/licenses/by/3.0/.

In a study by Mouazen *et al.*,[66] backpropagation NN (BPNN) was compared to the multivariate statistical techniques partial least squares regression (PLSR) and principal component regression (PCR) for the calibration of soil property accuracy measurements acquired using visible-near infrared (vis-NIR). Parameters analysed in this study included organic carbon, potassium, sodium, magnesium and phosphorous elemental contents in Belgian and French soil samples ($n = 168$). The dataset was randomly split into three replicates, with each sample dataset split 90%/10% for cross-validation and prediction sets, respectively.[66] Two forms of BPNN were created in this study, the first using input values consisting of the first five principal components (PCs) from PC analysis (PCA), and the second utilising the optimal number of latent variables (LVs) derived from PLSR. This study showed that the BPNN-LV method out-performed all of the methods undertaken, with BPNN-PC and PLSR both outperforming PCR. Therefore, the performance ranking for this study was shown to be BPNN-LV > BPNN-PC > PLSR > PCR in descending order of performance. The R^2 and residual prediction deviation (RPD) recorded for the BPNN-LV analysis were: organic carbon (0.84, RPD = 2.54), Mg (0.82, RPD = 2.54) and K, P and Na (0.68–0.74, RPD = 1.77–1.94). NN have also been used in conjunction with vis-NIR spectroscopy for the analysis of biodiesel properties such as biodiesel density, kinematic viscosity, methanol and water contents.[67] This study also compared the effectiveness of NN as a calibration method against the linear calibration methods multiple linear regression (MLR), PCR, PLSR and quasi non-linear methods, specifically poly-PLS and spline-PLS. Furthermore, a study by McKenzie *et al.*[68] characterised the total mineral contents of five different tea leaf varieties using ICP-AES, and analysed their dataset using probabilistic neural networks (PNN) and LDA models. The performance of the PNN and LDA models were shown to be 0.97 and 0.81 respectively.

Neural networks have been combined with the Monte Carlo Tree Search (MCTS) strategy, and can be applied for effective planning of chemical syntheses. In a double-blind AB test, it was found that the system generated was equivalent to the reported literature routes.[69] Segler's group were also able to narrow down the search for molecules for *de novo* drug design based on the properties and structures of those inputted. A study by Wei *et al.*[70] highlighted the viability of NN for the prediction of organic chemistry reactions. The NN was trained on a set of $n = 3400$ reactions, validated using k-fold cross-validation and tested on $n = 17\,000$ reactions. This NN returned a training accuracy of 86.0%, and a test accuracy of 85.7%.

Artificial neural networks have been used to predict age based on DNA methylation within forensic science. The study by Vidaki *et al.*[71] analysed blood DNA methylation profiles of $n = 1156$ participants, and successfully identified 23 age-associated CpG sites using stepwise regression. The NN was built upon 16 markers that accurately predicted forensic age with an R^2 value of 0.96 and a mean absolute deviation of 3.8 years. Further applications in forensic science include the use of NN in conjunction with thermal infrared imagery for determination of intoxicated individuals based on thermal imaging of participants' faces.[72]

Neural networks have also been applied in semi-supervised learning for classification and drift counteraction of artificial olfaction with coffee as a model. This study utilised the Pico-1 electronic-nose, a thin film semiconducting sensor device, for artificial olfaction analysis, and applied the 'Semi-Boost' algorithm with an adapted BPNN for classification purposes.[73]

5.3.3 *k* Nearest Neighbours

k nearest neighbours is a supervised learning technique that predicts input data based on its similarity within the full dataset: the technique takes a *k* number of datapoints closest to the new input data, summarises these values and then predicts an outcome classifier based upon the mean or median values, depending upon application, of the dataset. *K* refers to the number of datapoints to be analysed and is a variable chosen by the user. However, determination of the value of *k* can require some optimisation. This method tends to employ Euclidean distance measuring for determination of the nearest neighbour datapoints in the fields of chemistry, as datasets tend to consist of similar variables. Although other distance measures exist and can be more suitable for different data types, such as Manhattan distance measuring, which can be used for variables that are dissimilar. KNN is an advantageous technique for smaller datasets and becomes increasingly more complex with the accumulation of larger ones.

k nearest neighbour algorithms have seen application for the prediction of drug-target interactions through biochemical and physiochemical features such as modelling the adverse liver function effects of drugs,[74] the binding of endocrine-disrupting agents with androgen and oestrogen receptors,[75] and the prediction of adverse drug reactions in neurological drug treatment for conditions such as Alzheimer's disease.[76] Environmental chemistry applications of KNN have been utilised for the derivation of a forecast model of the daily river water temperature,[77] modelling of the half-life levels of chemical substances present in sediment,[78] and prediction of aquatic toxicity.[79]

Further applications of KNN include a method for the non-invasive detection of colon cancer using Raman spectroscopy;[80] the classification of iron ore acidity and alkalinity, which has been investigated by means of laser-induced breakdown spectroscopy;[81] analysis of the geographical origin of medicinal herbs using ICP-AES/MS and ^{1}H NMR analyses;[56] and age determination of bloodstains using spectrophotometry.[82] The application of KNN has also led to improved sensitivity, such as reducing the over-prediction rate for xenobiotic phase I and II metabolic pathway predictions for metabolites derived from the Meteor expert system.[83]

k nearest neighbours have also found applicability in conjunction with other machine learning techniques. A KNN molecular field analysis (KNN-MFA)-based quantitative structure-activity relationship (QSAR) method has seen wide use in the fields of medicinal chemistry, with applications including modelling the design of anticoagulant drugs,[84] anti-Alzheimer's agents[85] anti-microbial agents,[86] anti-hypertensives,[87] and anti-platelet agents,[88] in addition to hepatitis C inhibitors.[89] The KNN-MFA method can utilise different variable selection approaches, including simulated annealing, a simulation of a heating and cooling process distributed according to the Boltzmann distribution; stepwise variable selection, a step-by-step independent variable input method that inspects the model after each step; or a genetic algorithm, a stochastic technique that mimics the process of evolution and natural selection of genetic information and chromosome generation.[90] KNN has also been utilised in conjunction with fuzzy logic methods for determination of secondary and tertiary structures of outer membrane proteins from genomic sequences,[91] the classification of hybrid cancer systems,[92] and the identification of translation initiation sites in genomic studies.[93]

5.3.4 Decision Trees

Decision tree learning is a supervised machine learning technique in which new input data is classified by a tree of defined descriptor values derived from training data using a recursive algorithm. The DT begins with a root node consisting of a descriptor that allows for the largest splitting of a dataset. For example, in a population-based DT system, a common starting node would be gender, as data would split approximately 50 : 50. From these variables, additional descriptors are defined which further reduces the size of the dataset based on the probability of the next descriptor's output values. This process continues until each sub-set or branch is fully split into a single pure outcome value. A major disadvantage of DT learning is that the technique suffers from overfitting, in which the tree will continue to grow until each input data can be individually defined. This problematic situation can be avoided by restricting splitting based on statistical significance, or by pruning. Pruning is a technique which revisits the original tree formed from the training set, and determines whether any of the tree branches are underperforming or unutilised in future training samples. The DT is then reanalysed based on whether the decisions made would still be necessary without these branches, and if they are not required, they are removed from the tree in an effort to consolidate the branches. RF is an adaptation of DT learning that involves creating a number of different decision trees or k decision trees randomly from the original training dataset. Each tree will then split a sub-set, or attribute, and fully expand the node that separates the data; subsequently, data may be classified on the basis of the most common classification derived from all of the trees created.

The application of DT learning techniques has been applied within the fields of proteomic research for the prediction of tryptic cleavage sites using MS.[94] RF techniques have also been applied in proteomics for the identification of classical and non-classical secretory proteins in mammalian genomes,[50] as well as for improvement of scoring functions of protein-ligand binding data.[95] Moreover, applications of RF have been employed in the field of drug discovery for the selection of molecular descriptors,[96] and the prediction of *in vitro* drug sensitivity.[97] Figure 5.6 shows the algorithm upon which Riddick *et al.*97 built their RF model.

Within metabolomics, RF has been applied to phenotype discrimination and biomarker selection,[98] prediction of pharmaceutical toxicity,[99] prediabetic metabolite identification[100] and metabolic profile importance in non-alcoholic fatty liver disease,[101] for example. In a study conducted by Shah *et al.*,[102] the application of RF classification and regression trees (CARTs) was used to determine metabolite significance for separation of chronic kidney disease (CKD) stages. This RF approach randomly selected CKD stage 2 and stage 4 samples, then proceeded to randomly select the metabolites in which to build CARTs selection processes based on the chosen samples, and then further formulated 50 000 CARTs in order to classify the remaining samples that had not been selected between stages 2 and 4 of CKD.[102] RF strategies have also been used in conjunction with benchtop NMR analysis for the metabolomics investigation of diabetes in human urine samples.[103]

The field of environmental chemistry has also featured a diverse application of DT learning, for instance in the analysis of acid rain patterns in

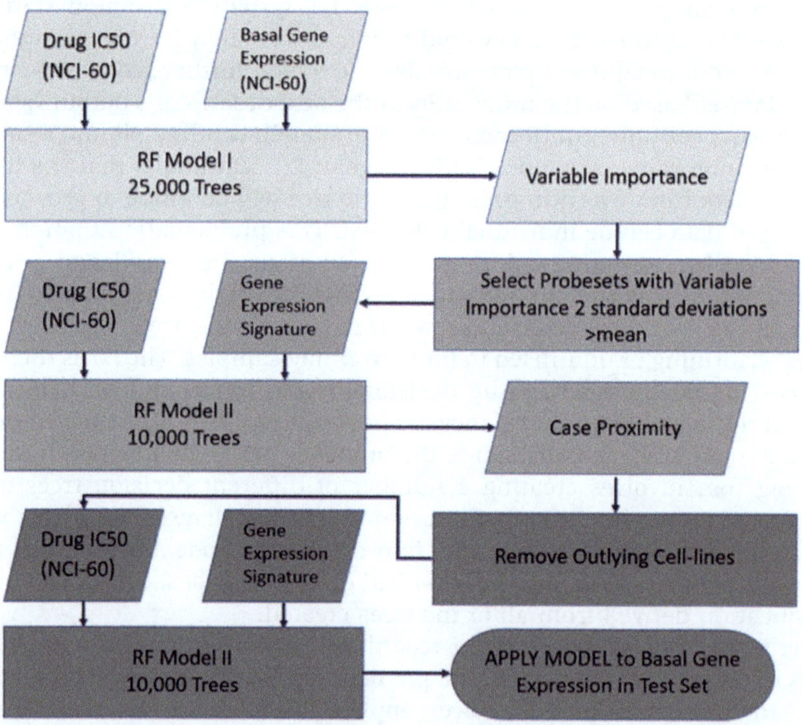

Figure 5.6 Flowchart for the algorithm building process for RF analysis in the model employed by Riddick *et al.*[97]
Reproduced from ref. 97 with permission from Oxford University Press, Copyright 2011.

China.[104] The initial DT node in this instance was the distance from the coastline. Branching from this node, the descriptors that were used include terrain elevation, month, daily meteorological factors such as atmospheric pressure, temperature, humidity, wind speed, wind direction and precursors of acid rain.

Furthermore, RF has shown significant diverse applications in areas such as the prediction of larval mosquito abundance in tree genera based on seasonal timings,[105] as well as classification of iron ore and elements present in steel using laser-induced breakdown spectroscopy.[106,107] RF techniques have been applied to the discovery and distinguishability of Heusler, inverse Heusler and non-Heusler compounds[108] (Heusler compounds are a form of inter-metallics). The descriptors used to build a training set for the determination of these compounds included electronegativity of metallic elements, s and p valence values, atom radius differential, electron number, group number, molar mass and period number. The data used to form the training sets were derived from crystallographic compound information from the Pearson's Crystal Data and the ASM Alloy Phase Diagram Database.

5.3.5 Naïve Bayes Classifiers

Naïve Bayes (NB) is a Bayesian network probability-based binary and multi-class classification machine learning technique. The technique consists of *n* number of features and classes, in which the probability of the feature occurring within each class is calculated and the most likely class is determined. NB uses the maximum *a posteriori* (MAP) decision rule for classification, in which the estimate is made of an unknown value equivalent to the mode of the posterior distribution. NB only considers the individual input frequency and assumes that input data is conditionally independent, and as such performs successfully with data-sets which uphold this assumption.

Naïve Bayes application has been used for classification models for predicting mutagenicity using the Ames method, both *in silico* and *in vitro*,[109,110] for classification of the carcinogenic properties of chemicals.[111] Furthermore, NB has been applied to *in silico* analysis of ligand libraries to determine protein ligand selection based on a set of known active compounds.[112]

Applications within drug discovery include prediction of butyrylcholinesterase inhibition for the treatment of Alzheimer's disease,[113] P-glycoprotein inhibition,[114] prediction of synergism between structural features and chemical-genetic interaction matrixes,[115] along with applications to large datasets; for example, prediction of adverse drug reaction effects using physiochemical, biological and phenotypic properties of drugs acquired from numerous databases including PubChem, Drug Bank, KEGG and SIDER.[116] NB was also used, amongst a number of previously discussed learning techniques, for a multi-classification approach for the prediction of acute oral toxicity *in silico.* A study of acute oral toxicity in rats collated data on 12 204 median lethal dose values (LD_{50}) of toxic compounds from the admetSAR database.[117] Upon sample pruning and removal of compounds containing inorganics, salts and organometallics, the total dataset consisted of 10 151 compounds. The data was then split in a 4:1 training:test set data ratio, and validated using an external dataset from the MDL Toxicity Database and Toxicity Estimation Software Tool. Samples were then classified according to four different toxicological categories defined in the United States Environmental Protection Agency guidelines. In this instance, Molecular ACCess System (MACCS) keys and FP4 fingerprints were utilised as substructure dictionary sources. The resulting q-value for NB applications within the study were 40, 42, and 25% for MACCS, and 60, 58, and 36% for FP4. NB was significantly outmatched by all other learning techniques used, and proved the least accurate method for classification purposes. The accuracy of the techniques used in this multiclassification ranks as $SVM_{OAO} > KNN > SVM_{BT} > RF > DT > NB$ for MACCS data, and $SVM_{BT} > SVM_{OAO} > KNN > RF > DT > NB$ for FP4 data.[117]

However, NB was later used as a classification technique for drug-induced human liver trauma.[118] In this study, NB was evaluated using 5-fold cross validation and presented an overall prediction accuracy of 94%. Interestingly, this study utilised a different fingerprint descriptor, that is, extended connectivity fingerprints (ECFP_6), followed by that used in the Xiao Li[117] study. However, a significantly smaller sample size was analysed ($n = 287$).[118]

5.3.6 Linear Regression

Linear regression is among the most well-known and simplest methods of statistical analysis; however, this technique also finds applications within the machine learning field. Such systems determine the regression value based upon a linear equation for a specific input value (x) and its predicted output one (y). In simple linear regression problems, there is a singular input and output value, although in higher dimensional data, in which multiple input values exist, the regression line is known as a hyperplane. When dealing with multi-dimensional data, as in multiple linear and/or polynomial regressions, an adaption of the ordinary least squares (OLS) method is used, and this approach estimates coefficient values by minimising the squared residuals, that is, the sum of all of the squared deviations from the regression line. The relationship between input (x_i) and output (y_i) variables within the OLS method is defined by eqn (5.5):

$$y_i = \alpha + \beta * x_i + \varepsilon_i \quad (5.5)$$

In which α is the intercept term, β is the regression coefficient, and ε is the residual term.

The residual (ε) refers to the difference between the output variable and the estimated influence of the input variable on the single output one. OLS can be modified to further reduce error and complexity within the model by regularisation techniques, such as Lasso (L1) or Ridge (L2) regressions. These techniques minimise the absolute sum of the coefficients and the squared absolute sum of the coefficients respectively.

Ordinary least squares has been utilised for the calibration of methods to detect essential and toxic metals present in mature ($n = 17$) and colostrum ($n = 12$) human breast milk, using inductively coupled plasma-optical emission spectroscopy (ICP-OES).[119] This study used OLS and weighted least squares (WLS) methods for the determination of the metal ions K, Na, Ba, Al, Cu, Mg, Mn, Cr, Fe, Ca, Ni, Co, Cd. Pb, Zn, Li, and V. The major difference between the two least squares methods is that OLS deals with data containing a constant uncertainty, whereas WLS considers data with a varying uncertainty. The Nascimento[119] study noted that all macrominerals and microminerals present in human milk showed differences between the two methods at a 95% confidence level, and that inappropriate choice of the LS method directly affects the estimated concentration of the sample, which led to a loss of precision. Figure 5.7 shows the application of linear regression through a comparison of the OLS and WLS techniques for metal ion analysis.

Ordinary least squares methods have seen further application in the fields of forensic science for the determination of age of death through periodontitis and dental root translucency.[120] The study by Schmitt *et al.*[120] utilised the Lamendin method for OLS analysis and showed that the model, consisting of a sample size of $n = 214$, had a tendency to overestimate the age of younger adults, and underestimate the age of older age groups, and gave R^2 values of only 0.288–0.404. Further work into the field of forensic age prediction included the application of OLS, WLS and quantile regression for the age prediction based on DNA methylation data.[121] Smeers *et al.*[121] concluded that all three techniques exhibited positive applications within this area of research, with a successful prediction percentage of 97.10%, 95.65% and 91.30% respectively, whilst also highlighting that WLS and quantile regression strategies allowed for wider prediction intervals when the prediction error increased with increasing age.

Similar to the Schmitt[120] study, Smeers[121] also noted that the successfulness of the model depended on the correct application of the regression model based on the characteristics of data acquired, and that there is not a

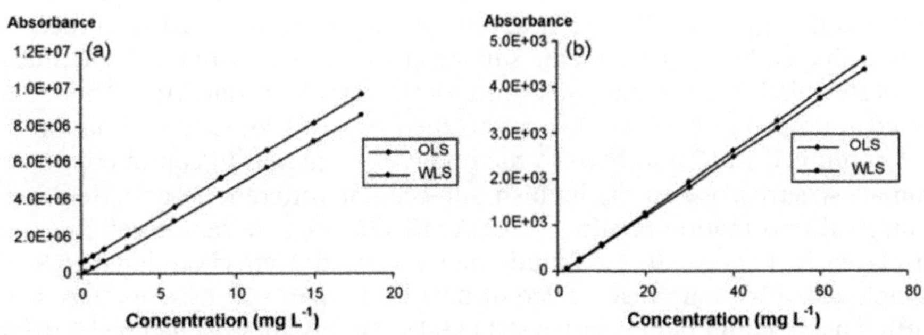

Figure 5.7 Highlights of the difference between the OLS and WLS analysed for K and Zn samples.
Reproduced from ref. 119 with permission from Elsevier, Copyright 2010.

simple best fit method for all analyses. OLS has been utilised in the fields of environmental modelling within a chemical mass balance model for source apportionment of airborne particulates,[122] within food chemistry for the analysis of stable hydrogen and oxygen isotope ratios present in samples obtained from fast-food chains and retail supermarket outlets,[123] as well as in the field of metabolomics for biomarker screening and classification using the Lasso regularisation technique.[124]

5.3.7 *k*-means Clustering

k-means clustering (KMC) is an unsupervised learning technique used to identify clusters or groupings within unlabelled data sets. This algorithm identifies the number of groups within datasets based on a pre-specified or unspecified number of *k* groups or centroids. The algorithm sorts data into specific clusterings based on feature similarities, and uses an iterative refinement process in order to produce the final labelled output. The centroid is recalculated during the training process, and is based upon the mean distance between datapoints and the centre of each clustering class. A *k* value is not predetermined, and requires some user trial and error to identify the optimal *k* value required for each dataset. Application of KMC has been utilised in the fields of environmental chemistry, such as in environmental risk management through zoning of chemical industrial areas,[125] and mapping human vulnerability to chemical hazards[126] in China.

A combined KMC multilayer perceptron approach was undertaken in the study by Ay and Kisi[127] for the modelling of chemical oxygen demand (COD) in samples obtained upstream of a Turkish wastewater remediation centre. This study analysed daily-measured water-suspended solids, pH, temperature, discharge and COD concentration. Furthermore, the effectiveness of a *k*-means clustering-multilayer perceptron (KMC-MLP) approach against established techniques of MLR, MLP, a radial-based neural network, a generalised regression neural network, and a subtractive clustering neuro-fuzzy inference system, as well as a grid partition neuro-fuzzy inference system were compared. The developmental method for the KMC-MLP approach is outlined in Figure 5.8. The accuracy of each learning and statistical method was evaluated using root mean square error, mean absolute error, mean absolute relative error and determination coefficient statistics. This combined approach of KMC-MLP outperformed its rivals for COD estimation.[127]

A combined KMC and PLS-DA method was employed in conjunction with Raman spectroscopy to distinguish sub-cellular differences, and yield new biological information relating to MDA-MB-435 isogenic cancer cell lines, as displayed in Figure 5.9. This study demonstrated a cross-validated PLS-DA sample classification success rate of 0.92.[128] Raman spectroscopy has seen further use in combination with KMC-PLS-DA for the study of chemical fixation methods, specifically formalin for the preservation of proteins, and Carnoys fixative and the methanol-acetic acid approach (a methanol-based fixative that preserves nucleic acids), for the analysis of cell constituents and processes.

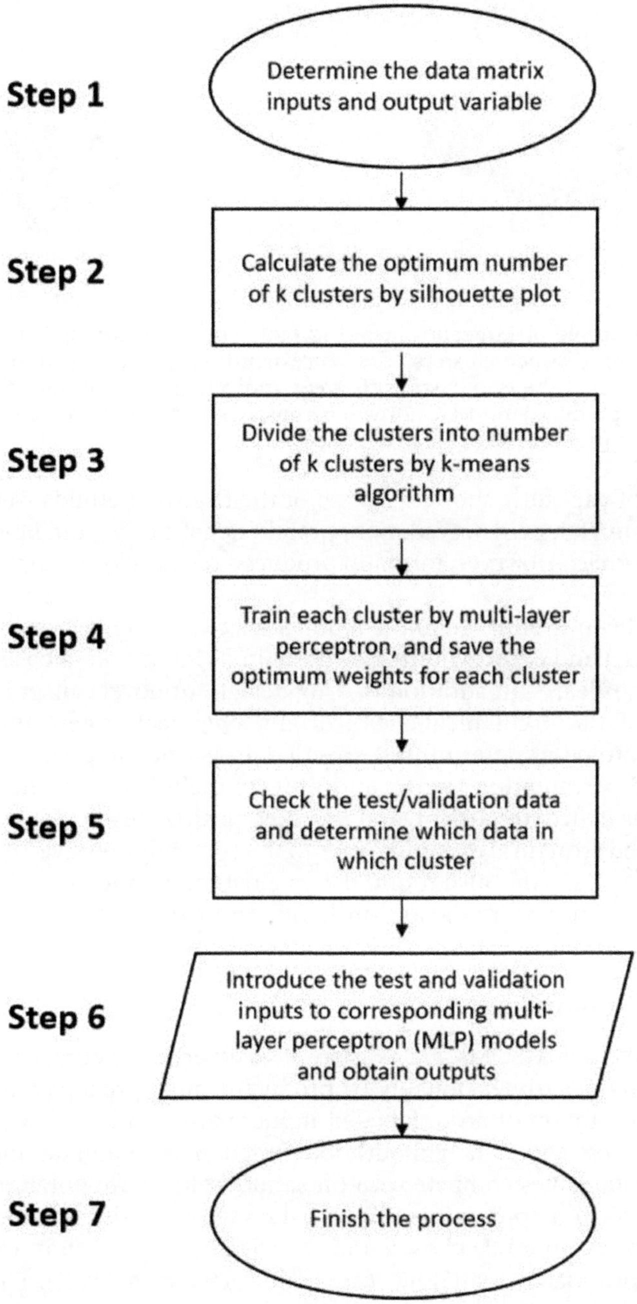

Figure 5.8 Flow chart of *k*-means cluster-multilayer perceptron method creation process.
Reproduced from ref. 127 with permission from Elsevier, Copyright 2014.

Figure 5.9 Example of *K*-means clustering plot. The algorithm split the data into four distinct clusters, the corresponding colours represent different parts of the cytoplasm with green indicating the nucleus.
Reproduced from ref. 128 with permission from American Chemical Society, Copyright 2010.

The results of the study showed that all of the fixative methods were capable of observing nucleic acid degradation, protein denaturation and lipid leaching to varying degrees. However, formalin produced results closest to that of a live cell.[129]

KMC has been used in genomic studies for clustering genome-wide polymerase (Pol II) and carboxyl-terminal domain (CTD) markers based on their ChIP-chip profiles,[130] in addition to drug design for novel cancer treatments through analysis of chemical-chemical and chemical-protein interactions. This study, aimed at determining effective drugs for the treatment of lung cancer, used permutation testing and KMC to exclude drugs with a low viability for lung cancer treatment, and returned positive results for compounds that possessed structural dissimilarities to the currently approved lung cancer treatments.[131] KMC has also found use in clustering approaches for the prediction of novel non-steroidal anti-inflammatory drugs.[132]

5.3.8 Self-organising Maps

Self-organising maps (SOMs) are a variety of unsupervised neural networks that are used to reduce dimensionality by producing an approximate low dimensional representation of an unlabelled input space, or map. SOMs use competitive learning and a neighbourhood function to maintain input space topology. Input values compete with one another for representation, and the SOMs proceed to map data based upon the vector weight, at which point a sample vector is randomly chosen and the current vector map is analysed for the most appropriately matching vector value. This vector is then 'rewarded', and as such gains a greater likelihood of random selection during the mapping process. Each vector weight will have nearby 'neighbour' vectors of a similar value, which are also 'rewarded' in this process, a process increasing their likelihood in the random selection component. The SOMs technique then continues this process of mapping as it grows.

Self-organising maps have been used alongside factor analysis for determination of the geochemical associations of elements from different anthropogenic sources, including smelters, ironworks and chemical industries.[133] SOMs have been used to model molecular descriptors of sesquiterpene lactones found in the flowering plant family *Asteraceae*. The chemotaxonomy approach used a dataset containing 1111 sesquiterpene lactones, extracted from 658 species, 161 genera, 63 subtribes and 15 tribes of *Asteraceae*. The input data descriptors for this study included constitutional functional groups, BCUT (also known as Burden Eigenvalues), auto-centering, 2D autocorrelations, topological, geometrical, resource development framework (RDF), 3D-molecule representation of structure based on electron diffraction (MoRSE), geometry, topology, and atom weights assembly (GETAWAY) and weighted holistic invariant molecular descriptors (WHIM) to separate the aforementioned botanical sources.[134]

Applications of SOMs have been seen in the field of environmental chemistry assessment models for the removal of chlorinated solvents in petrochemical wastewater systems. A study by Tobiszewski *et al.*[135] featured wastewater samples ($n = 72$) that were successfully clustered into a total of five groups. The first three groups featured samples that were collected from drainage water, process water and oiled rainwater treatment streams. The further two groups consisted of samples collected after biological treatment, and those collected after an unusual event such as a solvent spill. SOMs highlighted that the biological treatment significantly reduced the chlorine levels within the wastewater samples.[135] The variables considered in this study included: flow-rate, temperature, and dichloromethane, chloroform, carbon tetrachloride, 1,2-dichloroethane and perchloroethylene contents, along with the concentrations of chlorides, bromides, nitrates and sulphates. Further applications within the field of environmental chemistry include the monitoring of rainfall runoff in gauged basins,[136] the distribution of iron in coastal soil and sediment,[137] and coastal water quality.[138]

Self-organising maps have been used in conjunction with PCA and Raman spectroscopy for the spectral mapping and determination of histological features in a human colon polyp sample. The application of SOMs allowed Lloyd *et al.* to identify more subtle features present by increasing the spectral contrast in the polyp image, which allows for a higher level of bioinformation to be obtained.[139] SOMs have seen further application in fluorescence spectroscopy through excitation-emission matrices for the study of dissolved organic matter obtained from freshwater and marine sources,[140] and for the classification of lighter fuels using gas chromatography-mass spectroscopy (GC-MS).[141]

5.3.9 Hierarchical Clustering

Hierarchical clustering (HC) is an unsupervised clustering strategy that utilises an input variable hierarchal system in order to classify data. HC requires a data similarity matrix, commonly calculated using a cosine function, to form tree-like dendogram structures, and falls into one of two categories; agglomerative, involving a dendrogram that grows from the bottom-up, in which data begins as individual clusters and groups together similar clusters to form additional ones; and divisive, a dendrogram that grows from the top-down using clustering algorithms, such as KMC to divide each cluster down into subsequent smaller clustering formats.

Hierarchical clustering has been applied to the analysis of non-small-cell lung carcinomas in mouse models. The study reported by Westcott *et al.*[142] featured whole exome sequencing of adenomas obtained from three mouse models, which were introduced either through exposure to carcinogens, or through genetic activation.

Further examples of HC analysis are its use in parallel with PCA and PLS-DA for the classification of Moroccan olive oil samples using Fourier-transform infrared (FTIR) spectroscopy. A combinational approach using HC and PCA was used in the field of food chemistry for the classification of fruits and vegetables based on antioxidant activity *in vitro.* The study by Patras *et al.*[143] sampled six vegetables and eight fruits, and used descriptor variables of the global antioxidant activity, total phenolic, total anthocyanin and ascorbic acid contents, instrumental parameters, colour, and moisture levels. This process successfully split the data set into four separate clusters. The study found berries to possess the highest antioxidant activity and vegetables had the lowest.

Figure 5.10 represents a fairly simple application of HC, with four specific clusters being identified based upon antioxidant activities. A more complex example of SOM complementing HC is shown in Figure 5.11, in which a combinational application of HC and SOMs were applied to the analysis of protein structural conformations.[144]

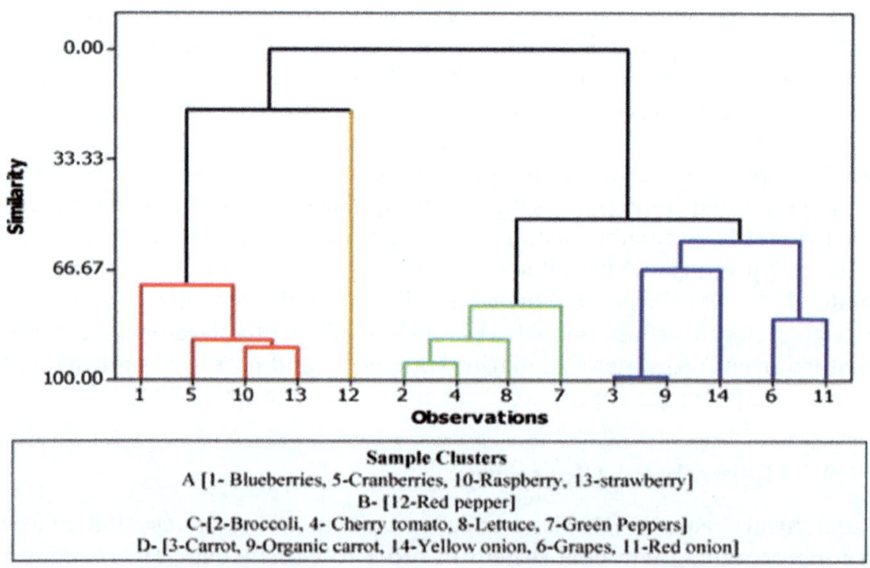

Figure 5.10 Simple annotated dendrogram for hierarchical classification of fruits and vegetables *in vitro*.
Reproduced from ref. 143 with permission from Elsevier, Copyright 2011.

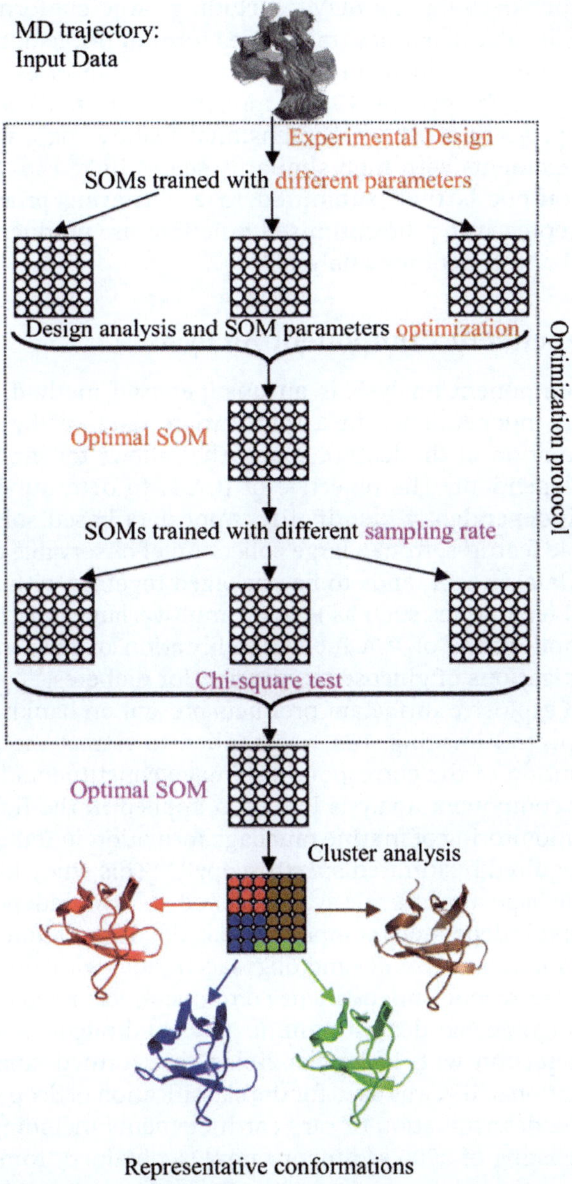

Figure 5.11 Structural ensemble analysis based on SOMs trained with different parameters, which are then optimised prior to verification through a Chi squared test to confirm the optimal SOM. These prototype vectors, which were trained with a molecular dynamics conformational ensemble were submitted to a HC process.
Reproduced from ref. 144 with permission from Springer Nature, Copyright 2011.

Here, the input data for the SOMs including large conformational sets output from molecular dynamics trajectories form multiple domains. These conformations are described in terms of Cartesian axes as opposed to Z-matrices, in which once the SOM is trained, the original conformational ensemble is a projection on a bi-dimensional feature map, with relationships between elements with high similarity scores linked to neighbouring neurons. The output is then submitted to a clustering process and the conformation representing the optimised functions are produced for further conformational and functional analysis.[144]

5.3.10 Independent Component Analysis

Independent component analysis is an unsupervised method that aims to maximise input independence in a multivariate data set by predicting a linear transformation of the feature space, that allows for input data to be statistically independent. The objective of ICA is to determine the hidden variables that independently identify the input data based solely upon the linear observable features, from a large collection of observable data, such as a multivariate dataset. ICA tends to be packaged together with a number of other statistical techniques such as PCA for multivariate statistical analyses. Multivariate applications of ICA include calibration of near-infrared spectroscopic investigations of glucose monitoring for diabetes,[145] as well as for the detection of explosive surfactant products present on banknotes through Raman hyperspectral imaging. ICA was applied to extract the pure spectra and the distribution of the corresponding image constituents.[146]

Independent component analysis has been applied in the field of marine biology for the monitoring of marine mucilage formation in Italian seas using spectral data acquired by infrared spectroscopy.[147] This study identified that the marine mucilage process always consisted of two independent components. The first independent component was the degradation of algal cells leading to the formation of mono- and oligosaccharides with amino acids and oligopeptides. The second independent component was related to the polymerisation of oligosaccharides with amino acids and oligopeptides, and the subsequent interaction with less polar lipids that formed supramolecules. A further application of ICA was used for the classification of deep sea sediment materials for the determination of rare earth elements including yttrium.[148] This study, consisting of 2000 sediment samples, obtained from 78 different deep-sea beds covering a large portion of the Pacific Ocean, was able to identify four independent components covering 91% of the total sample variance. However, for this investigation, the independent components should be noted as vectors or linear trends, rather than points with specific compositions.

One of the widely used applications of ICA is for exploratory analysis of functional magnetic resonance imaging (MRI) data. ICA decomposes functional MRI data into activity patterns that are statistically independent.[149] As an example, the denoising of MRI signals involves distinguishing the signal components that represent brain activity from the noisy ones, such as scanner artefacts or motion. Using ICA, it is possible to differentiate between these two components and therefore remove the noise from the MRI scans. The Salimi (2014)[149] study utilised an ICA-based X-noiseifier which achieved an overall accuracy of 95% from single run datasets, and 99% classification accuracy from high-quality rfMRI data obtained from the *Human Connectome Project*. Other studies have used ICA for MRI applications, including classification of brain tissue analysis using spectral clustering ICA,[150] as well for auditory resting-state connectivity analysis in tinnitus.[151]

5.3.11 Deep Learning

In 2015, Geoffrey Hinton reported his seminal work on the concept of deep learning (DL), which has revolutionised the use of ML techniques in a multitude of scientific research fields. Fundamentally, DL allows for multiple processing layers to learn data representation through multiple levels of abstraction. DL is capable of determining structures within large sets of data by utilising backpropagation to alter the internal parameters of a system, which are then used to compute data within each layer from that in the previous layer.[24] A large number of DL models are conducted within a NN system. Deep neural networks (DNN) process data by passing the input value through a series of hierarchal algorithm layers and applying non-linear transformations to create a statistical output model. The network learns through various iterations until a suitable accuracy level has been acquired.

One of the many benefits of DNN are that they are unsupervised, and this learning process occurs without the need for supervised feature extraction. However, the major disadvantage of DNN and DL are that the models require large amounts of training data. Certain studies have attempted to combat this limitation, such as the Altae-Tran[152] approach. In this study, 'one-shot' learning was applied as a means of significantly reducing the large training data requirement for prediction models within drug discovery models.[152]

With the development of DNN, networks have been tested and compared against other learning techniques that had previously been more applicable to the task at hand; one such field is QSAR. In a study by Ma *et al.*,[154] the viability for DNN for QSAR analysis was compared to that of a RF model. This study showed that an unoptimized DNN, that is one that consisted of only a single set of parameters, outperformed RF in the majority of datasets analysed from the Merck drug discovery system.[154] Comparison studies for QSAR approaches between NN and DNN have also been able to tackle the issue of activity cliffs within QSAR datasets. A study by Winkler and Le[153] compared the capabilities of a NN and DNN system for prediction in large drug datasets, using the Kaggle drug data set as a testing model. A graphical representation of the differences between a NN and DNN system is shown in Figure 5.12.

A study by Lenselink *et al.*[155] employed DNN as a comparative technique against learning approaches: SVM, RF, NB, and logistic regression; all commonly applied to the analysis of QSAR and proteochemometrics data. Lenselink acquired a dataset from ChEMBL which was standardised, using a Boltzmann

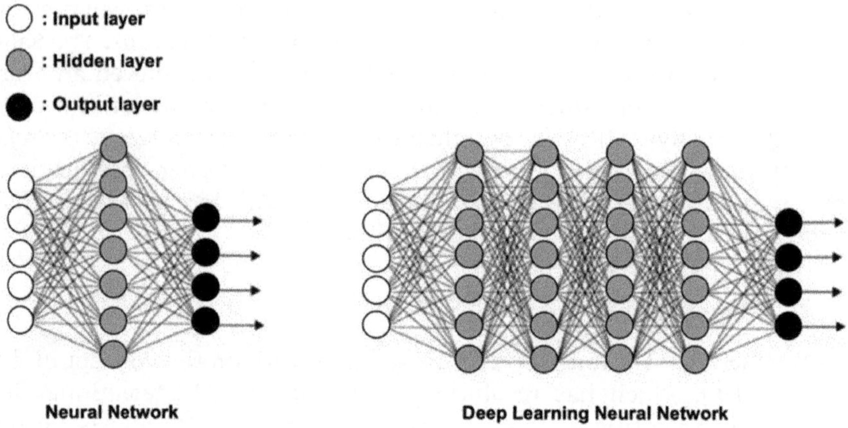

Figure 5.12 Graphical visualisation of the differences between a deep neural network and a non-deep neural network. The deep neural net consists of multiple hidden layers that are used to classify datasets, as opposed to the singular algorithm layer in traditional neural network systems.[153]

evaluation discrimination receiver operating characteristic (BEDROC) and MCC, and validated using random split validation and temporal validation methods. DNN was recorded as the top classifying method, with an enhanced performance 27% greater than the other methods involved in this study. The most effective method was shown to be a deep neural network-proteochemometric (DNN-PCM) strategy, and this exhibited an accuracy of one standard deviation higher than the mean performance.[155] Similar findings were previously recorded by Unterthiner *et al.*[156] for drug target prediction, for which the DNN outperformed the SVMs, binary kernel discrimination, logistic regression and kNN, in addition to commercially-available drug targeting products, such as the Parzen–Rosenblatt kernel density estimation (KDE) based model, NB utilising pipeline pilot Bayesian classifiers and the similarity ensemble approach (SEA). The DNN outperformed all other methods and possessed a performance prediction value with an AUC value of 0.830, followed by SVM (AUC = 0.816). The method with the weakest performance in this study was SEA with an AUC value of 0.699.[156]

In more recent years, DNN themselves have been developed through the use of convolutional neural networks (CNNs). CNNs are a sub-group of DNN commonly applied to image analysis studies. CNNs techniques apply bias to various areas, aspects, or objects within an image of which the network can differentiate between. The main benefit of CNNs are that they require significantly less image pre-processing and hence are notably faster. CNNs have seen use in QSAR and quantitative structure–property relationships (QSPR) studies, such as for the prediction of chemical properties based purely upon the 2D chemical structure.[157] Originally, the CNN model in this study outperformed multi-layered perceptron DNN for determination of activities and solvations, but underperformed in its prediction of toxicity. However, advancements have been made by Goh *et al.* (2018),[158] and the new Aug-Chemception platform now outperforms the original model for toxicity, activity, and solvation. Further models have been created for the prediction of small molecule properties,[159] as well as quantum interactions in molecules,[160–162] and the prediction of complex organic chemistry reactions.[163]

Drug discovery is a field that has offered significant benefit and development with the application of DNN for structural analysis. Refining the computational process of drug discovery can greatly reduce the cost and time investment required for the screening of novel chemotypes and experimental assays.[164] DNN and CNNs have been used for the prediction of bioactivity in structure-based drug discovery by the Heifets group's AtomNet project,[165] and comprehensive 3D representation of protein-ligand interactions,[164] respectively. DNN have the capacity to inversely design photonic structures, and this process overcomes a fundamental issue of photonic structure design by effectively training datasets on non-unique electromagnetic scattering instances. This is achieved by combining forward modelling and inverse design in tandem architecture.[166] A DNN method developed by Putin *et al.*[167] termed reinforced adversarial neural computer (RANC) was applied for the *de novo* design of small-molecule organic structures. This technique was developed based upon a generative adversarial network paradigm, and reinforcement learning and was shown to possess promising applications for the development of novel molecules with activity against different biological targets and pathways in the field of drug discovery.[167] Furthermore, creation of evaluation frameworks for *de novo* design such as the GuacaMol benchmarking framework, presented by Brown *et al.*,[168] have been designed for validation and optimisation of DNN-based *de novo* design models.

Deep neural networks have seen significant applications in fields that utilise imaging instrumentation or image analysis. Examples include the study of lung segmentation CT imaging,[169] neuroimaging for brain extraction using MRI,[170] along with brain tumour segmentation, specifically targeting glioblastomas with MRI.[171] A study by Gibson *et al.*[172] showed the applicability of DL for segmentation, regression, image generation and representation using the NiftyNet infrastructure. The study generated simulated ultrasound images, predicted computed tomography through image regression of brain MRI images, and segmentation of abdominal organs through computed tomography.[172] However, its application is not limited to medical imaging fields; indeed, CNNs have been utilised for crop composition and 'in-field' biomass *via* semantic segmentation in RGB images. This study was able to distinguish between oil, radish, barley, weed, stump, soil, equipment, and unknown features within farmland images.[173]

The fields of genomics and proteomics are ideal areas of application for DL approaches. DNN and CNNs have been employed for the quantification of non-coding DNA sequencing regions *de novo*;[174] application with the nanopore MinION device, employed in base calling for infectious disease detection and custom target enrichment during DNA sequencing,[175] as well as the prediction of DNA- and RNA-protein binding sequence specificities.[176] In proteomics, research DL networks have been constructed for the determination of mammalian malonylation sites in conjunction with a RF classifier,[177] for nitration and nitrosylation prediction.[178] Further applications have included the prediction of protein backbone angles and structures using sparse auto-encoder DNN,[179] and protein-protein interactions from protein sequences, also using stacked sparse auto-encoder DNN.[180]

Catalytic optimisation was also explored in a recent study by Zhai *et al.*[181] for modelling platinum clusters. This study used a seven layer multidimensional network combining limited step density functional theory for geometric optimisation and utilised a bond length distributional algorithm to sample the configuration space and create random initial catalytic structures.[181] DNN have been applied to lattice Boltzmann flow simulations to improve computational dynamic fluid simulation predictions. The Lat-Net program created in the Hennigh[182] study is a method capable of application for general compression of other lattice Boltzmann simulations, such as electromagnetism.[182]

5.3.12 Quantum Computing and Applications in Chemical Analysis

Although the field of quantum computing has seen a dramatic upsurge owing to innovations such as harnessing the capabilities of the NV qubit in diamond, there have been numerous applications to quantum chemistry and chemical analysis. The application of neuromorphic models, drawn from the biological sciences, have been employed to enhance the capability of computing to solve real-world problems. For example, neuromorphic models have been employed in order to act similarly to the well-parametered, learning models inherent in NN.[183] Employing the input to such algorithms as a quantum state, and producing such an output is the basis of modern quantum learning, which has far reaching possibilities in enhancing quantum computers and solving highly non-trivial computational problems.[184] Berlinguette, Aspuru-Guzik and coworkers have recently reported the implementation of a ML algorithm in the high-dimensional optimization of a self-driving modular robotic laboratory.[185] This platform is capable of learning from experiments and is aimed at the acceleration of materials development and optimization. The computational methodology employs a Bayesian optimization algorithm for new experimental design,[186] which has a built-in bias parameter, leading experimental design in a more explorative manner alternating with exploitation-based approaches. This has been demonstrated to improve on alternatives such a random and systematic searches.[187] Whilst this approach is currently employed for the discovery of thin-film materials, it is not difficult to imagine the fields this could encompass in the future. This platform also includes UV-Vis-near infrared (NIR) spectroscopic equipment in order to allow for a diagnostic quantity to be established for the stratification of new materials.

In a similar approach, semantically-constrained graphs have been applied to the *de-novo* design of new molecules. It is now possible to generate both images and sound, as well as text through the implementation of deep generative models,[167] which has a direct influence on the design of chemical systems from pharmacology to materials science. The SELFIES program (self-referencing embedded strings), can be employed to input into DL models, for example in molecular evolution techniques; methods which are used for the design of new molecules and precursors.[188] The SELFIES program is based on the requirement in the natural sciences to consider and develop complex structures or models, which can often be described in graph form. For example, a More O'Ferral–Jencks diagram will represent the multiple reaction coordinates in 2D for chemical reactions with changes in two degrees of freedom. Should constraints not be applied to this representation, then highly unphysical structures/complexes would arise. The key is therefore to derive such constrained, or semantically constrained graphs, with deep generative models for these; a task implemented in SELFIES. Most remarkably, this technique has been implemented in the reverse design of molecules, in which property prediction becomes an inherent part of the problem. Solubility was calculated as logP using the QM9 molecular dataset[189] being split into training and validation molecules. The SELFIES algorithm correctly predicted the property of previously unseen molecules with a validity of 95.2% *cf.* SMILES 71.9% all based on the best performing variational autoencoder architectures.[188]

With this growth in the use of deep generative models in chemistry and materials science, there comes the need for benchmarking platforms. The molecular sets (MOSES) model incorporates a number of molecular generation models, whilst combining a method of assessing the spread and viability of generated chemical systems.[190] The molecular dataset employed within the model is based on the ZINC biological molecule database, and contains over 4.5 million molecules with weights between 250 and 350 Daltons.

Within the field of medicinal chemistry and lead identification, a number of filtering algorithms allow candidates with unstable, reactive or biotransformative moieties to be discounted from the compound list. MOSES incorporates a number of filters with bespoke designs for alkyl halides, epoxides, aldehydes and numerous other functions. The action of these discriminatory filters leads to a dataset of more than 1.9 million molecules, which is then used to represent the systems as either SMILES strings or molecular graphs. MOSES therefore provides comparative tools in order to assess new generative models.

In an effort to develop basic computational frameworks to aid in the understanding of potential energy surfaces of organic reactions, heuristically aided quantum chemistry approaches have been developed.[191]

5.3.13 Particle Swarm Optimisation

Within the last few years, DL algorithms have attracted tremendous traction in the ML community in view of their success in image recognition and natural language processing. It is notable, however, that there are yet significant pipelines of ML techniques which are responsible for the recent success of DL and other novel algorithms. For this broad range of ML algorithms applied in industry and academia for image and pattern recognition, or other inductive learning purposes, the reliability of the resultant model in providing valid and efficient generalised performance for new configurations of the existing factors, depends on good theoretical knowledge of the underlying spatial distribution of the feature space, and the identification of the most informative configurations of this.

Finding an appropriate set of features from data of high dimensionality for building an accurate classification model is a well-known NP-hard computational problem.[183] This complexity is amplified by the consideration that modern techniques of knowledge abstraction are characterized by a huge compass of data volume, and instances which also involve uninformative features. The feature selection avenue therefore represents one of the most active research areas in ML because of its significance in model building. Liu and Yu[184] have reported several algorithms for selecting informative features for pattern recognition and other ML tasks, the intuition involved generally consisting of the following steps:

a. generation strategy to generate the next configurations of vector;
b. performance metric to evaluate each configuration generated;
c. stopping criterion to decide when to stop exploring the search space; and
d. validation procedure to assess the quality of the result.

In the first stage, multiple copies of feature vectors are generated for evaluation. Depending on the specific choice of algorithm, the initialisation may commence with an empty cluster and then repeatedly populate or exclude features. It may also initialise with all the features and then iteratively include or remove features on the basis of their evaluation report. Yet another initialisation strategy is the random addition/exclusion of subsets with the support of their evaluation information. The evaluation stage measures the quality of each feature configuration presented to it, replacing the pre-existing best configuration with a new found one. Evaluation is a very crucial phase in the selection which requires the installation of a typical choice of classifier to guide the search towards the global optimum. The stopping criterion controls the iteration by determining whether predefined conditions have been attained. As the basis for every ML task, the validation procedure has been incorporated in order to assess the quality of the final configuration realised from the algorithm with existing benchmark results. These four phases for selecting features have been presented in Figure 5.13. Advances in research and software developments have led to vast improvements in model performances and interpretability in the literature for the selection of informative features with the prediction accuracy devised as the primary measure. The approaches of the informative feature selection can be categorised into four main classes: filter-based, wrapper-based, embedded-based, and hybrid based. Among these classes, the review in this section focuses on the wrapper-based method of feature selection implemented via an evolutionary computation algorithm known as the particle swarm optimisation (PSO). PSO is a

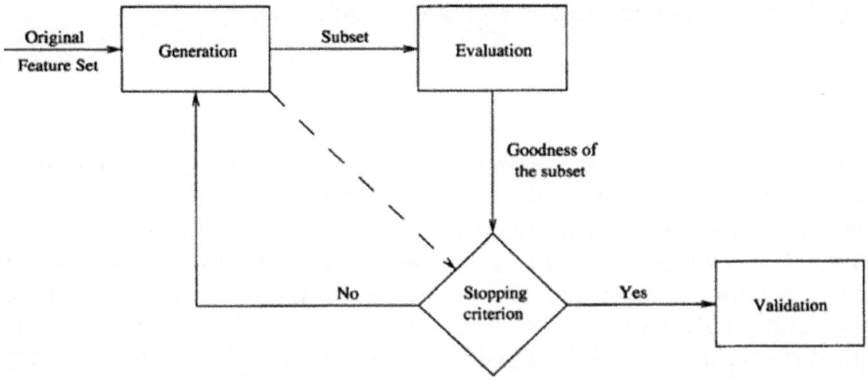

Figure 5.13 Diagrammatic representation of different stages of selecting an informative feature configuration from the original dimension of the data.

nature inspired metaheuristic which iteratively adapts the social behaviour of swarms, such as flocking birds and schooling fish to optimise non-linear problems. Originally developed by Kennedy and Eberhart in 1995.[185] The underlying phenomenon of PSO is that knowledge is optimized by social interaction in the population in which thinking is not only personal, but also social.[186] Like many evolutionary computations, the particle swarm optimisation algorithm as a global metaheuristic does not guarantee it will converge to an optimal solution for non-linear problems and they usually generate a weak local convergence ability, although their solutions are often near optimal. The major benefit of evolutionary computations such as PSO over classic optimisation routines such as the quasi-newton method is that the former does not require the problem space to be convex or differentiable. The PSO also involves simple implementation and is practically insensitive to the scaling of design variables.

In PSO parlance, each individual in the population is regarded as a particle in the search space, and in turn described as a possible solution of the optimisation problems for which fitness is measured by a classifier. The movement of each particle, in the problem space, is governed by its current global best *pk* and local best *pi* positions, respectively. Consequently, the updating policy of each particle for an optimal solution is based on the individual position and the position of neighbouring particles with some incorporation of randomness. The classic algorithm proposed by Kennedy and Eberhart in 1995 is as follows:[185]

x_k^i – Particle position;
v_k^i – Particle velocity;
p_k^i – Historical individual best particle position;
p_k^g – Historical swarm or global best particle position;
c_1, c_2 – Cognitive and social parameters;
r_1, r_2 – Random numbers between 0 and 1.

In which the updating rule is evaluated as follows:

$$v_{k+1}^{i} = v_k^i + c_1 r_1 (p_k^i - x_k^i) + c_2 r_2 (p_k^g - x_k^i)$$

$$v_{k+1}^{i} = x_k^i + v_{k+}^i$$

The position of each particle in the swarm is the vector configuration of features encoded as zeros and ones, in which 0 represents an instance of exclusion and 1 represents an instance of inclusion. Hence, a design model using PSO is a wrapper-based feature selection algorithm which consists of two components, a classifier that evaluates the fitness of each particle in the swarm and a PSO optimiser which searches for the optimal particles of the feature subset in the high dimensional search space. The pseudo code for the PSO-based feature selection is presented in Box 5.1.

Box 5.1 PSO-based Feature Selection Algorithm.

Input: $0 < r_1, r_2 < 1$ (random weights); set constants c_1, c_2 (cognitive and social parameters); $V_{\max}$ = maximum velocity; D = dimension of the search space; |Swarm| = population size; $K_{\max}$ = maximum number of iterations.

Output: Swarm best = global best feature configuration measured by fitness value.

```
begin
    randomly initialise the position and velocity of each particle;
    while K_max or other the stopping criterion is not met do
        evaluate fitness of each particle;
        for i = 1 to |Swarm| do
            if fitness of x^i is better than that of an existing best, i.e. p_k^i then
                p_k^i = x^i;   // update the local best particle position
            if fitness of p_k^i is better than that of swarm best, i.e. p_k^g then
                p_k^g = p_k^i;   // update the global best particle position

        for i = 1 to |Swarm| do
            for d = 1 to D do

            update the velocity of particle i according to Equation 1;
            update the position of particle i according to Equation 1;
    return swarm best position i.e., p_k^g and its fitness value;
```

5.4 Conclusions

Throughout the years, chemical and biochemical analysis has worked in tandem with ML techniques to improve the overall accuracy and reliability of data models. With the recent innovative development of DL, the scientific community has seen a surge in its application and most likely will continue to do so for some time. DL has proved to be a very suitable, successful and accurate ML technique for data models, but is limited by the requirement for a large amount of training data. Studies are currently underway in order to counteract this limitation, for example the one-shot learning approach noted above. SVMs have also shown a significant aptitude for classification purposes, and have been shown in numerous instances to outperform many other ML techniques. This chapter has highlighted some of the more widely used ML techniques that have had many previous applications, and may therefore also be applied to future analysis in a multitude of chemistry disciplines. Based upon their specific field, the reader may take a more informed approach to ML techniques that show the greatest degree of accuracy for their particular need or application, and potentially explore alternative ML concepts that have demonstrated levels of success within that field.

List of Abbreviations

3D-MoRSE	Molecule Representation of Structure based on Electron diffraction
AI	Artificial Intelligence
ANN	Artificial Neural Network
AUC	Area Underneath Curve
AUROC	Area Underneath Curve Receiver Operating Characteristic
BEDROC	Boltzmann-Evaluation Discrimination Receiver Operating Characteristics
BIC	Bayesian Information Criteria
BPNN	Backpropagation Neural Network
CART	Classification and Regression Trees
CKD	Chronic Kidney Disease
CNN	Convolutional Neural Network
COD	Chemical Oxygen Demand
CTD	Connective Tissue Disease
CpG	Cytosine Guanine linked nucleotide site
CT	Computed Tomography
DARPA	Defence Advanced Research Projects Agency
DENDRAL	Dendritic Algorithm
DL	Deep Learning
DNA	Deoxyribose Nucleic Acid
DNN	Deep Neural Network
DSERS	Dynamic Surface-Enhanced Raman Spectroscopy
DT	Decision Tree

ECFP-6	Extended Connectivity Fingerprint
FN	False Negative
FP	False Positive
FTIR	Fourier-Transform Infrared
GC-MS	Gas Chromatography-Mass Spectroscopy
GETAWAY	GEometry, Topology, and Atom Weights AssemblY
GMM	Gaussian Mixture Model
HC	Hierarchical Clustering
HIV	Human Immunodeficiency Virus
HSA	Human Serum Albumin
ICA	Independent Component Analysis
ICP-AES	Inductively Coupled Plasma – Atomic Emission Spectroscopy
ICP-MS	Inductively Coupled Plasma – Mass Spectroscopy
ICP-OES	Inductively Coupled Plasma – Optical Emission Spectroscopy
KDE	Kernel Density Estimation
KEGG	Kyoto Encyclopaedia Of Genes And Genomes
KMC	K Means Clustering
KMC (MLP)	K Means Clustering-Multilayer Perceptron
KNN	K Nearest Neighbour
KNN-MFA	K Nearest Neighbour Molecular Field Analysis
L1	Lasso Linear Regression
L2	Ridge Linear Regression
LD	Lethal Dose
LDA	Linear Discrimination Analysis
LOOCV	Leave One Out Cross Validation
LV	Latent Variable
MACCS	Molecular Access System
MAP	Maximum a Posteriori
MCC	Matthews Correlation Coefficient
MCTS	Monte Carlo Tree Search
MDL	Molecular Design Limited
MDMA	3,4-Methylenedioxy-methamphetamine
ML	Machine Learning
MLP	Multilayer Perceptron
MLR	Multiple Linear Regression
MRI	Magnetic resonance imaging
MS	Mass Spectrometry
MT-4	Metallothionein 4
NB	Naïve Bayes
NMR	Nuclear Magnetic Resonance
NN	Neural Networks
OLS	Ordinary Least Square
PC	Principal Component
PCA	Principal Component Analysis

PCR	Principal Component Regression
PLS-DA	Partial Least Square-Discriminant Analysis
PLSR	Partial Least Square Regression
PNN	Probabilistic Neural Network
POL II	Polymerase II
QSAR	Quantitative Structure Analysis Relationship
RBF	Radial Basis Function
RDF	Resource Development Framework
RF	Random Forest
RGB	Red/Green/Blue colour spectrum
ROC	Receiver Operating Characteristic
RPD	Residual Prediction Deviation
SEA	Similarity Ensemble Approach
SNARC	Stochastic Neural Analog Reinforcement Calculator
SOM	Self-organising map
SIDER	Side Effect Response Resource
SSP	Shortest Spinning Path
SVM	Support Vector Machines
SVM (BT)	Binary Tree Support Vector Machines
SVM (DR)	Dense Regions Approach Support Vector Machines
SVM (OAO)	One Against One Support Vector Machines
SVM (RS)	Retrosynthetic Approach Support Vector Machines
TEP	Tennessee Extraction Process
TIBO	Thiobenzimidazolone
TN	True Negative
TP	True Positive
Vis-NIR	Visible-Near Infrared
WHIM	Weighted Holistic Invariant Molecular descriptors
WLS	Weighted Least Squares

Acknowledgements

BCP would like to acknowledge De Montfort University for her fee waiver for her PhD studies.

References

1. H. A. Simon, *Information-processing Theory of Human Problem Solving*, vol. 5, 1978.
2. M. Coray, B. W. Millis, S. Strack and S. D. Olson, *Homer's Iliad: the Basel commentary, Book XVIII.*
3. A. Long and D. Sedley, *Plato: Meno and Phaedo*, Cambridge University Press, 2010.
4. T. Hobbes, *Leviathan or the Matter, Forme, and Power of a Commonwealth Ecclesiasticall and Civil*, London, Malmesbury, 1651.

5. E. Knobloch and W. Berlin, *The Mathematical Studies of G.W Leibniz on Combinatorics*, vol. 1, 1974.
6. R. Descartes, *The Philosophical Writings of Descartes*, Cambridge University Press, vol. 1, 1900.
7. A. G. Bromley and C. Babbage, *Charles Babbage's Analytical Engine*, 1838, 1982.
8. L. F. Menabrea and A. A. Lovelace, *Sketch of the Analytical Engine Invented by Charles Babbage*, 1843.
9. G. Boole, An Investigation of the Laws of Thought: On Which are Founded the Mathematical Theories of Logic and Probabilities, Dover Publications, 1854.
10. C. E. Shannon and W. Weaver, *The Mathematical Theory of Communication*, 1949.
11. A. M. Turing, K. Ford, C. Glymour and P. Hayes, *Computing Machinery and Intelligence*, 1950.
12. G. O'Regan, *Giants of Computing*, Springer, 2013, pp. 193–195.
13. W. S. Mcculloch and W. Pitts, *A Logical Calculus of the Ideas Immanent in Nervous Activity*, vol. 5, 1943.
14. D. O. Hebb, *The Organization of Behavior A Neuropsychological Theory*, 1949.
15. J. Mccarthy, M. L. Minsky, N. Rochester and C. E. Shannon, *A Proposal for the Dartmouth Summer Research Project*, AIMAG.V2714, 2006.
16. F. Rosenblatt, *Psychol. Rev.*, 1958, **65**, 386–408.
17. M. Minsky and S. A. Papert, *Perceptrons: An Introduction to Computational Geometry*, MIT press, 2017.
18. H. A. Simon and A. Newell, *Am. Psychol.*, 1971, **26**, 145–159.
19. A. Newell, J. C. Shaw and H. A. Simon.
20. K. Fukushima, *Biological Cybernetics Neocognitron: A Self-organizing Neural Network Model for a Mechanism of Pattern Recognition Unaffected by Shift in Position*, vol. 36, 1980.
21. J. J. Hopfield, *Neural Networks and Physical Systems With Emergent Collective Computational Abilities*, vol. 79, 1982.
22. C. J. C. H. Watkins and P. Dayan, *Q-Learning*, vol. 8, 1992.
23. D. E. Rumelhart, G. E. Hinton and R. J. Williams, *Nature*, 1986, **323**, 533–536.
24. Y. Lecun, Y. Bengio and G. Hinton, *Nature*, 2015, **521**, 436–444.
25. R. K. Lindsay, B. G. Buchanan, E. A. Feigenbaum, J. Lederberg and R. K. Lindsay, *DENDRAL: A Case Study of the First Expert System for Scientific Hypothesis Formation**, 1993.
26. T. H. Pierce and B. A. Hohne, 1986.
27. L. B. Sybrandt and S. P. Perone, *Anal. Chem.*, 1971, **43**, 382–388.
28. P. C. Jurs, B. R. Kowalski, T. L. Isenhour and C. N. Reilley, *Anal. Chem.*, 1970, **42**, 1387–1394.
29. P. C. Jurs, L. E. Wangen, N. M. Frew, T. L. Isenhour, P. C. Jurs, B. R. Kowalski and C. A. Reilly, *The Fourier Transform and Its Applications*, McGraw-Hill Book Co, vol. 42, 1970.

30. P. C. Jurs, *Anal. Chem.*, 1971, **43**, 22–26.
31. K. L. Ting, R. C. Lee, G. W. Milne, M. Shapiro and A. M. Guarino, *Science*, 1973, **180**, 417–420.
32. B. R. Kowalski and C. F. Bender, *Anal. Chem.*, 1972, **44**, 1405–1411.
33. S. R. Heller, L. Chin Chang and K. C. Chu, *Anal. Chem.*, 1974, **46**, 951–952.
34. K. C. Chu, *Anal. Chem.*, 1974, **46**, 1181–1187.
35. B. R. Kowalski, T. F. Schatzki and F. H. Stress, *Anal. Chem.*, 1972, **44**, 2176–2180.
36. E. H. Shortliffe, *MYCIN: A. Knowledge-Based Computer Program Applied to Infectious Diseases*, 1977.
37. I. G. Maglogiannis, *Emerging Artificial Intelligence Applications in Computer Engineering: Real World AI Systems with Applications in eHealth, HCI, Information Retrieval and Pervasive Technologies*, IOS Press, 2007.
38. E. Alpaydin, *Introduction to Machine Learning*, 2009.
39. Y. Bengio, A. Courville and P. Vincent, *Unsupervised Feature Learning and Deep Learning: A Review and New Perspectives*, 2012.
40. B. Schölkopf, C. J. Burges and A. J. Smola, *Advances in Kernel Methods: Support Vector Learning*, MIT Press, 1999.
41. A. Lavecchia, *Drug Discovery Today*, 2015, **20**, 318–331.
42. J. M. Granda, L. Donina, V. Dragone, D.-L. Long and L. Cronin, *Nature*, 2018, **559**, 377–381.
43. Y. Podolyan, M. A. Walters and G. Karypis, *J. Chem. Inf. Model.*, 2010, **50**, 979–991.
44. S. Vilar, G. Cozza and S. Moro, *Curr. Top. Med. Chem.*
45. W. (Walter) Karcher and J. Devillers, *Practical applications of quantitative structure-activity relationships (QSAR) in environmental chemistry and toxicology*, Kluwer Academic Publishers, 1990.
46. C. Hansch, D. Hoekman, A. Leo, L. Zhang and P. Li, *Toxicol. Lett.*, 1995, **79**, 45–53.
47. R. Darnag, E. L. Mostapha Mazouz, A. Schmitzer, D. Villemin, A. Jarid and D. Cherqaoui, *Eur. J. Med. Chem.*, 2010, **45**, 1590–1597.
48. S. Xuan, Y. Wu, X. Chen, J. Liu and A. Yan, *Bioorg. Med. Chem. Lett.*, 2013, **23**, 1648–1655.
49. L. A. Cysique, J. M. Murray, M. Dunbar, V. Jeyakumar, B. J. Brew and S. Wales, *HIV Med*, 2010, **11**, 642–649.
50. K. K. Kandaswamy, G. Pugalenthi, S. Möller, E. Hartmann, K.-U. Kalies, P. N. Suganthan and T. Martinetz, *Prediction of Apoptosis Protein Locations with Genetic Algorithms and Support Vector Machines Through a New Mode of Pseudo Amino Acid Composition*, vol. 17, 2010.
51. S. Lise, D. Buchan, M. Pontil and D. T. Jones, *PLoS One*, 2011, **6**, 16774.
52. R. Kumar, A. Srivastava, B. Kumari and M. Kumar, *J. Theor. Biol.*, 2014, **365**, 96–103.
53. C. Xu, H. Qu, G. Wang, B. Xie, Y. Shi, Y. Yang, Z. Zhao, L. Hu, X. Fang, J. Yan and L. Feng, *Nat. Publ. Gr.*, 2015, **5**, 17788.

54. F. Santos, P. Guyomarc'h and J. Bruzek, *Forensic Sci. Int.*, 2014, **245**, 204.e1–204.e8.
55. C. Gambino, P. Mclaughlin, L. Kuo, F. Kammerman, P. Shenkin, P. Diaczuk, N. Petraco, J. Hamby and N. D. K. Petraco, *Scanning*, 2011, **33**, 272–278.
56. Y.-K. Kwon, Y.-S. Bong, K.-S. Lee and G.-S. Hwang, *Food Chem.*, 2014, **161**, 168–175.
57. R. Dong, S. Weng, L. Yang and J. Liu, *Anal. Chem.*, 2015, **87**, 2937–2944.
58. P. Van Renterghem, P.-E. Sottas, M. Saugy and P. Van Eenoo, *Anal. Chim. Acta*, 2013, **768**, 41–48.
59. F. Zsila, Z. Bikadi, D. Malik, P. Hari, I. Pechan, A. Berces and E. Hazai, *Bioinformatics*, 2011, **27**, 1806–1813.
60. I. Monroy, R. Benitez, G. Escudero and M. Graells, *Comput. Chem. Eng.*, 2010, **34**, 631–642.
61. L. Qian, E. Winfree and J. Bruck, *Nature*, 2011, **475**, 368–372.
62. Y. Shen and A. Bax, *J. Biomol. NMR*, 2010, **48**, 13–22.
63. Y. Shen and A. Bax, *J. Biomol. NMR*, 2013, **56**, 227–241.
64. Y. Shen and A. Bax, Springer, New York, NY, 2015, pp. 17–32.
65. D. F. Brougham, G. Ivanova, M. Gottschalk, D. M. Collins, A. J. Eustace, R. O'Connor and J. Havel, *J. Biomed. Biotechnol.*, 2011, **2011**, 158094.
66. A. M. Mouazen, B. Kuang, J. De Baerdemaeker and H. Ramon, *Geoderma*, 2010, **158**, 23–31.
67. R. M. Balabin, E. I. Lomakina and R. Z. Safieva, *Fuel*, 2011, **90**, 2007–2015.
68. J. S. McKenzie, J. M. Jurado and F. de Pablos, *Food Chem.*, 2010, **123**, 859–864.
69. M. H. S. Segler, M. Preuss and M. P. Waller, *Nature*, 2018, **555**, 604–610.
70. J. N. Wei, D. Duvenaud and A. Aspuru-Guzik, *ACS Cent. Sci.*, 2016, **2**, 725–732.
71. A. Vidaki, D. Ballard, A. Aliferi, T. H. Miller, L. P. Barron and D. Syndercombe Court, *Forensic Sci. Int.: Genet.*, 2017, **28**, 225–236.
72. G. Koukiou and V. Anastassopoulos, *Forensic Sci. Int.*, 2015, **252**, 69–76.
73. S. De Vito, G. Fattoruso, M. Pardo, F. Tortorella and G. Di Francia, *IEEE Sens. J.*, 2012, **12**, 3215–3224.
74. A. D. Rodgers, H. Zhu, D. Fourches, I. Rusyn and A. Tropsha, *Chem. Res. Toxicol.*, 2010, **23**, 724–732.
75. Y. Chen, F. Cheng, L. Sun, W. Li, G. Liu and Y. Tang, *Ecotoxicol. Environ. Saf.*, 2014, **110**, 280–287.
76. S. Jamal, S. Goyal, A. Shanker and A. Grover, *Sci. Rep.*, 2017, **7**, 872.
77. A. St-Hilaire, T. B. M. J. Ouarda, Z. Bargaoui, A. Daigle and L. Bilodeau, *Hydrol. Processes*, 2012, **26**, 1302–1310.
78. A. Manganaro, F. Pizzo, A. Lombardo, A. Pogliaghi and E. Benfenati, *Chemosphere*, 2016, **144**, 1624–1630.
79. M. Cassotti, D. Ballabio, V. Consonni, A. Mauri, I. V. Tetko and R. Todeschini, *Altern. Lab. Anim.*, 2014, **42**, 31–41.

80. X. Li, T. Yang, S. Li, D. Wang, Y. Song and S. Zhang, *Laser Phys.*, 2016, **26**, 035702.
81. C. Yan, Z. Wang, F. Ruan, J. Ma, T. Zhang, H. Tang and H. Li, *Anal. Methods*, 2016, **8**, 6216–6221.
82. T. Bergmann, F. Heinke and D. Labudde, *Forensic Sci. Int.*, 2017, **278**, 1–8.
83. C. A. Marchant, E. M. Rosser and J. D. Vessey, *Mol. Inf.*, 2017, **36**, 1600105.
84. P. B. Choudhari, M. S. Bhatia and N. M. Bhatia, *Med. Chem. Res.*, 2013, **22**, 976–985.
85. K. S. Bhadoriya, M. C. Sharma, S. Sharma, S. V. Jain and M. H. Avchar, *Arabian J. Chem.*, 2014, **7**, 924–935.
86. L. Owen, K. Laird and P. B. Wilson, *Mol. Cell. Probes*, 2018, **38**, 25–30.
87. M. C. Sharma, S. Sharma, N. K. Sahu and D. V. Kohli, *J. Saudi Chem. Soc.*, 2013, **17**, 167–176.
88. P. B. Choudhari, M. S. Bhatia and S. D. Jadhav, *Sci. Pharm.*, 2012, **80**, 283–294.
89. V. M. Patil, S. P. Gupta, S. Samanta and N. Masand, *Med. Chem. Res.*, 2011, **20**, 1616–1621.
90. S. Ajmani, K. Jadhav and S. A. Kulkarni, *J. Chem. Inf. Model.*, 2006, **46**, 24–31.
91. M. Hayat and A. Khan, *Protein Pept. Lett.*, 2012, **19**, 411–421.
92. A. Sungheetha and R. Rajesh Sharma, *J. Med. Imaging Heal. Informatics*, 2016, **6**, 1652–1656.
93. W. Chen, P.-M. Feng, E.-Z. Deng, H. Lin and K.-C. Chou, *Anal. Biochem.*, 2014, **462**, 76–83.
94. T. Fannes, E. Vandermarliere, L. Schietgat, S. Degroeve, L. Martens and J. Ramon, *J. Proteome Res.*, 2013, **12**, 2253–2259.
95. C. Wang and Y. Zhang, *J. Comput. Chem.*, 2017, **38**, 169–177.
96. G. Cano, J. Garcia-Rodriguez, A. Garcia-Garcia, H. Perez-Sanchez, J. A. Benediktsson, A. Thapa and A. Barr, *Expert Syst. Appl.*, 2017, **72**, 151–159.
97. G. Riddick, H. Song, S. Ahn, J. Walling, D. Borges-Rivera, W. Zhang and H. A. Fine, *Bioinformatics*, 2011, **27**, 220–224.
98. T. Chen, Y. Cao, Y. Zhang, J. Liu, Y. Bao, C. Wang, W. Jia and A. Zhao, *J. Evidence-Based Complementary Altern. Med.*, 2013, **2013**, 298183.
99. P. R. West, A. M. Weir, A. M. Smith, E. L. R. Donley and G. G. Cezar, *Toxicol. Appl. Pharmacol.*, 2010, **247**, 18–27.
100. R. Wang-Sattler, Z. Yu, C. Herder, A. C. Messias, A. Floegel, Y. He, K. Heim, M. Campillos, C. Holzapfel, B. Thorand, H. Grallert, T. Xu, E. Bader, C. Huth, K. Mittelstrass, A. Döring, C. Meisinger, C. Gieger, C. Prehn, W. Roemisch-Margl, M. Carstensen, L. Xie, H. Yamanaka-Okumura, G. Xing, U. Ceglarek, J. Thiery, G. Giani, H. Lickert, X. Lin, Y. Li, H. Boeing, H.-G. Joost, M. H. de Angelis, W. Rathmann, K. Suhre, H. Prokisch, A. Peters, T. Meitinger, M. Roden, H.-E. Wichmann, T. Pischon, J. Adamski and T. Illig, *Mol. Syst. Biol.*, 2012, **8**, 615.

101. S. C. Kalhan, L. Guo, J. Edmison, S. Dasarathy, A. J. McCullough, R. W. Hanson and M. Milburn, *Metabolism*, 2011, **60**, 404–413.
102. V. O. Shah, R. R. Townsend, H. I. Feldman, K. L. Pappan, E. Kensicki and D. L. Vander Jagt, *Clin. J. Am. Soc. Nephrol.*, 2013, **8**, 363–370.
103. B. C. Percival, M. Grootveld, M. Gibson, Y. Osman, M. Molinari, F. Jafari, T. Sahota, M. Martin, F. Casanova, M. L. Mather, M. Edgar, J. Masania and P. B. Wilson, *High-Throughput*, 2018, **8**, 2.
104. X. Zhang, H. Jiang, J. Jin, X. Xu and Q. Zhang, *Atmos. Environ.*, 2012, **46**, 590–596.
105. A. M. Gardner, T. K. Anderson, G. L. Hamer, D. E. Johnson, K. E. Varela, E. D. Walker and M. O. Ruiz, *Parasites Vectors*, 2013, **6**, 9.
106. L. Sheng, T. Zhang, G. Niu, K. Wang, H. Tang, Y. Duan and H. Li, *J. Anal. At. Spectrom.*, 2015, **30**, 453–458.
107. T. Zhang, L. Liang, K. Wang, H. Tang, X. Yang, Y. Duan and H. Li, *J. Anal. At. Spectrom.*, 2014, **29**, 2323–2329.
108. A. O. Oliynyk, E. Antono, T. D. Sparks, L. Ghadbeigi, M. W. Gaultois, B. Meredig and A. Mar, *Chem. Mater.*, 2016, **28**, 7324–7331.
109. C. Xu, F. Cheng, L. Chen, Z. Du, W. Li, G. Liu, P. W. Lee and Y. Tang, *J. Chem. Inf. Model.*, 2012, **52**, 2840–2847.
110. H. Zhang, Y.-L. Kang, Y.-Y. Zhu, K.-X. Zhao, J.-Y. Liang, L. Ding, T.-G. Zhang and J. Zhang, *Toxicol. In Vitro*, 2017, **41**, 56–63.
111. H. Zhang, Z.-X. Cao, M. Li, Y.-Z. Li and C. Peng, *Food Chem. Toxicol.*, 2016, **97**, 141–149.
112. A. Bender, ed. J. Bajorath, *Bayesian Methods in Virtual Screening and Chemical Biology*, Humana Press, Totowa, NJ, 2010, pp. 175–196.
113. J. Fang, R. Yang, L. Gao, D. Zhou, S. Yang, A. Liu and G. Du, *J. Chem. Inf. Model.*, 2013, **53**, 3009–3020.
114. L. Chen, Y. Li, Q. Zhao, H. Peng and T. Hou, *Mol. Pharmaceutics*, 2011, **8**, 889–900.
115. J. Wildenhain, M. Spitzer, S. Dolma, N. Jarvik, R. White, M. Roy, E. Griffiths, D. S. Bellows, G. D. Wright and M. Tyers, *Cell Syst.*, 2015, **1**, 383–395.
116. M. Liu, Y. Wu, Y. Chen, J. Sun, Z. Zhao, X. Chen, M. E. Matheny and H. Xu, *J. Am. Med. Informatics Assoc.*, 2012, **19**, e28–e35.
117. X. Li, L. Chen, F. Cheng, Z. Wu, H. Bian, C. Xu, W. Li, G. Liu, X. Shen and Y. Tang, *J. Chem. Inf. Model.*, 2014, **54**, 1061–1069.
118. H. Zhang, L. Ding, Y. Zou, S.-Q. Hu, H.-G. Huang, W.-B. Kong and J. Zhang, *J. Comput.-Aided Mol. Des.*, 2016, **30**, 889–898.
119. R. S. Nascimento, R. E. Froes, N. O. e Silva, R. L. Naveira, D. B. Mendes, W. B. Neto and J. B. B. Silva, *Talanta*, 2010, **80**, 1102–1109.
120. A. Schmitt, B. Saliba-Serre, M. Tremblay and L. Martrille, *J. Forensic Sci*, 2010, **55**, 590–596.
121. I. Smeers, R. Decorte, W. Van de Voorde and B. Bekaert, *Forensic Sci. Int.: Genet.*, 2018, **34**, 128–133.
122. G. Argyropoulos and C. Samara, *Environ. Model. Software*, 2011, **26**, 469–481.

123. L. A. Chesson, D. W. Podlesak, B. R. Erkkila, T. E. Cerling and J. R. Ehleringer, *Food Chem.*, 2010, **119**, 1250–1256.
124. G.-H. Fu, B.-Y. Zhang, H.-D. Kou and L.-Z. Yi, *Chemom. Intell. Lab. Syst.*, 2017, **160**, 22–31.
125. W. Shi and W. Zeng, *Front. Environ. Sci. Eng.*, 2014, **8**, 117–127.
126. W. Shi, W. Zeng, W. Shi and W. Zeng, *Int. J. Environ. Res. Public Health*, 2013, **10**, 2578–2595.
127. M. Ay and O. Kisi, *J. Hydrol.*, 2014, **511**, 279–289.
128. M. Hedegaard, C. Krafft, H. J. Ditzel, L. E. Johansen, S. Hassing and J. Popp, *Anal. Chem.*, 2010, **82**, 2797–2802.
129. A. D. Meade, C. Clarke, F. Draux, G. D. Sockalingum, M. Manfait, F. M. Lyng and H. J. Byrne, *Anal. Bioanal. Chem.*, 2010, **396**, 1781–1791.
130. J. R. Tietjen, D. W. Zhang, J. B. Rodríguez-Molina, B. E. White, M. S. Akhtar, M. Heidemann, X. Li, R. D. Chapman, K. Shokat, S. Keles, D. Eick and A. Z. Ansari, *Nat. Struct. Mol. Biol.*, 2010, **17**, 1154–1161.
131. J. Lu, L. Chen, J. Yin, T. Huang, Y. Bi, X. Kong, M. Zheng and Y.-D. Cai, *J. Biomol. Struct. Dyn.*, 2016, **34**, 906–917.
132. R. Bartzatt, *Anti-Inflammatory Anti-Allergy Agents Med. Chem.*, 2012, **11**, 151–160.
133. G. Žibret and R. Šajn, *Math. Geosci.*, 2010, **42**, 681–703.
134. M. T. Scotti, V. Emerenciano, M. J. P. Ferreira, L. Scotti, R. Stefani, M. S. da Silva, F. J. B. M. Junior, M. T. Scotti, V. Emerenciano, M. J. P. Ferreira, L. Scotti, R. Stefani, M. S. Da Silva and F. J. B. M. Junior, *Molecules*, 2012, **17**, 4684–4702.
135. M. Tobiszewski, S. Tsakovski, V. Simeonov and J. Namieśnik, *Chemosphere*, 2012, **87**, 962–968.
136. A. J. Adeloye and R. Rustum, *Hydrol. Res*, 2012, **43**, 603–617.
137. S. C. Löhr, M. Grigorescu, J. H. Hodgkinson, M. E. Cox and S. J. Fraser, *Geoderma*, 2010, **156**, 253–266.
138. T. Li, G. Sun, C. Yang, K. Liang, S. Ma and L. Huang, *Sci. Total Environ.*, 2018, **628–629**, 1446–1459.
139. G. R. Lloyd, J. Wood, C. Kendall, T. Cook, N. Shepherd and N. Stone, *Vib. Spectrosc.*, 2012, **60**, 43–49.
140. E. Ejarque-Gonzalez and A. Butturini, *PLoS One*, 2014, **9**, e99618.
141. W. N. S. M. Desa, N. N. Daéid, D. Ismail and K. Savage, *Anal. Chem.*, 2010, **82**, 6395–6400.
142. P. M. K. Westcott, K. D. Halliwill, M. D. To, M. Rashid, A. G. Rust, T. M. Keane, R. Delrosario, K.-Y. Jen, K. E. Gurley, C. J. Kemp, E. Fredlund, D. A. Quigley, D. J. Adams and A. Balmain, *Nature*, 2015, **517**, 489–492.
143. A. Patras, N. P. Brunton, G. Downey, A. Rawson, K. Warriner and G. Gernigon, *J. Food Compos. Anal.*, 2011, **24**, 250–256.
144. D. Fraccalvieri, A. Pandini and F. Stella, *et al.*, Conformational and functional analysis of molecular dynamics trajectories by Self-Organising Maps, *BMC Bioinf.*, 2011, **12**, 158.

145. M. Goodarzi, S. Sharma, H. Ramon and W. Saeys, *TrAC, Trends Anal. Chem.*, 2015, **67**, 147–158.
146. M. R. Almeida, D. N. Correa, J. J. Zacca, L. P. L. Logrado and R. J. Poppi, *Anal. Chim. Acta*, 2015, **860**, 15–22.
147. M. Mecozzi, M. Pietroletti, M. Scarpiniti, R. Acquistucci and M. E. Conti, *Environ. Monit. Assess.*, 2012, **184**, 6025–6036.
148. Y. Kato, K. Fujinaga, K. Nakamura, Y. Takaya, K. Kitamura, J. Ohta, R. Toda, T. Nakashima and H. Iwamori, *Nat. Geosci.*, 2011, **4**, 535–539.
149. G. Salimi-Khorshidi, G. Douaud, C. F. Beckmann, M. F. Glasser, L. Griffanti and S. M. Smith, *NeuroImage*, 2014, **90**, 449–468.
150. S. S. Kumar and K. Balakrishnan, *Biomed. Signal Process. Control*, 2013, **8**, 667–674.
151. A. Maudoux, P. Lefebvre, J.-E. Cabay, A. Demertzi, A. Vanhaudenhuyse, S. Laureys and A. Soddu, *PLoS One*, 2012, **7**, e36222.
152. H. Altae-Tran, B. Ramsundar, A. S. Pappu and V. Pande, *ACS Cent. Sci.*, 2017, **3**, 283–293.
153. D. A. Winkler and T. C. Le, *Mol. Inform.*, 2017, **36**, 1600118.
154. J. Ma, R. P. Sheridan, A. Liaw, G. E. Dahl and V. Svetnik, *J. Chem. Inf. Model.*, 2015, **55**, 263–274.
155. E. B. Lenselink, N. ten Dijke, B. Bongers, G. Papadatos, H. W. T. van Vlijmen, W. Kowalczyk, A. P. IJzerman and G. J. P. van Westen, *J. Cheminform.*, 2017, **9**, 45.
156. T. Unterthiner, A. Mayr, G. Klambauer, M. Steijaert, J. K. Wegner and H. Ceulemans, 2015.
157. G. B. Goh, C. Siegel, A. Vishnu, N. O. Hodas and N. Baker.
158. G. B. Goh, C. Siegel, A. Vishnu, N. Hodas and N. Baker, in *2018 IEEE Winter Conference on Applications of Computer Vision (WACV)*, IEEE, 2018, pp. 1340–1349.
159. G. B. Goh, C. M. Siegel, A. Vishnu and N. O. Hodas, 2017.
160. K. T. Schütt, H. E. Sauceda, P.-J. Kindermans, A. Tkatchenko and K.-R. Müller, *J. Chem. Phys.*, 2018, **148**, 241722.
161. N. Lubbers, J. S. Smith and K. Barros, *J. Chem. Phys.*, 2018, **148**, 241715.
162. J. S. Smith, O. Isayev and A. E. Roitberg, *Chem. Sci.*, 2017, **8**, 3192–3203.
163. P. Schwaller, T. Gaudin, D. Lányi, C. Bekas and T. Laino, *Chem. Sci.*, 2018, **9**, 6091–6098.
164. M. Ragoza, J. Hochuli, E. Idrobo, J. Sunseri and D. R. Koes, *J. Chem. Inf. Model.*, 2017, **57**, 942–957.
165. I. Wallach, M. Dzamba and A. Heifets, AtomNet: A Deep Convolutional Neural Network for Bioactivity Prediction in Structure-based Drug Discovery, arXiv preprint, 2015, arXiv:1510.02855.
166. D. Liu, Y. Tan, E. Khoram and Z. Yu, *ACS Photonics*, 2018, **5**, 1365–1369.
167. E. Putin, A. Asadulaev, Y. Ivanenkov, V. Aladinskiy, B. Sanchez-Lengeling, A. Aspuru-Guzik and A. Zhavoronkov, *J. Chem. Inf. Model.*, 2018, **58**, 1194–1204.
168. N. Brown, M. Fiscato, M. H. S. Segler and A. C. Vaucher, *J. Chem. Inf. Model.*, 2019, **59**, 1096–1108.

169. B. Ait Skourt, A. El Hassani and A. Majda, *Procedia Computer Sci.*, 2018, **127**, 109–113.
170. J. Kleesiek, G. Urban, A. Hubert, D. Schwarz, K. Maier-hein, M. Bendszus and A. Biller, *NeuroImage*, 2016, **129**, 460–469.
171. M. Havaei, A. Davy, D. Warde-farley, A. Biard, A. Courville, Y. Bengio, C. Pal, P. Jodoin and H. Larochelle, *Med. Image Anal.*, 2017, **35**, 18–31.
172. E. Gibson, W. Li, C. Sudre, L. Fidon, D. I. Shakir, G. Wang, Z. Eaton-rosen, R. Gray, T. Doel, Y. Hu, T. Whyntie, P. Nachev, M. Modat, D. C. Barratt, S. Ourselin, M. J. Cardoso and T. Vercauteren, *Computer Methods Programs Biomed.*, 2018, **158**, 113–122.
173. A. K. Mortensen, M. Dyrmann, H. Karstoft, R. N. Jørgensen and R. Gislum, CIGR-AgEng Conf. 26–29 June 2016, Aarhus, Denmark. Abstr. Full Pap., 2016, 1–6.
174. D. Quang and X. Xie, *Nucleic Acids Res.*, 2016, **44**, e107.
175. V. Boža, B. Brejová and T. Vinař, *PLoS One*, 2017, **12**, e0178751.
176. B. Alipanahi, A. Delong, M. T. Weirauch and B. J. Frey, *Nat. Biotechnol.*, 2015, **33**, 831–838.
177. Z. Chen, N. He, Y. Huang, W. T. Qin, X. Liu and L. Li, *Genomics, Proteomics Bioinf*, 2018, **16**, 451–459.
178. Y. Xie, X. Luo, Y. Li, L. Chen, W. Ma, J. Huang, J. Cui, Y. Zhao, Y. Xue, Z. Zuo and J. Ren, *Genomics. Proteomics Bioinformatics*, 2018, **16**, 294–306.
179. J. Lyons, A. Dehzangi, R. Heffernan, A. Sharma, K. Paliwal, A. Sattar, Y. Zhou and Y. Yang, *J. Comput. Chem.*, 2014, **35**, 2040–2046.
180. Y.-B. Wang, Z.-H. You, X. Li, T.-H. Jiang, X. Chen, X. Zhou and L. Wang, *Mol. BioSyst.*, 2017, **13**, 1336–1344.
181. H. Zhai and A. N. Alexandrova, *J. Chem. Theory Comput.*, 2016, **12**, 6213–6226.
182. O. Hennigh, arXiv:1705.09036.
183. S. Fong, X.-S. Yang and S. Deb, in *2013 IEEE 16th International Conference on Computational Science and Engineering*, IEEE, 2013, pp. 902–909.
184. H. Liu and L. Yu, *IEEE Trans. Knowl. Data Eng*, 2005, 491–502.
185. B. P. MacLeod, F. G. L. Parlane, T. D. Morrissey, F. Häse, L. M. Roch, K. E. Dettelbach, R. Moreira, L. P. E. Yunker, M. B. Rooney, J. R. Deeth, V. Lai, G. J. Ng, H. Situ, R. H. Zhang, M. S. Elliott, T. H. Haley, D. J. Dvorak, A. Aspuru-Guzik, J. E. Hein and C. P. Berlinguette, *Sci. Adv.*, 2020, **6**, eaaz8867.
186. B. Xue, M. Zhang and W. N. Browne, *IEEE Trans. Cybern.*, 2012, **43**, 1656–1671.

CHAPTER 6

Computational Spectroscopy and Photophysics in Complex Biological Systems: Towards an In Silico *Photobiology*

ANTONIO FRANCÉS-MONERRIS,*[a,b] MARCO MARAZZI,*[c,d] VANESSA BESANCENOT,[e] STÉPHANIE GRANDEMANGE,[e] XAVIER ASSFELD[a] AND ANTONIO MONARI*[a]

[a] Université de Lorraine and CNRS, LPCT UMR 7019, F-54000, Nancy, France; [b] Departament de Química Física, C/Dr. Moliner 50, 46100, Burjassot, Spain; [c] Universidad de Alcalá, Departamento de Química Analítica, Química Física e Ingeniería Química, E-28871, Alcalá de Henares, Spain; [d] Universidad de Alcalá, Chemical Research Institute "Andrés M. del Río" (IQAR), E-28871, Alcalá de Henares, Spain; [e] Université de Lorraine and CNRS, CRAN UMR7039, F-54000, Nancy, France
*Emails: antonio.frances@univ-lorraine.fr; marco.marazzi@uah.es; antonio.monari@univ-lorraine.fr

6.1 Introduction

The interaction of light with biological systems gives rise to a number of complex, yet crucial, phenomena that to a large extent have contributed to shape the way life on Earth has evolved.[1] To cite some important, but non-exhaustive examples, one could highlight photosynthesis,[2] for example the possibility of exploiting photochemical reactions for the production of

Theoretical and Computational Chemistry Series No. 20
Computational Techniques for Analytical Chemistry and Bioanalysis
Edited by Philippe B. Wilson and Martin Grootveld

Published by the Royal Society of Chemistry, www.rsc.org

energy,[3] but also the use of light as a signal, such as circadian rhythm control[4–6] or in vision.[7–9] On the other hand, the absorption of the high energy density contained in visible or ultraviolet radiation may also lead to unwanted side-effects, triggering harmful chemical reactions that should be properly controlled by living organisms, limiting their impact.[10–13]

In living cells, and even more so in complex organisms, the interaction with light induces a rather complex cascade of events whose precise regulation will ultimately determine the final biological outcome. In many respects, such a cascade will involve the activation of different signalling pathways, the recruitment of enzyme complexes, or even the up- or down-regulation of gene expressions.[14–16] As a secondary consequence, the time-scale involved in the response to light activation spans a rather large domain. Furthermore, such a complexity in the global response phenomena, precludes the use of a totally reductionist approach that would neglect the precise study of the cross talks occurring at the cellular and multi-cellular level as a consequence of the application of the light stimulus.

However, even keeping in mind the previous considerations, one should not forget that, at least for what concerns the very first steps, light absorption will induce some elementary photochemical reaction the bases of which should be properly characterized at a molecular and electronic level.[17–20] Hence, in this sense, the domain of chemical biology is fundamental to understand and analyse photobiology, that is the interaction of light and biological matter.[21]

Traditionally, the domain of photobiology, as studied from the point of view of chemistry, has comprised the use of different analytical techniques to understand the different pathways and characterize the distribution of products and reactants, together with the eventual labile intermediates, and hence possibly drawing realistic kinetic models. It is obvious in the case of the study of the interaction between light and matter that the techniques of choice have mostly been various spectroscopic techniques. Indeed, spectroscopic determinations performed either on isolated model systems or in cellular environments are common and allow researchers to gain sufficient information to build realistic models of the various photochemical pathways.[22] This has in most cases comprised the use of rather standard spectroscopic techniques such as UV/Vis absorption and fluorescence measurements or infrared spectroscopy, as well as more advanced protocols allowing an ultra-fast time-resolution to be achieved, such as in the case of pump-probe spectroscopies.[23] More recently, spectroscopic techniques have also been complemented by the use of fluorescence live cell imaging,[24] including super resolution microscopy[25] or single-molecule approach,[26,27] leading to assessment at an unprecedented level of detail. In parallel, the development of other non-linear spectroscopies such as 2D electronic spectroscopy and its extension to the domain of the visible and ultraviolet spectra provides an increased capacity for disentangling the phenomena taking place in complex multi-chromophoric systems, and also takes into account the conformational complexity and flexibility of biological macromolecules.[28–32]

This last aspect also highlights a further consideration and underlines the crucial role played by the complex, and crowded, environment that chromophores embedded in biological systems are obliged to face. Indeed, in many cases the presence of the molecular environment, such as proteins, nucleic acids, or lipid membranes, totally alters the spectroscopic and photophysical properties of the original chromophore,[33–35] *via* either a modification of its potential energy landscape in the ground or in the excited states, or through the opening of novel photochemical and photophysical reactive pathways *via* preferential interactions with other chromophores happening in an organized environment. The latter phenomena may involve energy- or electron-transfer or even novel chemical reaction possibilities.[36–39]

In other words, the macromolecular environment is far from innocent and instead behaves like a real actor in shaping and tuning the photochemical landscape and hence in directing the reactivity and the final outcome. Once again, and as a non-exhaustive example, one can take into account the crucial role of the rhodopsin environment[40,41] in increasing the efficiency of the retinal isomerization, both in terms of the characteristic time and the quantum yield, or the increase in the photostability of the DNA nucleobases in the double-helical arrangement as compared to the isolated monomers.[42,43]

Although fascinating from a scientific point of view, and also potentially opening a route to the design of efficient bio-inspired or biomimetic devices, this scenario also induces a supplementary and considerable difficulty in the fine interpretation of the experimental results. Indeed, even if powerful spectroscopic experiments usually give a global statistic picture, they lack one-to-one mapping with the molecular and electronic description of the various processes.

In this respect the use of molecular modelling and simulation to complement and interpret experimental determinations in complex biological systems may represent the key to achieving a correct and global description of the different phenomena taking place, of the interactions between the different chromophores, and finally of the complex outcomes. However, such a possibility has been made possible by the impressive development of computational techniques observed in recent years and complemented by the parallel impressive growth of the computational power and efficiency provided by modern supercomputers. Indeed, today the proper use of molecular modelling[44] allows precise solving of the electronic structure, and hence provides an energy profile of the ground and excited states of complex chromophores in the presence of complex and inhomogeneous environments.[45] Molecular simulations, on the other hand, permit the inclusion of vibrational and dynamic degrees of freedom and also to explore the free energy landscape of complex systems.[46,47] The excited state time evolution can also be modelled *via* the use of either quantum dynamics or semiclassical molecular dynamics techniques, offering a mapping of the results of the ultrafast time-resolved spectroscopies in terms of the electronic state population and evolution of the energies.[48]

Globally, this unprecedented surge has also been made possible thanks to the development of hybrid quantum mechanics/molecular mechanics

(QM/MM) methods that allow the electronic and geometric effects of complex surroundings to be precisely taken into account.[49]

In this chapter we aim to provide a review on the use of computational modelling and simulations for the study of photobiological processes, and its complementarity with experimental spectroscopies and analytical techniques. In this respect, after an introduction on the methodological bases of QM/MM methods for excited states, we will present how those same methods can be used to offer a finer interpretation of spectroscopic experiments, in particular of linear absorption and emission, also taking into account the effects of the vibrational and conformational movement and the complex surroundings. Afterwards, and to go a step further, we will also show how photophysical and photochemical landscapes may be studied *via* computational methods, in some cases providing a guide for future experiments. At first we will consider a static approach based on the determination of potential energy surfaces connecting critical points, which are essential to understand the photoreactivity. Subsequently, we will show that (non-)adiabatic molecular dynamics may also be used to fully characterize the evolution of the excited states and hence access crucial quantities such as time-scales and quantum yields.

All in all, we aim to provide a useful guide detailing the capacities of modern computational techniques in providing a veritable *in silico* photobiological approach and how this approach should be properly used with constant relation to the experimental determinations.

6.2 Computational Modelling and QM/MM Techniques

As already mentioned, computer simulations offer a complementary tool to understand the interaction of light and biological matter. In this chapter, we are not interested in the actual interaction of light with matter, but instead on the state matter is pushed into after having interacted with light and its time-evolution. By state we mean electronic state, that is the arrangement, allowed by the laws of quantum mechanics, of the electrons within the system under study. For theoretical chemists, matter is composed of two types of particle, electrons and nuclei. The position in space of the nuclei defines the geometry of the system. Solving, if possible, the electronic Schrödinger equation for a given geometry will provide several values of the energy, each of them corresponding to a state. The lowest energy corresponds to the ground state, the others to the so-called excited states. It is indeed the absorption or emission of light that allows transitions from one state to another. In general, the energy of a state corresponds to the sum of the kinetic energy of the electrons and the coulombic interaction between the charged particles (electrons, nuclei). Knowing the energy of a given state for all possible geometries defines a potential energy (hyper)surface (PES) (see Figure 6.1 below). For two geometrical parameters these PESs look like a landscape presenting valleys, mountains, passes, ridges and so on. Stable structures correspond to the bottom of valleys, transition states to mountain

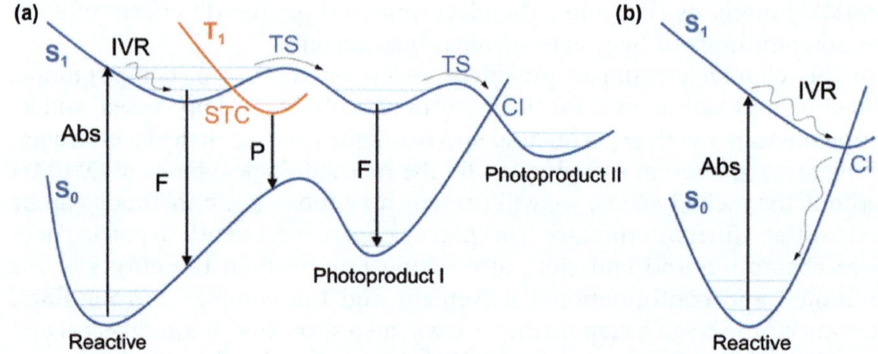

Figure 6.1 Photophysical and photochemical processes. Panel (a) shows the general scheme of the excited-state reactivity toward photoproduct I (adiabatic photoreactivity) and II (non-adiabatic photoreactivity), as well as the light emission processes of fluorescence and phosphorescence. Panel (b) shows the ultrafast, non-radiative and non-reactive decay of S_1 to the ground state. Abs = light absorption; F = fluorescence; P = phosphorescence; IVR = internal vibrational relaxation; TS = transition state; CI = conical intersection; STC = singlet-triplet crossing.

passes. One can remark that until now the kinetic energy of the nuclei was not considered. It is this component, depending on the temperature, that allows movement on the PES, that is going from one valley to another. To understand what light does to matter one then needs to know three main features: (i) what PES looks like for several accessible electronic states; (ii) how the geometry can evolve; and (iii) how the system can pass from one state to another. In this section, emphasis will be placed on the first one. The third feature corresponds to the amount of coupling between the different electronic states, and hence the different PESs. In many instances the crossing probability can be found by solving the Schrödinger equation *via* different methods such as time-dependent density-functional theory (TD-DFT), complete active space self-consistent field (CASSCF), or complete active-space second-order perturbation theory (CASPT2) (see below). In this chapter we will not provide detailed information on how to solve the electronic structure problem for ground or excited states. The curious reader can find technical details of the different methods in reviews and books[50,51] detailing the progresses made in recent years, as well as the future perspectives.[44] The second aspect is handled by means of molecular dynamics (MD) techniques that solve the Newtonian equations of motion. Each nucleus possesses a velocity and evolves in time owing to the forces, that is the derivatives of the PES with respect to nuclei displacements, applied to it. Several types of MD exist and these will not be discussed in detail here. The interested reader can find all of these explanations in Andrew Leach's book.[52]

As previously stated, in this section we will explain how to obtain the electronic energy of a biological molecular system at a given geometry.

There is indeed a crucial particularity of biological systems as compared to most chemical ones: a relevant biological (macro)molecule generally contains hundreds or thousands of atoms (DNA or proteins) and is surrounded by either tons of water molecules or of lipid molecules in membranes. For systems of more than approximately one hundred atoms, theoretical chemists can no longer use accurate quantum mechanics tools as the required computational resources are unattainable, both from the output storage point of view and from the required computational time. Facing this situation, two options are available. Either simplify the molecular system or simplify the theoretical model. Chemical phenomena (*e.g.* chemical reactions or electronic transitions) are generally localized in space and involve few atoms. It is then tempting to simplify the molecular system by removing the atoms not directly involved in the chemical process. This is, however, generally not a good option as the removed atoms can have a very strong influence on the chemically relevant atoms. It could be a simple mechanical influence, for example imposing a peptide secondary structure, and/or an electrostatic one, as the different solvation effects on different electronic states. Then, the only way to solve the problem is to simplify the theoretical model.

A common solution is to divide the whole system into two distinct regions. The region in which the chemical process takes place is called the region of interest (RI). The remaining part of the system is called the surrounding (S). Obviously, RI needs to be described with the help of quantum mechanics (QM) as we are interested in the motion of the electrons and the changes brought about in the electronic distribution when passing to an excited state. Thus, the simplification should be performed on S. One way to describe S at a lower level is to forget the detailed description of matter with electrons and nuclei. One can instead consider that S is composed of atoms linked to one another by means of chemical bonds. Then, the equations are simplified as classical mechanics instead of quantum mechanics, and the total number of particles drops down dramatically, making the numerical procedure possible. This idea is at the origin of molecular mechanics (MM) in which solely atoms, bonded or not, are considered instead of electrons. Joining together QM for RI and MM for S gives rise to the well-known QM/MM methods at the heart of the 2013 Nobel Prize in Chemistry.

Let's have a closer look at the MM region. In addition to the list of atoms and their positions in space, one initially needs also the connectivity, for example which atoms are bound to which atoms (the chemical bonds mentioned above in S). The bond between the two atoms, A and B, is represented by a spring and thus the corresponding energy is simply expressed as $k_{AB}(R_{AB}-R_{AB}{}^{0})^2$, in which k_{AB} is the force constant (up to the factor one half), $R_{AB}{}^{0}$ is the equilibrium bond length and R_{AB} the actual interatomic distance. Angular deformation between two geminal bonds, AB and BC, are treated the same way with a harmonic potential $k_{ABC}(\alpha_{ABC}-\alpha_{ABC}{}^{0})^2$. Dihedral torsions around a bond are expressed with a combination of cosine functions to reproduce the rotation profile around the central bond. In addition to these bonded terms, each atom possesses an atomic point charge with which it interacts with other charged atoms, and van der Waals parameters to account for the remaining (induction, dispersion, ...) interatomic interactions. All force constants, equilibrium parameters, atomic point charges and van der Waals parameters (k_{AB}, k_{ABC}, $R_{AB}{}^0$, q_A, ...) form what is called a force field. A plethora of force fields exist, but the most widely used in the applications mentioned in this chapter is AMBER, as it was specifically developed to describe biological systems.[53,54]

Apart from the trivial distinction between levels of theory and force fields, the QM/MM methods differ by the way they treat the frontier between the two regions and how RI feels S.

6.2.1 QM/MM Frontier

If the frontier between the MM and QM worlds goes through intermolecular interactions, like a solute in a solvent, then no special care has to be paid to its description. If, on the contrary, the frontier goes through covalent bonds (*i.e.* QM/MM boundary cuts) then more care is required. Formally, if a covalent bond is cut, a valency is created, generating a so called dangling bond. Many solutions have been proposed to deal with these dangling bonds and they can be cast in three families, all together referred to as covalent embedding.

The first one includes the link atom (LA) method. It simply proposes to saturate the valency with a monovalent atom, typically a hydrogen atom. Albeit very simple at first glance, these methods suffer from the artificial nature of the additional atom. Does it interact with the other atoms? How does one treat the degrees of freedom? There are no definitive answers and all of them contain arbitrariness.

The second family is termed the connection atom (CA) methods. It moves the frontier at the end of the covalent bond and replaces the actual terminating atom by an atom for which specific parameters need to be defined. With respect to LA methods it has the advantage that the two troubling questions do not need to be posed. One drawback, however, is that one needs to redefine new parameters for each QM level of theory and for each force field, meaning that there is no generality.

Finally, the third type of method uses electron density, generally frozen but this is not compulsory. This method was originally proposed in a seminal paper by Warshel back in 1976,[55] and further developed by Assfeld and Rivail.[56] It is based on the same concept of force fields, that is, transferability. Indeed, in most of the cases the electronic density is built using localized molecular orbitals. These methods, as with the CA methods, do not need to answer the two unphysical questions raised by LA methods. In addition, they are much less sensible to the level of theory and completely insensitive to the nature of the force field. Hence, they are more general. In a recent review selected examples have been discussed and compared.[49]

6.2.2 QM/MM Embedding

The second important difference between the two QM/MM methods is the embedding scheme, that is how RI interacts with S. For all QM/MM methods one can write the total energy as a sum of three terms. The first one is the pure, unperturbed, QM energy of RI. The second is the MM energy of S. The last one is the interaction between both parts.

$$E_{\mathrm{QM/MM}}^{\mathrm{RI+S}} = E_{\mathrm{QM}}^{\mathrm{RI}} + E_{\mathrm{MM}}^{\mathrm{S}} + E_{\mathrm{QM/MM}}^{\mathrm{RI/S}}$$

Depending on the last term RI will not experience the same effect owing to S. The simpler embedding scheme is called mechanical embedding. RI only feels the steric hindrance of S either *via* the bonded terms (stretch, angular deformations...) or by the non-bonded van der Waals interaction. This embedding is meaningful when RI is apolar as well as S. Otherwise, one should go to the next level, that is electrostatic embedding, by which RI feels the Coulombic interaction of the classical atomic point charges brought by the MM atoms. As a result, the electronic wave function of RI is polarized by S. One question that may arise is whether S should be polarized in return. As the atomic point charges of the force fields are derived most of the time from experimental data on condensed phase systems, one considers that the MM point charges are implicitly polarized and do not need to be back polarized by the induced charges in RI. This approximation is fairly correct for most situations. However, there are some cases in which it breaks down. Typically, when the charge distribution of RI varies drastically between the initial and the final state of the chemical process. This could happen for chemical reactions, such as the Menshutkin reaction, but also for electronic transitions involving large charge redistribution, for example charge transfer states. In such situations, one has to use the polarizable embedding refinement in which RI is polarized by S and S by RI, until convergence is reached. This can be achieved either by using an explicitly polarizable force field or by using the so-called universal electronic response of the surroundings (ERS) method, which represents the electrons of the MM part by a polarizable continuum. For a review see the works of Monari *et al.*[49]

6.3 Linear Spectroscopy in Chemical and Biological Systems

The outstanding development of computational tools in recent years has provided a vast and growing variety of computational methods, and the corresponding acronyms, that can be overwhelming for non-expert users.[44,57] The method(s) of choice will always depend on the chemical problem to be solved, in particular, on the nature and the size of the system under study, and also on the properties that one needs to compute. As a starting point, a common practice is to consider a system model composed solely of a single molecule (chromophore) isolated *in vacuo*. Even though this model is an oversimplification, as detailed in Section 6.2, the use of electronic structure methods properly calibrated on such a system provides valuable information about the geometries, energy, and nature of the electronic excitations, dipole moments, and so forth that represent the first step towards an understanding of the full complex system. Later, the second step involves the proper inclusion of the environment and the molecular surroundings. For molecules in solution, the environment can be considered homogenous and often an implicit method is sufficient to represent the solvation, whereas for photobiological systems the heterogeneous biological environments often require a full atomistic description of the system. Indeed, the excitation can take place in one or several chromophores embedded in a complex biological structure, in which the environment can play an active role in the deactivation pathways, and therefore the QM/MM approach is necessary to capture all these important phenomena, as described in Section 6.2. In this scenario of growing complexity, one has to find a suitable method that provides a satisfactory answer to the more or less intricate chemical, biochemical or biological problem under consideration, always at a reasonable computational cost.

6.3.1 Analysing the Excited States PES: A Computational Perspective

The steady-state absorption and emission of light are fundamental processes of light-matter interaction that constitute the basis of countless protocols in analytical chemistry and biochemistry laboratories and is used to identify and/or quantify chemical species. From a theoretical point of view, the essentials of computational photophysics and photochemistry rely on the determination of the ground- and excited-state PESs. PESs are the representation of the states' energies *versus* the molecular coordinates of a given system, as molecular motions change the total (electronic and nuclear) potential energy of the molecule. Thereby, the energy and shape of the PESs will determine the molecular absorption properties, as well as the possible photochemical paths of deactivation after light absorption, for example light emission. Three types of regions are of special relevance, the minima, maxima and crossing points. Minima are related to light absorption and emission processes; maxima are ascribed to transition states connecting two minima and the crossing points are related to non-adiabatic processes.

6.3.1.1 PES Minima

The critical points of minimum energy are relevant because they represent energy traps for the systems. The ground-state minima of a stable molecule (see the reactants area, Figure 6.1a and b) represent a region in which the system will remain indefinitely in the absence of any external perturbation. Therefore, light absorption (Abs) will take place from this area, transferring the system *vertically* from the electronic ground state to the excited state. The fact that light absorption occurs *via* vertical transition can be ascribed to the fact that the motion of electrons is orders of magnitude faster than that of the nuclei (Born–Oppenheimer approximation), hence electron density rearrangement will take place before any nuclear rearrangement. Drawing a parallelism with spectroscopy experiments, most of the molecules of the sample will be vibrating around this ground-state minimum when photons are absorbed by each of the molecules that constitute the sample. On average, the vertical absorption energy (or energy gap between the excited and the ground state) solely at the ground-state equilibrium geometry will provide a good estimation of the experimental absorption maxima, given the statistically high probability of finding most of the molecules in this area. This fact is, of course, reflected in the experimental record as the highest intensity of absorption. This approximation already provides information on the nature of the excitation, dipolar moments of the states, and so forth and has the advantage of a relatively low computational cost and the simplicity of the analysis. However, vibrational effects are neglected, and they can be important in many cases by producing significant energy shifts. Further characterization of the spectral width, caused by the molecular vibrations not being taken into account when solely computing the vertical absorption energy at the ground-state minimum of a molecule (see Figure 6.1a), requires a sampling of those relevant regions of the PES. The convolution of the ensemble of the excitation energies at different geometries, randomly generated around the vibrational modes close to the ground-state minima, allows generation of the electronic spectrum of a given system, including the effects of the vibrational degrees of freedom. Usually, the nuclear density distribution is sampled by means of (approximated) quantum Wigner distributions or *via* MD simulations of the system, following well tested protocols that work in many situations.[58–63] The choice of the first or second option depends on the nature of the system and the inclusion or not of the environment by means of the QM/MM approach, which hampers, in general, the Wigner distribution option. It should be noted that significant improvements of these protocols have been recently reported by some of the authors of the present book chapter.[64]

On the other hand, minima regions in excited-state PESs are associated with areas in which light emission takes place if no crossing with the ground state is accessible. Trapped in these zones, the molecules will emit a photon as the only way to repopulate the ground state, which generally lies at a much lower energy. Fluorescence emission is characterized by the same spin multiplicity between the excited and the ground state (typically, singlet→ singlet transitions), whereas phosphorescence emission involves states of different multiplicities (typically, triplet→singlet transitions). The system can either return to the initial structure (reactant) or to a photoproduct (photoproduct I in Figure 6.1a), the fate will depend on the particular PES landscape of the molecule. Light emission usually takes place from the S_1 and T_1 states (*i.e.* the lowest excited states), following Kasha's rule.[65] In general, long excited-state lifetimes are ascribed to fluorescence (10^{-9}–10^{-6} s) and even longer ones to phosphorescence (10^{-3}–10^{2} s) emissions. The vertical emission energy, that is the energy difference between the excited and ground states at the excited-state equilibrium geometry, is associated with the emission maximum. Analogously to the absorption spectrum, a sampling of the excited singlet or triplet state minima can be performed in order to compute the spectral band width, even though this practice is in general less common with respect to the absorption spectrum owing to the large computational costs associated to the calculation of excited-state frequencies.[59]

6.3.1.2 PES Maxima

As in thermal reactions, energy barriers in the excited state will provide valuable information about the excited-state reactivity, which is in general different than that of the ground state, allowing the study of the competition between different decay channels and the feasibility of a given photochemical pathway. The true maxima of excited PESs are often very difficult to obtain since, as previously mentioned, frequency calculations in the excited state can be a tedious task. For this reason, alternative protocols such as coordinate interpolations are usually used to provide upper bounds for the energy barriers. As shall be mentioned below, the study of energy barriers is of crucial importance when performing a photodynamic study, as the presence of any significant energy barrier will strongly increase the timescale of the event and consequently the computational cost, often making the dynamic study prohibitive.

6.3.1.3 PES Crossing Points

There are certain molecular coordinates in which the energies of two or more electronic states are degenerate. At these specific geometries, the probability of the population transfer from one electronic state to another is maximal, and hence those structures give rise to the so-called non-adiabatic events, for example allowing the system to jump from one PES to another. Structures that have energy degeneracy between two states of the same spin multiplicity are called conical intersections (CIs), whereas structures of energy degeneracy between singlet and triplets states are called singlet-triplet crossings (STCs), as displayed in Figure 6.1a.

Conical intersections allow a particular case of non-adiabatic events named internal conversions, processes that deactivate the excited state in a non-radiative way. In the specific situation in which the CI is directly accessed through a barrierless path from the Franck–Condon region (ground-state minimum), as displayed in Figure 6.1b, the decay to the ground state is ultrafast (10^{-15}–10^{-12} s) and the initial structure of the reactant may be recovered after the decay. This type of excited-state decay represents photostability, however, other topologies around the CIs can lead to a different GS minimum and hence will induce photoreactivity. CIs that can be reached by a barrierless path are particularly suitable to be studied by means of dynamic methods given the ultrashort excited-state lifetime.

On the other hand, STCs enable intersystem crossings between singlet and triplet states, and represent a way to transfer the population from the singlet manifold to the triplet one (and *vice versa*), and therefore being a prerequisite to phosphorescence emission as depicted in Figure 6.1a. Intersystem crossings are spin-forbidden processes and are thus much slower than internal conversions, actually the intersystem crossing efficiency depends on many factors such as the spin-orbit coupling (SOC) between the involved singlet and triplet states.[66] Many photochemical, photobiological and nanotechnological phenomena are mediated by triplet states, we will highlight in the following subsections the particular case of DNA photosensitization through some recent studies reported by our research group.

6.3.2 Practical Guidelines to Model Light Absorption and Emission in Complex Environments

Being aware that the theoretical description of photophysics and photochemistry can be a complicated task, especially in complex biological systems, here we provide some practical guidelines with the aim to help the reader choose a computational protocol according to the chemical, biochemical, or biological problem under study. Specific examples reported in the literature will be provided in order to illustrate the advice.

(i) *Start with the Chromophore in the Gas Phase*

The first step in a photochemical characterization is always the study of the chromophore in the gas phase. Thereby, the intrinsic molecular properties can provide very insightful information at a reasonable computational cost. Indeed, the study of isolated molecules by means of semiempirical methods started in the 1950s–1970s.[67–70] In spite of the strong computational limitations present in that period, calculations allowed helpful analyses of structural parameters, orbital energies and many other properties of molecules of great interest, paving the way for the development of the computational chemistry field and subsequently the understanding of fundamental chemical problems. Pioneering studies reported in those years demonstrated that certain research problems do not require sophisticated computations to be solved, and that relatively simple calculations in the gas phase can provide useful results if properly used. Moreover, it should be noted that the chromophore in the gas phase constitutes the simplest system to benchmark the performance of different electronic structure methods, of crucial relevance for the quality of the results, simplifying the analysis as no external perturbation to the system is included.

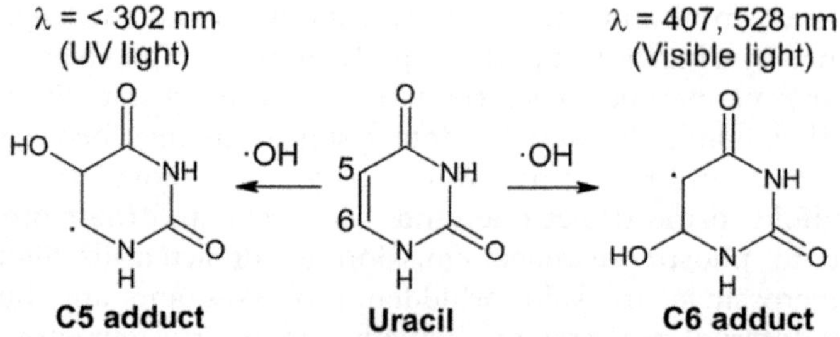

Figure 6.2 OH radical addition to the C5 and C6 positions of uracil. The CASPT2 vertical absorption wavelengths of the radical adducts are also shown.[71]

Nowadays, strongly correlated methods can be routinely applied to molecules in the gas phase, notably increasing the accuracy of the predictions, in some cases being sufficient to solve certain spectroscopic problems.[44] An illustrative example can be found in the resolution of the experimental transient absorption spectra of the hydroxyl radical adducts of the DNA/RNA pyrimidinic nucleobases,[71,72] which were reported in the early 1970s.[73] OH, a highly reactive species produced in oxidative stress conditions, can attack either the C5 or C6 position of the pyrimidinic C=C double bond, showing a clear preference for the regioselectivity at C5 (Figure 6.2).[72,74] The spectra, characterized by broad bands peaking at 380 nm,[73] were misinterpreted for many decades: the signals were ascribed to the C5 adducts as they are the most favourable regioisomers.[75,76] However, highly accurate CASPT2 calculations later evidenced clear differences between the absorption wavelengths of both C5 and C6 adducts, unambiguously showing that the former species were not absorbing at the recorded wavelengths. Thus, the computational results showed that the recorded spectra must be due to the absorption of the C6 adducts.

A similar protocol was applied to the more complex reaction of ·OH with adenine, providing a reinterpretation of the transient spectra of the species and thus the overall mechanism involved in the OH addition with purine derivatives.[77] These conclusions are in fact of crucial relevance in the design of chemical analysis protocols to detect these and other radical species of high biological relevance, given their paramount participation in DNA/RNA oxidative damage.

(ii) *Check the Performance of Implicit Solvation Methods*

Unfortunately, in many cases the information provided by the electronic spectra of isolated molecules in the gas phase is not sufficient to provide a satisfactory resolution for the absorption spectrum. This is the case if the environment has a strong influence on the properties to be computed, for instance, when the ground and excited electronic wave functions have large differences in their respective dipolar moments, making the excitation energies highly sensitive to solvatochromic effects. A typical case is the presence of a charge-transfer state characterized by a large dipolar moment owing to the charge separation, which will be strongly stabilized by the solvent.

In these situations, the correct reproduction of the shift requires the study of solvent effects. An extremely popular way to do this, given its very low computational cost and its user-friendly implementations, is to use implicit solvation methods based on a continuum model such as the polarizable continuum model (PCM)[78] or the conductor-like screening model (COSMO),[79] implemented in most common quantum-chemistry codes. Within this approach, the electrostatic effect of the solvent is represented by means of a homogenous cavity surrounding the chromophore and mimicking the average electrostatic influence of the solvent on the electronic wave functions. Even though the molecules of the solvent are not explicitly included in the calculation, the most important electrostatic solute-solvent interactions are in most cases successfully captured. Thus, the user only needs to specify the dielectric constant of the solvent to calculate the absorption and/or emission spectra in solution. This approach is, however, less appropriate to describe heterogeneous biological environments, in this case the explicit representation of the environment's molecules by means of the QM/MM approach is the recommended option.

We highlight here an example in which the strong solvatochromic effect observed in the electronic excitations of the betaine **B30** (see Figure 6.3a),[80] a molecule used to establish a scale for solvatochromic effects, was studied by means of TD-DFT and the PCM model.[81] The electronic excited states have a strong long-range intramolecular charge-transfer nature, as the electron density is redistributed from the phenolate ring to the pyridinium fragment (see red arrow in Figure 6.3a), as revealed by the analysis of the natural transition orbitals (NTOs),[82,83] *i.e.* of the orbital couples that better describe the electronic density rearrangements, and hence the transitions. Figure 6.3b shows the molar transition energies, which are closely related to the excitation energies, computed using the TD-DFT/PCM protocol for different aprotic solvents (top red data). It is clearly shown that the blue shift registered in the experiments (bottom blue data) is well reproduced by the theoretical calculations.[81] This solvent effect is not surprising as the charge-transfer excitation is actually neutralizing the positive charge localized in the quaternary ammonium cation and the negative charge of the phenolate fragment. Thus, the blue shift in **B30** is explained on the basis of the much larger dipole moment module of the ground state (20–24 D), making it more

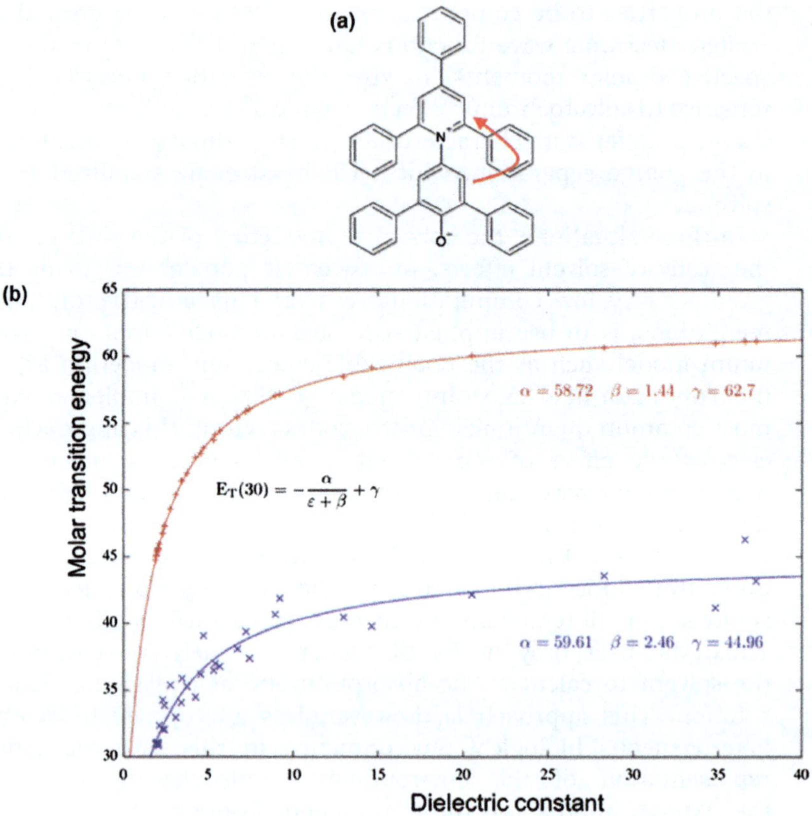

Figure 6.3 (a) Molecular structure of **B30**. The red arrow indicates, roughly, the electron density redistribution in the excited state. (b) Representation of the molar transition energy, a magnitude proportional to the absorption energy, *versus* the dielectric constant of aprotic solvents. Top line (red) corresponds to the theoretical data, whereas the bottom line (blue) corresponds to the experimental records.
Part (b) reproduced from ref. 81 with permission from Elsevier, Copyright 2014.

sensitive to solvent stabilization, with respect to that of the excited state (~3 D). Those effects together lead to an increase of the excitation energy as the dielectric constant of the solvent increases.[81] Given the clear relationship between the solvatochromic effects and the ground- and excited-state dipolar moment modules, chemical intuition may often allow a reasonable prevision of either blue or red shifts prior to the actual quantification.

(iii) *Sample the Ground State Equilibrium Area*

So far, the excitation properties have been studied using only one structure, obtaining the so-called vertical absorption energy that can be related to the experimental absorption maximum. As mentioned previously, a more accurate calculation of the absorption spectrum

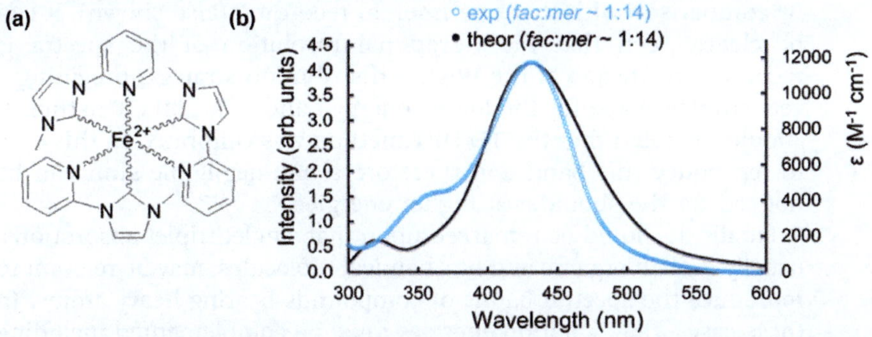

Figure 6.4 (a) Molecular structure of the Fe–NHC complex. (b) Theoretical spectrum for the *fac* and *mer* (1 : 14) mixture of Fe–NHC obtained as a convolution of the excitation energies of 20 structures (Wigner distribution) for each isomer.
Reproduced from ref. 84 with permission from American Chemical Society, Copyright 2018.

requires the sampling of the ground-state configurational space *via* a semiclassical Wigner distribution and/or molecular dynamics simulations.[64] These computational strategies are actually the only way to capture the vibrational effects in a physically meaningful manner. This is of special relevance in those cases in which the vibrational distortions cause energy shifts, or even the appearance of new shoulders or peaks, features that cannot be resolved by the determination of the vertical absorption energy from a single geometry. Indeed, in general, the convolution of the vertical absorptions of a single structure is not recommended as it does not allow proper recovery of the band shape.

As an example, we can cite recent studies performed by some of the authors of the present work regarding the photophysics of the Fe(II)-based complex displayed in Figure 6.4.[84] The rationalization of the absorption properties, as well as the excited-state decay mechanisms of these compounds, are crucial for use in optoelectronic and other technological devices.[85] Their interest relies on the abundance and relatively low toxicity of iron as compared to the currently used complexes based on heavy and rare metals such as ruthenium.[86] Light absorption of N-heterocyclic carbene (NHC) Fe(II) complexes in the visible range of the spectrum leads to the population of metal-to-ligand (MLCT) excited states, usually the recorded band is the result of several excitations given the high density of states placed at those wavelengths. Figure 6.4 shows the theoretical spectrum built as a convolution of the excitation energies and oscillator strengths of 20 structures obtained *via* a Wigner distribution around the ground-state minimum. The method of choice was TD-DFT in combination with the PCM method to describe the solvent (acetonitrile).

By comparison with the experimental recording (also shown), it can be clearly seen that the vibrational resolution of the spectra is recovered by means of the Wigner distribution strategy, providing a very similar shape for the lowest-energy band of an MLCT nature. It should be noted that the TD-DFT method was calibrated in this case to reproduce this band and therefore a non-negligible shift can be noticed for the shoulder at higher energies.

Finally, it should be remarked upon that singlet-triplet absorptions, usually dark in organic and biochemical molecules, may be relevant to reproduce the spectral bands of compounds bearing heavy atoms. In these cases, the excitation energies must be complemented including the calculations of both triplet state manifolds and the SOC effects to allow the estimation of the probability of singlet→triplet transitions. Indeed, SOC elements will act in mixing the singlet and triplet states, and hence while breaking the degeneracy of the triplet manifold they will also break down the spin-conservation selection rule. Recent examples can be found in the photophysics of mercury halides, of interest in atmospheric chemistry,[87] or of Rhenium(I) carbonyl diamine complexes, used to study long-range electron transfer processes in proteins.[88]

(iv) *Using the QM/MM Approach to Describe the Environment*

Dealing with photochemical or photobiological problems often requires the use of more sophisticated methods to describe the environment, because in the vast majority of cases biochemical and biological surroundings are not homogenous and thus the chromophore-environment interactions are not equal in all directions of the space. Moreover, and owing to this heterogeneity, the interactions may change over time given the spatial degrees-of-freedom of the environment that may affect the chromophore-environment interactions. For this reason, a single configuration is usually not sufficient to represent the absorption or emission spectrum of the chromophore and hence the configurational space, that is the phase space, has to be sampled either using classical molecular dynamics or QM/MM simulations.

As an example, we can cite the UV–vis spectrum of the **BMEMC** dication (see Figure 6.5) interacting with DNA.[89] This molecule can photosensitize DNA and exhibits two-photon absorption capacities, making it a good candidate to improve photodynamic therapy.[90–92] In this case, the only way to properly resolve the absorption spectrum of the molecule interacting with DNA is using the QM/MM approach.[93] Figure 6.5b shows a representative snapshot of the intercalation mode of **BMEMC** inside a double-stranded DNA, whereas Figure 6.5c shows one- and two-photon absorption spectra obtained as the convolution of the vertical absorptions computed, at QM/MM, for 100 snapshots obtained from a 100 ns classical molecular dynamics run. Using this computational protocol the authors were able

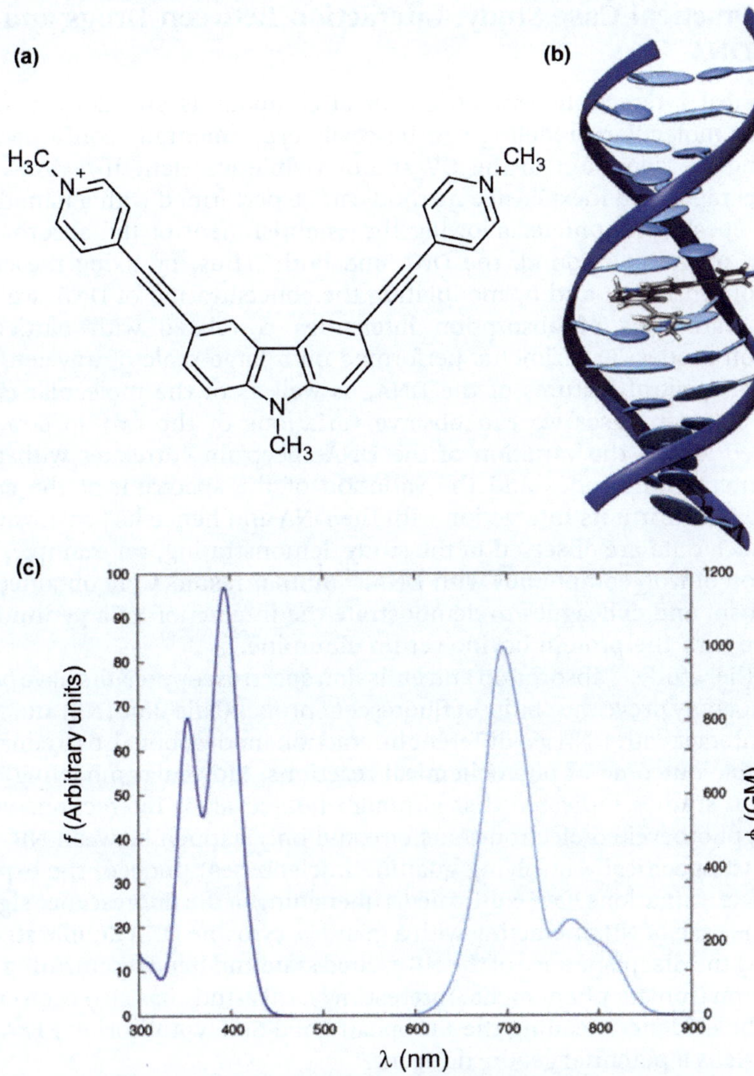

Figure 6.5 (a) Molecular structure of the **BMEMC**. (b) Representative snapshot of the intercalation of **BMEMC** within a DNA double strand. (c) One- (left) and two-photon (right) spectrum of the intercalated **BMEMC** as shown in panel (b).
Reproduced from ref. 93 with permission from the PCCP Owner Societies.

to reproduce the impact of the DNA in the absorption spectrum of the chromophore with a reasonable agreement with the experimental results.[89,93]

6.3.3 Practical Case Study: Interaction Between Drugs and DNA

The physical interaction between chemical compounds and DNA demonstrated by molecular modeling can be easily experimentally confirmed by measuring the spectrum in the UV and/or visible wavelength regions. Indeed, this rapid and inexpensive method can be performed with a nanodrop or other spectrophotometer allowing the establishment of the absorbance spectrum of the compound, the DNA and both. Thus, by fixing the compound concentration and by modulating the concentration of DNA, we can measure variations of absorption intensities correlated with particular interaction modes. Experiments performed on a large scale of wavelengths can reveal spectral features of the DNA, as well as of the molecular compounds. In such cases we can observe variations of the two spectra. As mentioned above, the variation of the DNA spectrum correlates with particular interaction modes and the variation of the spectrum of the compounds can confirm its interaction with the DNA and hence its sensitization action. Such data are observed in the study demonstrating, for example, the interaction of iron compounds with DNA.[94] Similar results were obtained by Suryawanshi and colleagues to demonstrate the interaction of a pyrimidine derivative with the protein bovine serum albumine.[95]

In another study,[96] absorption and emission spectroscopy results have been used to not only prove the ability of fluorescent probes (Nile Blue (NB) and Nile Red) to interact with DNA *via* different interaction modes, but also to confirm the possible outcome of photochemical reactions. Molecular modeling and simulation studies indicated that although non-covalent interactions were possible, photoexcited electron-transfer could only happen between NB and DNA (most specifically involving guanine nucleobases). Indeed, the experimental determinations have confirmed a quenching of the fluorescence signal only in the case of NB interacting with a guanine-cytosine DNA double strand indicating the disappearance of the NB excited state and hence confirming the fast electron transfer phenomena. Interestingly, this study has also been cited among the evidences leading the European Food Safety Authority (EFSA) to declare NB as a potential genotoxic agent.[97]

6.4 Modelling Circular Dichroism Spectroscopy

One of the technically simpler, but extremely powerful techniques allowing researchers to gain insights into the structure of biological macromolecules in solution, is certainly electronic circular dichroism (CD). The CD signal is represented by the difference in the absorption, namely in the extinction coefficient, between left and right circularly polarized light exhibited by chiral chromophores. Indeed, the circular polarization of the electromagnetic field induces a chiral environment that interacts differentially with other chiral counterparts, here the chromophore. The power of CD as an analytic technique relies on the fact that on the one hand most biological components are enantiomeric pure compounds, such as the naturally occurring amino acids, and on the other hand on the fact that the self-organization of biological macromolecules produces chiral arrangements such as helices or β-sheets. Furthermore, the CD signal is extremely sensitive to modification in the secondary structure of self-organized systems and hence can be used to prove the structural modification of biological macromolecules upon interaction with external compounds or as a response to an external stimulus. However, the rich density of information contained in the CD signal makes the fine interpretation of the experimental results, and in particular the relationship between the change in the CD intensities and the specific structural modification, quite complicated. In this context the role of molecular modelling and simulation in providing a clear atomistic picture of the structural evolution of biological macromolecules is fundamental.[159] However, to complete the picture one should not only obtain structural information, but also have access to a one-to-one mapping between experimental and computed CD spectra.

The calculations of CD spectra of small-size chiral molecules are relatively simple and can be performed in a straightforward manner by using the same techniques (CASPT2, TD-DFT) as the ones detailed in the previous examples, however the situation is more complicated when the CD signal is derived from a chiral supramolecular arrangement such as in a DNA double-helix. Indeed, in this case the CD spectra is the clear manifestation of the interaction of multichromophores and hence special computational techniques should be used to tackle this more complicated phenomenon. Different approaches have been developed among different groups, in particular some direct calculations of the multichromophore system are performed thanks to the complex polarization propagator (CPP) approach that allows researchers to restrict the calculations to only the excited states lying in a pre-defined excitation energy window. As such, this technique avoids wasting computational efforts on the less relevant parts of the spectrum and hence solves the problem related to the high density of excited states present in the case of multichromophoric aggregates. The technique developed mainly by the groups of Linares and Norman has been successfully applied in a number of applications,[160] including the study of CD of naked and nucleosomal DNA.[161–163]

Despite the clear success of the CPP approach alternatives have been developed, explicitly taking into account the physical principle at the basis of the induction of a CD signal in the aggregates of chromophores. Indeed, in the case of close-by and strongly-interacting chromophores the excited states manifold is altered by the formation of excitons, that is by the superposition of the excitations centred in the single monomers to give rise to delocalized states. Hence, the CD signal, excitation energies and signal intensities, can be recovered by building and diagonalizing a phenomenological semi-empirical Hamiltonian, taking into account the energies of the excitations of the monomers (diagonal elements) and the coupling between the different states. The advantage of this approach is that the expensive calculation of the excitation energies is performed on a relatively small partition (each monomer), hence strongly limiting the overall computational cost, even if the excited state calculation needs to be repeated for every monomer. While in principle every quantum chemistry method providing an excited state can be used, in most of the cases the exciton approach has been used at the affordable TD-DFT or semiempirical levels. Owing to the partition of the systems in relatively small QM regions, and hence the reduction of the overall computational cost, it is also possible to repeat the calculations for the CD spectra for a statistically meaningful ensemble of snapshots extracted, for instance, from a MD simulation, hence providing the effects of the thermal motion and vibrations on the shape of the CD signal and facilitating mapping with the experimental results. Obviously, the inclusion of environmental effects is straightforward *via* the use of hybrid QM/MM techniques.

Depending on the way the Hamiltonian off-diagonal elements, the couplings, have been calculated one can have slightly different variations of the methods. The simplest approach is to consider the coupling as dominated by the dipolar interactions, in this way one relies on the original Frenkel approach and the matrix elements are obtained simply as the scalar product of the transition dipole moments between different excited states belonging to different monomers weighted by the distance between the monomers.[164] In other cases, for instance as in Mennucci's approach,[165–167] the coupling is calculated by explicitly taking into account the density matrices of the monomer. Although the former approach is conceptually simpler and less expensive, the latter is more adapted in the case of strong coupling between the monomers, and generally reduces the errors owing to the use of a semiempirical Hamiltonian.

The use of the Frenkel based protocol, coupled with extensive sampling provided by MD has allowed unravelling of important characteristics of the native DNA circular dichroism signal, in particular the important sequence-dependent differences in the spectrum have been highlighted, as well as the crucial role of the more flexible terminal bases, especially in the case of short DNA fragments.[168] The protocol has also been applied to non-canonic DNA structures such as G-quadruplexes (G4).[169] Once again, important differences in the CD signal of different G4 conformers (parallel, hybrid, and antiparallel) have been evidenced and the results match satisfactorily with the experimental spectra, confirming the possibility of using such a protocol to interpret and assign unknown CD spectra.

Finally, in a more challenging application, the protocol has proven capable of reproducing the experimentally observed change in the CD spectrum of a peptide under the influence of a covalently bound photoswitch.[170] Indeed, MD simulations have proven that the secondary structure of the peptide strongly depends on the photoswitch isomer. In particular, although in the case of the *trans* arrangement the peptide was almost entirely folded in a α-helix, the transition to the *cis* isomer induces a breaking of the helix owing to the sterical constraints. The calculated CD spectra for the two isomers show important changes in the α-helix signature region coherently compared with the one observed experimentally, clearly confirming the loss of helicity owing to the photoswitch isomerisation. Interestingly, the importance of a wise choice of monomer, that is the QM partition, in order to include all the important interactions, such as specific hydrogen bonds patterns, has also been underlined.

6.5 Photochemistry and Photobiology in Complex Environments

The absorption process described in the previous sections represents the first step of a photoinduced process. From a theoretical standpoint, the events that take place after, for example the photochemical mechanisms by which the excited state decays to the ground state, need the characterization of the ground and excited PESs in order to elucidate the main photochemical properties of a given system (radiationless decay or light emission, photostability or photoreactivity, *etc.*). Indeed, several pathways are usually operative and one has to evaluate the competition between them. Ideally, once the static description is achieved, the study should be completed with molecular dynamics calculations in order to capture the dynamic effects and determine the time scales and the quantum yields of the photo processes, although this latter step is not always achievable owing to the large computational overload.

Whereas DFT and its TD-DFT extension are usually able to reproduce the absorption and emission properties, they tend to fail in the description of crossing points. Multiconfigurational approaches such as the CASPT2 or the multireference configuration interaction (MRCI) methods, among others,[98] are deemed more appropriate for these purposes, however at much higher computational costs in most of the cases precluding dynamic studies. There are, however, situations in which the TD-DFT description is sufficiently good to describe the PESs, as illustrated in the following.

6.5.1 Computing Potential Energy Surfaces in Complex Environments

Quantifying the impact of complex environments in the PESs landscape is a challenging task that adds complexity to the already non-trivial determination of the relevant photochemical pathways of a system. In an analogous way to the absorption and emission properties detailed previously, one can start with the system *in vacuo* and assess in a second step the effect of the solvent by means of continuum model methods. This can be illustrated by a recent example reported by our research group.[99] The work focused on the determination of the triplet photosensitization mechanism of thymine exerted by an oxidized nucleobase, named 5-formyl-uracil (**ForU**), a common oxidative lesion of DNA formed under oxidative stress conditions. It was found that this molecule absorbs UV light and induces photolesions in model experiments, as well as in plasmid DNA.[100] The full triplet photosensitization mechanism was studied *in vacuo* and in water solution (through the PCM method) using both CASPT2 and TD-DFT approaches (see Figure 6.6). It was shown that both methods provide a coherent description

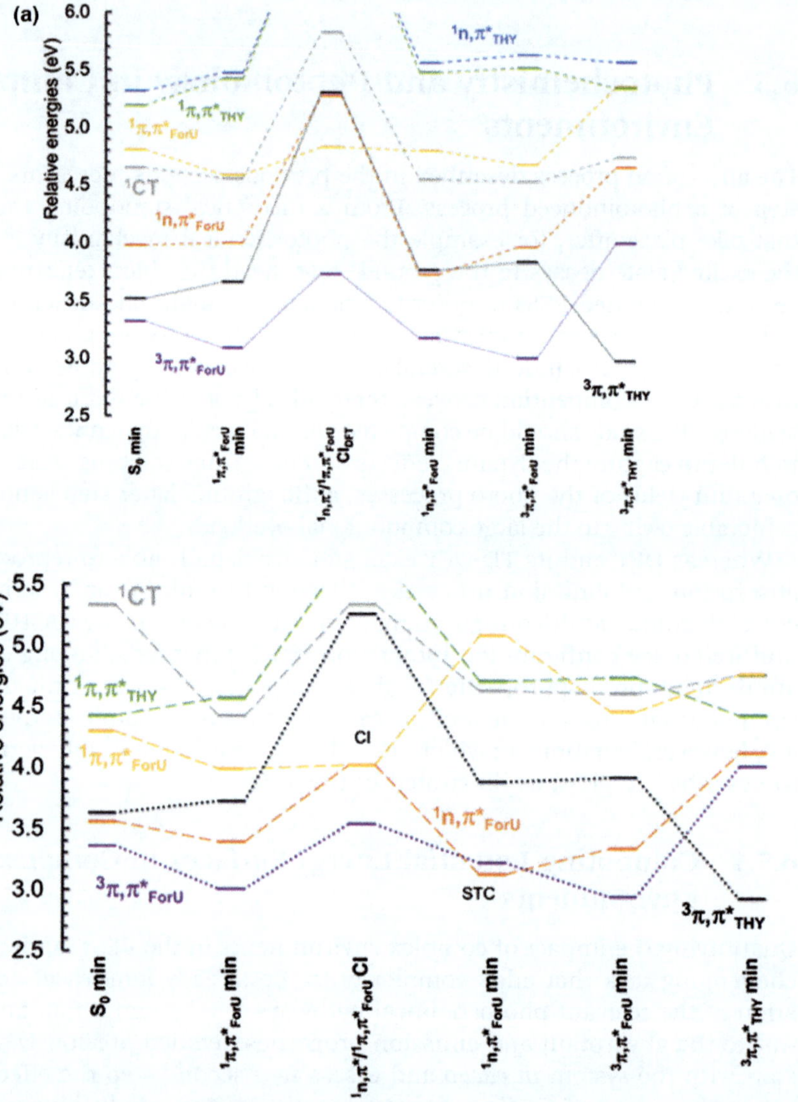

Figure 6.6 (a) CASPT2 energies of the most relevant singlet and triplet states using the triplet photosensitization mechanism. (b) TD-DFT/PCM energies of the most relevant singlet and triplet states using the triplet photosensitization mechanism. The dotted lines denote singlet states and dashed lines correspond to triplet states.
Reproduced from ref. 99 with permission from the PCCP Owner Societies.

of the triplet manifold population, mediated by the $^1(n,\pi^*)$ state localized in **ForU**, whereas the PCM results revealed a limited effect for water solvation. Small energy barriers were found between the equilibrium geometries of the $^3(\pi,\pi^*)$ state localized in **ForU** and thymine, respectively, proving the feasibility of the mechanism.[99]

Another example recently reported in the literature uses the QM/MM approach to determine the singlet π,π^* PESs of two stacked thymine molecules embedded in a complex DNA double strand.[101] The problem of the excited-state decay of nucleic acids has been extensively studied and discussed in recent decades given its relevance for the comprehension of the evolution of life and the insurgence of DNA lesions that can trigger a variety of serious diseases.[102,103] The topic has led to a vast amount of literature and a great number of experiments dealing with isolated nucleobases, oligonucleotides and mono- and double-stranded systems, as well as a formidable number of theoretical studies using a wide variety of methods.[44,57,103–106]

The presence of long-lived decay components, up to several hundreds of picoseconds in double stranded DNA, measured using time-resolved spectroscopy setups,[107,108] has challenged the scientific community as nucleobases in solution mostly decay in the subpicosecond regime[44,104,105] even though long-lived signals were also registered for pyrimidine nucleobases, nucleosides and nucleotides in solution.[109] The QM/MM PESs revealed that the π-stacking between the thymine molecules hampers one of the possible ring-puckering distortions that leads to the $\pi,\pi^*/S_0$ CI, giving rise to a barrier estimated at 0.5 eV (see Figure 6.7).[101] This channel was thus ascribed to the long-lived excited-state component measured in the experiments.

6.5.2 Photochemistry and Photobiology of Complex Systems Under the Dynamic Approach

Time evolution of the excited states is crucial to determining the spectroscopic and photochemical properties that depend on time, such as time-resolved spectra, quantum yields and the time scales of the photo processes. The outstanding development of more and more sophisticated experimental setups of increasing accuracy and resolution is accompanied by an astonishing development of theoretical methodologies able to study time-dependent properties. The synergy between experiments and computations has led to impressive results in recent years, providing an unprecedented understanding of a variety of chemical and biochemical events of utmost importance.

6.5.2.1 Adiabatic Molecular Dynamics Simulations

The PESs described in the previous section are a consequence of the Born–Oppenheimer approximation, which provides the so-called 'adiabatic description'. In this picture, the labelling of the states is based on an energy

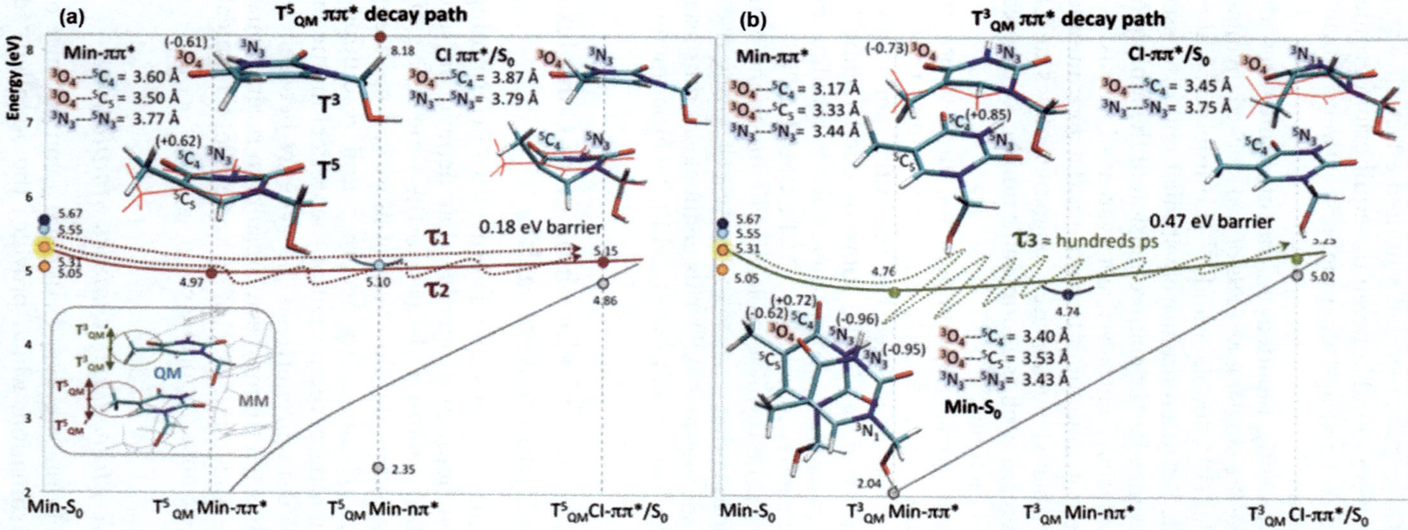

Figure 6.7 π,π^* decay pathways computed using the CASPT2//CASSCF/MM protocol for two different thymine-thymine stacking arrangements embedded in a DNA structure, exhibiting energy barriers of (a) 0.18 and (b) 0.47 eV, respectively. Reproduced from ref. 101 with permission from American Chemical Society, Copyright 2018.

criterion provided by the resolution of the time independent Schrödinger equation (Section 6.1). The determination of the nature of the states, which allows the study of the electronic structure of the system, must be done *a posteriori* by analysis of the linear expansion of the electronic configurations that describe the state of interest. This task is not trivial and recent advances in the field will be depicted in the next subsection.

Dynamic methods applied to medium/large systems in chemistry and biochemistry are usually based on a semi-classical approach, that is the nuclear displacements are computed classically, whereas the electronic gradients are computed on-the-fly using a QM method that solves the time-independent Schrodinger equation. In the QM/MM scheme, the overall nuclear displacements are obtained by solving the classic Newton equations of motion with the force given as the derivative of the QM energy with respect to the nuclear displacement for the QM region and *via* the force field for the MM partition, also taking into account their mutual coupling.[110]

The quantum behaviour of the nuclei can be partially recovered by collecting statistical data from a set of individual trajectories that start from different initial conditions, *i.e.* atomic coordinates and momenta (velocities). From a practical standpoint, the initial conditions can be generated through a Wigner distribution of the ground-state minimum or from classical, QM or QM/MM dynamic simulations covering the majority of configurations of the system, as previously mentioned. Thereby, the excited-state dynamics can be propagated on an adiabatic excited-state surface revealing the dynamic properties related to that state. As the time propagation is based on an adiabatic description, the trajectories are forced to stay in the same excited state and thus non-adiabatic events such as state hops are *a priori* excluded. This approach is useful to study the dynamics of the lowest-lying singlet or triplet excited states being able to quantify the impact of the excited state in certain properties, for instance geometrical distortions.

The adiabatic QM/MM simulations are less costly than the non-adiabatic scheme, however, they present two important drawbacks. The first one is the crucial choice of the excited state, which relies on the user. The photochemical relevance of the results will strongly depend on this selection and a strong knowledge of the system, for instance *via* a full static description of the PESs is required prior the dynamic study. The second one is the inclusion of artificial adiabatic trapping as state hops are not allowed. The choice of systems with a low density of PES crossings, and the monitoring of the excited state energies can minimize the latter problem.

An example of an adiabatic QM/MM MD study concerning the excited-state proton transfer in DNA was recently reported by our research group and will be briefly discussed here in order to illustrate the use of this method in a specific photochemical phenomenon of a complex system.[43] Excited-state hydrogen transfer (ESHT) was proposed about 15 years ago as an efficient channel to dissipate the excess energy of photoexcited DNA.[111–113] The mechanism is localized in the guanine-cytosine base pairs and temporarily alters the canonical Watson–Crick (WC) hydrogen bonding pattern, promoting the motion of the H1 proton, coupled with an electron transfer, back and forth between the guanine and cytosine nucleobases (see Figure 6.8a for the atom numbering). Even though from the static point of view other possibilities of proton movements that could lead to tautomeric structures are possible,[114] the aforementioned H1 transfer that recovers the WC form remained the preferred one as reported in time-resolved IR spectroscopy studies on single guanine-cytosine base pairs dissolved in chloroform,[115] in GC/GC miniduplexes[116] and in larger duplexes.[117] Theoretical studies supported this finding, even though the reasons for the preference of the pathway recovering the WC structure and ensuring the 'correct' DNA structure were not unveiled.[114,118,119] Thus, we performed QM/MM adiabatic simulations in the S_1 state to analyse the effects of the energy released during the proton transfer step.[43] This particular problem was especially suitable to be studied through adiabatic dynamic simulations as the PESs of the system are known in significant detail and a low density of states is predicted along the proton transfer coordinate.[112,114,118,119] The results of the S_1 state dynamics revealed not only out-of-plane distortions that were boosting photostability (see Figure 6.8b for an illustrative snapshot), explaining the experimental outcomes, but also a novel proton transfer mechanism named four proton transfer (FPT) in which more than one GC base pair is involved in the ground-state decay, as schematised in Figure 6.8. Thanks to the full atomistic picture provided by the QM/MM method, it was possible to conclude that the latter mechanism is indeed influenced by the presence of nearby counter ions in solution (Na^+ in this case), constituting a clear example of the photochemical event strongly modulated by a complex environment.

6.5.2.2 *Non-adiabatic Molecular Dynamics Simulations*

Non-adiabatic molecular dynamics simulations allow more than one potential energy surface to be taken into account, hence making it possible to study complex photochemical phenomena. In particular, important macroscopic observables such as quantum yields, absorption and emission intensities can be predicted with accuracy. Also, a direct comparison with ultrafast (*e.g.* pump-probe) spectroscopy is at hand.

Hence, in this respect, non-adiabatic molecular dynamics present obvious advantages over the adiabatic scheme, as the user does not have to choose in advance the only relevant excited state of interest in which the whole trajectory will run, avoiding artificial adiabatic trapping. On the other hand, the non-adiabatic scheme will strongly depend on the method selected to switch from a PES to another one. Hence, also in this case a previous static study of the mechanism (or mechanisms) involved can be highly beneficial, as it helps to foresee the type of expected photoproducts and to estimate the eventual energy barriers to reach them, finally ascertaining the feasibility of the different possible processes. Also, intersections between different PESs

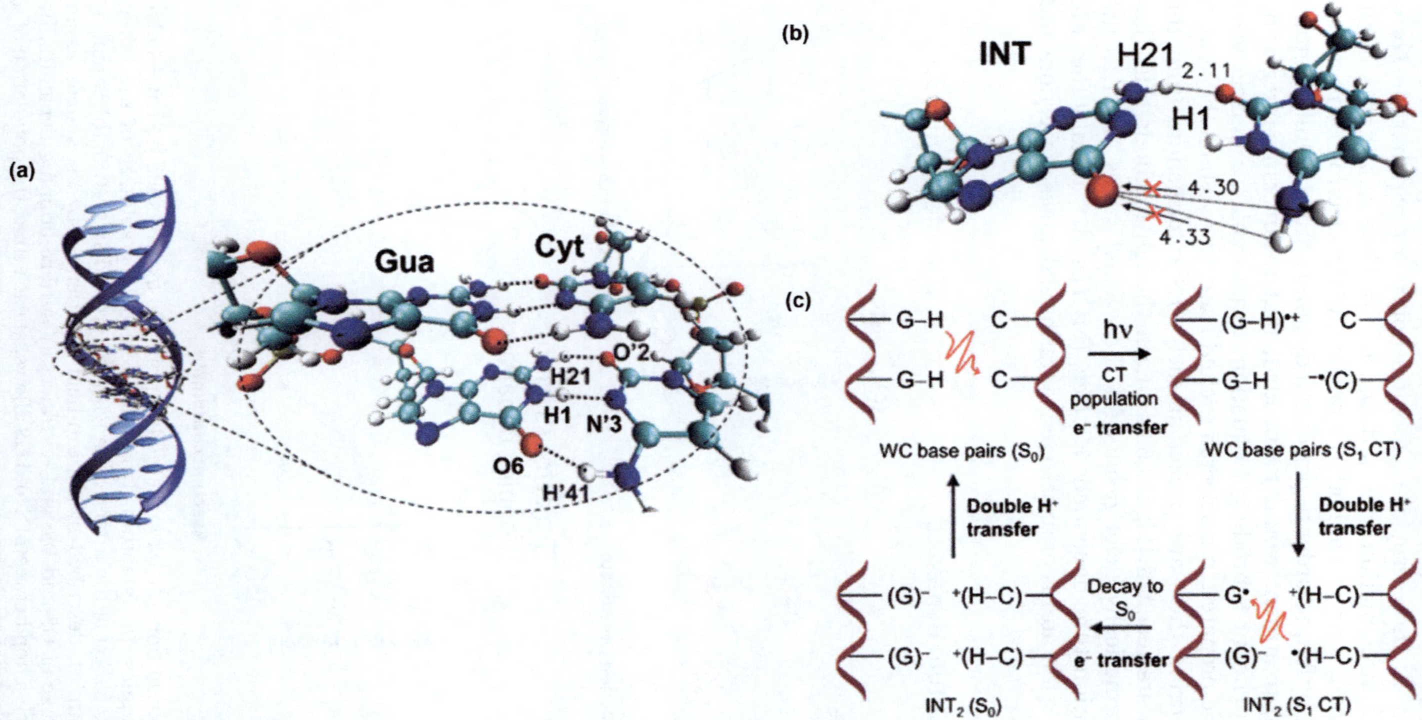

Figure 6.8 (a) Atom numbering and QM/MM partition of two guanine-cytosine base pairs embedded in a (dG)·(dC) homopolymer. (b) Illustrative snapshot of an adiabatic QM/MM trajectory that shows the out-of-plane distortions induced by the energy release during the proton transfer step. Only one guanine-cytosine base pair was included in the QM partition. (c) Schematic description of the FPT mechanism.
Reproduced from ref. 43 with permission from the Royal Society of Chemistry.

can be found *a priori* using a static approach, determining if a CI or a SCT can be reached. At these crucial topological points (or regions) the Born–Oppenheimer approximation breaks down, in principle requiring a quantum mechanical description of both the nuclei and the electrons. In practice this would mean that, in order to perform a dynamics study, a full quantum dynamics (QD) treatment would be needed, that means time-integration of the time-dependent Schrödinger equation. In particular, a nuclear wave packet has to be prepared in a selected excited state, from which the system evolves by splitting the wave packet at each PES intersection, until the photoproducts are reached (Figure 6.9a). Even though methods exist that can provide numerically exact results for the excited states (such as the multiconfiguration time-dependent Hartree, (MCTDH))[120] usually they are limited to only a few atoms or to selected vibrational degrees of freedom and

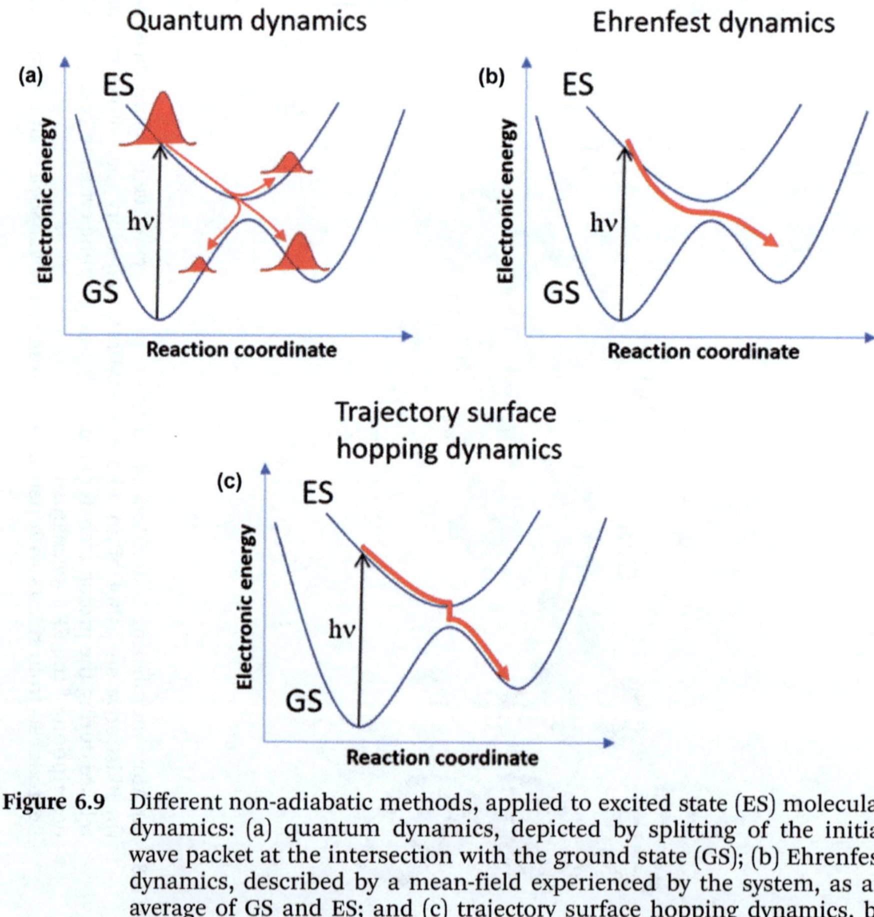

Figure 6.9 Different non-adiabatic methods, applied to excited state (ES) molecular dynamics: (a) quantum dynamics, depicted by splitting of the initial wave packet at the intersection with the ground state (GS); (b) Ehrenfest dynamics, described by a mean-field experienced by the system, as an average of GS and ES; and (c) trajectory surface hopping dynamics, by which a hop between ES and GS allows, at each time, the population of a single adiabatic surface.

hence in general they are less used in photobiological applications, even though some important progress has allowed the study of complex organometallic systems.[121] An interesting alternative approach was recently applied that couples QD to an environment treated classically, hence resulting in the following QD/MD scheme,[122] the excited state PES of an isolated chromophore is calculated at a high level of theory. Then, the chromophore is considered as part of an MM model, in order to perform classical MD. As we have remarked in this chapter, a MD trajectory can serve as a sampling strategy, as MD snapshots can be extracted and QM/MM calculations performed, by which an environmental potential can be derived and further added to the initially calculated PES of the isolated system. Using this procedure, QD can be performed by including the atomistic surrounding of biological interest. A valuable example is the RNA environment surrounding an uracil base. In this way, the photorelaxation of uracil can be studied in its native biological environment, resulting in an increased excited state lifetime compared to isolated uracil. Hence, it was shown that uracil photostability, predicted to be highly efficient when considered as an isolated base in solution,[123] can be partially inhibited by the RNA environment, thus increasing the probability of photodamage initiated by photoexcited uracil.[122]

Different methods of mixing quantum and classical methods to study dynamically excited states are Ehrenfest and trajectory surface hopping dynamics (Figure 6.9).[124,125] Both are semiclassical methods, since the nuclei are treated classically. The main difference between these two methods relies on the concept of trajectory, using the Ehrenfest method, trajectories evolve on a time-dependent average of all the considered electronic states, constituting a mean-field approach, while using surface hopping dynamics each trajectory is always located on a single electronic state, and a probability calculated at each time step will dictate if the trajectory will hop from one electronic state to another.

Among the two mentioned methods, the trajectory surface hopping method is much more popular, most probably because of the easier comparison of the calculated results with the experimental ones, in terms of the calculations of exchange rates between the population of different electronic states, a concept that is almost lost by a mean-field approach, but easily represented by non-adiabatic surface hopping.

Nevertheless, different versions of the Ehrenfest method were successfully applied to the description of biological systems on the excited state, especially when long-range events spanning DNA or protein environments are the target study. Indeed, starting from the Ehrenfest theorem, a method called semiclassical electron-radiation-ion dynamics (SERID) was developed and initially applied to describe photoisomerization.[126] Recently, SERID was applied to a DNA model, focusing on the bonded excimer formed by stacked cytosine bases in particular.[127]

Thanks, in particular to the implementation in a QM/MM scheme, accurate photobiological mechanisms were studied using the Ehrenfest method. As an example, we cite the long-range charge transport in Escherichia coli DNA photolyase, a photoactivated process in which a chain of tryptophan residues allows the electron to migrate between the protein surface and the flavin adenine dinucleotide (FAD) cofactor.[128]

On the other hand, as aforementioned, trajectory surface hopping is the method of choice when spectroscopical predictions are required. As photobiological processes usually involve more than one excited state pathway, and especially owing to the classical description of the nuclei, several trajectories should be performed, in order to reach a statistically representative ensemble. The number of trajectories that should be run depends largely on the computational cost, which should remain affordable. Nevertheless, some statistical criteria should be met, as the surface hopping method allows one to unequivocally assign the system under study to a specific PES (hence a specific electronic state) at each time step until a hop is experienced, the most valuable prediction is the decay time between different PESs. This event can be described in terms of the average time (among all calculated trajectories) and the standard deviation from the average. Once the standard deviation is statistically meaningful, we could have reached the right number of trajectories. Indeed, the standard deviation can be converted into the margin of error, which depends on the confidence interval selected by the user.

Concerning the levels of theory that can be used for a quantum mechanical description, several possibilities arise, all of them being coupled to classical MM: *ab initio* multiconfigurational methods (CASSCF,[129] CASPT2[130]) can be applied to chromophores of a limited size, mainly due to the limitations imposed by the selection of the active space. TD-DFT offers reasonable accuracy at a comparatively lower cost, being therefore applicable to QM regions of more than 100 atoms. The main limitation relies in the approximate description of the crossing between the electronic excited and ground state. Nevertheless, different TD-DFT non-adiabatic MD techniques have been implemented.[131–133] Even more interestingly for photobiological applications, some semiempirical methods especially adapted to the description of excited states are available for the study of even larger systems, thus including in the QM region a considerable chromophore environment. In particular, long-range corrected tight-binding TD-DFT(B),[134] orthogonalization-corrected OM2 and OM3 Hamiltonians in combination with a multireference configuration interaction (MRCI) treatment.[135]

Another important issue in trajectory surface hopping dynamics is the calculation of the hopping probability between adiabatic states, which depends on the coupling between the states and, usually, is significantly high only when the energy difference between adiabatic energies is small (lower than 10 kcal mol^{-1}). The technique most widely applied is Tully's fewest-switches surface hopping algorithm,[136] because of its implementation feasibility and ability to reproduce experimental results. It should be noted that a decoherence correction needs to be applied to Tully's original algorithm, as surface hopping can develop non-physical coherences between the quantum coefficients of the crossing states, especially when large simulation times are required.[137]

Regarding applications to photobiology, one of the most studied systems is rhodopsin, a transmembrane protein that initiates the process of visual phototransduction. In particular, the photoisomerization of a retinal chromophore within the opsin pocket of rhodopsin, covalently bound to a lysine residue *via* a protonated Schiff base linkage, is the ultrafast event that triggers the process of vision. A pioneering non-adiabatic QM/MM molecular dynamics study was performed by Frutos *et al.* in 2007,[40] by calculating a single excited state trajectory until decay to the ground state by a conical intersection. The QM region comprises the retinal chromophore at the CASSCF level of theory. Interestingly, the qualitative CASSCF energies along the trajectory were scaled in order to represent the quantitative CASPT2 energies, unaffordable for systems the size of the retinal chromophore. Such energy scaling makes it possible to scale the simulation time accordingly at the more reliable CASPT2 level. Moreover, valuable information was collected concerning the complex geometrical change of the retinal chromophore while isomerization occurred in a space-conserving environment, such as the opsin pocket. Indeed, an asynchronous bicycle-pedal (or crankshaft) motion was initiated at *ca.* 60 fs and continued until the conical intersection with the ground state was located, on a 110 fs time scale (Figure 6.10).

In the following years, increasing computational power and further implementations in quantum chemistry software packages have allowed extension of the study to a statistical number of non-adiabatic QM/MM dynamics. In particular, in 2010 Polli *et al.* applied the same QM/MM approach, by scaling the CASSCF energies within the QM region, but this time preparing a thermal sample of 38 initial conditions at 300 K.[138] The non-adiabatic MD trajectories were compared with ultrafast pump-probe experimental spectroscopy, finally evidencing the existence and importance of the S_1/S_0 conical intersection in the visual pigment. Indeed, $S_1 \rightarrow S_0$ stimulated emission before reaching the conical intersection and $S_0 \rightarrow S_1$ photoinduced absorption after hopping on the ground state, were detected with a sub-20 fs time resolution, this is in agreement with the theoretical study (Figure 6.10).

Apart from visual rhodopsin, several other types of rhodopsin exist in nature, extracted from prokaryotes and algae. The monomeric form of all of them is composed of seven α-helices, nevertheless modifications in the sequence of the amino acids lead to different light-induced biological functions, such as phototaxis (*i.e.* movement as response to light),[139] ion-pumps,[38,140] and ion-channels.[141]

Channelrhodopsins, in particular, are a class of ion-channels that allowed the discovery and further development of optogenetics, that is the control of neuronal activity by light-pulses.[142] From the point of view of the retinal chromophore, the main difference as compared to visual rhodopsin relies on the photoisomerization process: in the case of the visual rhodopsin it is an 11-*cis* → all-*trans* isomerization, while in channelrhodopsin it is an all-*trans* → 13-*cis* pathway. This, coupled to a different electrostatic

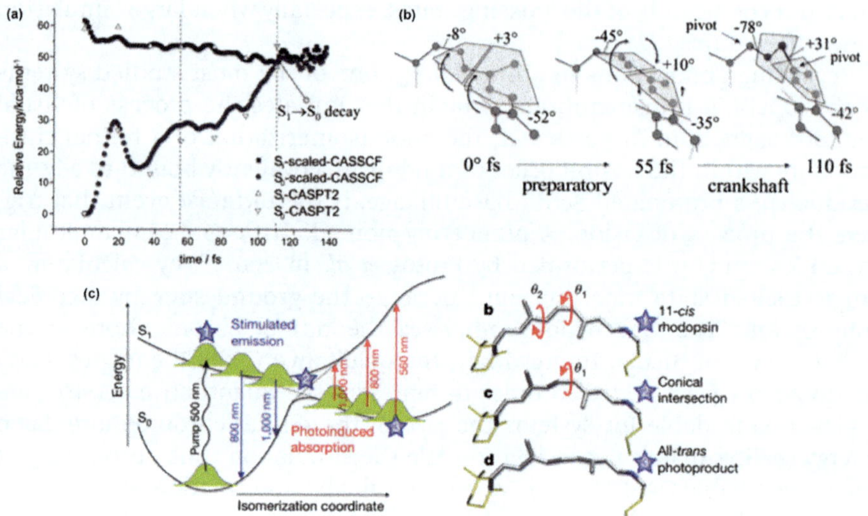

Figure 6.10 Excited state photoisomerization of the retinal chromophore within visual rhodopsin. (a) Non-adiabatic trajectory shown by energy as a function of time. No energy barrier is found on the S_1 electronic state. (b) Retinal motion during the trajectory: a preparatory step of 55 fs reorients the chromophore and allows bond length alternation before starting the crankshaft motion, by which a space-conserving photoisomerization can start. (c) Overall process of the 11-*cis* to all-*trans* photoisomerization.
Parts (a) and (b) were reproduced from ref. 40. Copyright (2007) National Academy of Science, USA. Part (c) is reproduced from ref. 138 with permission from Springer Nature, Copyright 2010.

environment, results in a slowed excited state lifetime for channelrhodopsin (*ca.* 400 fs), as a more flat S_1 potential energy surface was found. Moreover, a higher flexibility of the ground state in channelrhodopsin results in different hydrogen-bonding patterns, hence envisaging a more complex scenario.[143]

In particular, the hybrid C1C2 structure was studied using non-adiabatic MD. C1C2 corresponds to a hybrid composed by channelrhodopsin 1 and 2,[144] as it was the only X-ray structure available until 2017, when channelrhodopsin 2 was finally crystallized.[145] A first mechanistic study proposed two different paths after photon absorption, indicating both clockwise and counter-clockwise retinal rotation (Figure 6.11).[146] The two paths require different S_1 energy barriers to be overcome, owing to the asymmetric opsin pocket. Indeed, after absorption an S_1 minimum region is reached in both cases (as revealed by CASSCF/MM molecular dynamics). This first part of the dynamics corresponds to the vibrational relaxation of the chromophore, described by the partial bond length alternation between the conjugated single and double bonds. Both paths lead to a conical intersection with the ground state, but only one of them leads to the desired photoproduct (13-*cis*-retinal), while the other allows only internal conversion back to the initial all-*trans* retinal. Recently, surface hopping

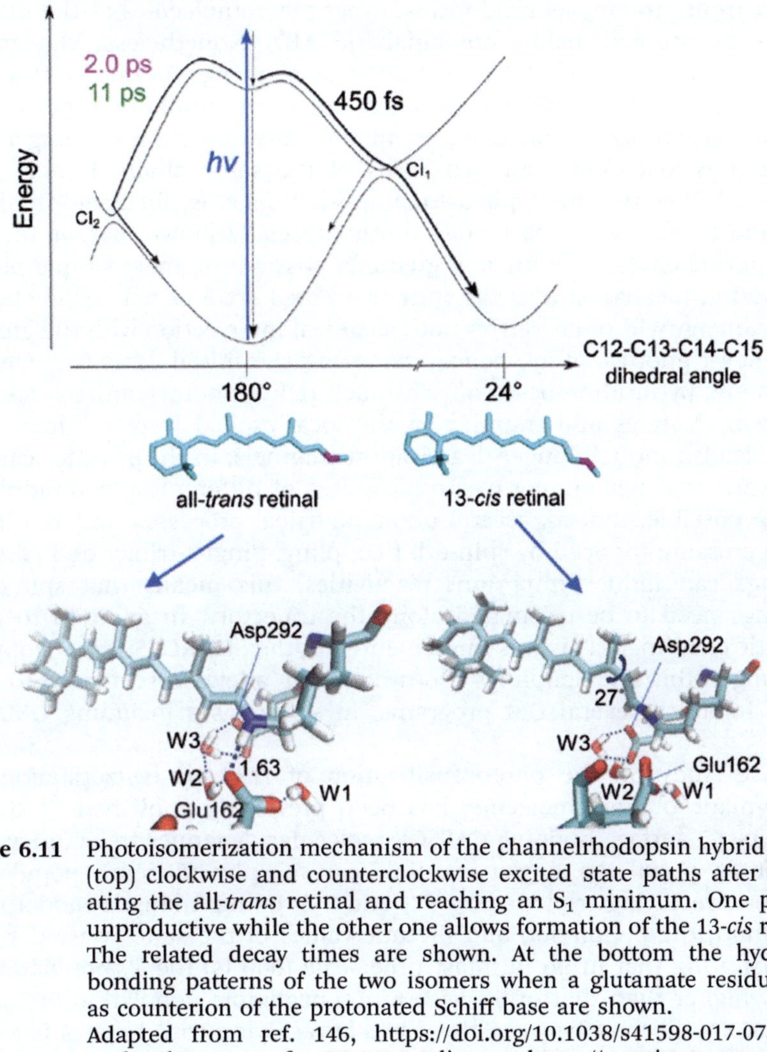

Figure 6.11 Photoisomerization mechanism of the channelrhodopsin hybrid C1C2: (top) clockwise and counterclockwise excited state paths after irradiating the all-*trans* retinal and reaching an S_1 minimum. One path is unproductive while the other one allows formation of the 13-*cis* retinal. The related decay times are shown. At the bottom the hydrogen bonding patterns of the two isomers when a glutamate residue acts as counterion of the protonated Schiff base are shown.
Adapted from ref. 146, https://doi.org/10.1038/s41598-017-07363-w, under the terms of a CC BY 4.0 license, https://creativecommons.org/licenses/by/4.0/.

dynamics were performed on the same C1C2 system, applying a semiempirical method to describe the retinal (OM3/MRCI) and the rest calculated at the MM level.[147] Being a semiempirical method it is much less computationally expensive (although less accurate) than the *ab initio* multiconfigurational CASSCF method, a statistical number of trajectories (1208) was reached. The results confirm that both clockwise and counterclockwise retinal photoisomerization processes are possible, including productive and unproductive isomerization events (Figure 6.11). Moreover, when aspartate is the retinal counterion, exclusive clockwise torsion is predicted, while the torsion is bi-directional when glutamate acts as a counterion.

Apart from proteins as rhodopsins, other macromolecules as the nucleic acids were studied using non-adiabatic MD. Nonetheless, the multi-chromophoric nature of DNA is an additional challenge for double stranded DNA, compared to proteins containing a single chromophore. Indeed, each DNA base is hydrogen bonded to a complementary base (thus forming a base pair) and is π-stacked with two bases of the same chain. Hence, non-adiabatic MD is usually applied to simplified systems, focusing on single DNA (and RNA) bases in particular. It was especially shown that, on the one hand, purine bases, adenine and guanine, possess the most simple photo-deactivation mechanism, as the initially excited state of π,π* character relaxes in energy without a barrier until a conical intersection with the ground state allows photostability, hence recovering the initial structure. On the other hand, pyrimidine bases offer a much richer photochemistry, also including n,π* states and trapping in the local excited state minima. This usually leads to much longer deactivation channels, in the ps time scale.[148] Moreover, crossings among electronic states of different spin multiplicity become possible. Indeed, several photobiological processes undergo intersystem crossing induced by spin-orbit coupling. Singlet-triplet excited state crossings can induce important reactivities, this means that spin-orbit couplings need to be calculated along the trajectory, in addition to non-adiabatic couplings. This was implemented in the SHARC (Surface Hopping including ARbitrary Couplings) software, that allows researchers to efficiently interface several QM programs, in some cases including QM/MM suites.[149,150]

A case study on the photosensitization of DNA by benzophenone, a paradigmatic organic molecule, has been previously published.[151] It was shown by 65 surface hopping CASSCF molecular dynamics trajectories that the initially populated singlet manifold can relax by efficiently populating the two lowest-lying triplet states (T_1 and T_2). Using a kinetic model built *a posteriori*, the population and lifetime values of the states were derived, demonstrating that in an ultrafast time scale (600 fs) the T_1 population is already higher than that of S_1, with a non-negligible population of T_2. Indeed, an equilibrium between these two lowest-lying triplet states is established just after the intersystem crossing, at least in the ps time scale, owing to the T_1–T_2 energy degeneracy (Figure 6.12). These two populated triplet states are responsible for the photosensitization towards thymine, through Dexter-type triplet–triplet energy transfer. Indeed, thymine is the DNA base with the lowest triplet energy, hence making it possible to accept energy from the triplet benzophenone. Moreover, benzophenone is in contact with DNA owing to the formation of stable DNA binding modes, namely groove binding and double insertion modes.[152]

Hence, not only complexes including heavy metals, but also simple organic molecules, can play an important role in opening a channel to triplet population after excitation. This is a fundamental aspect that has to be taken into account when considering photosensitization processes in biological media.

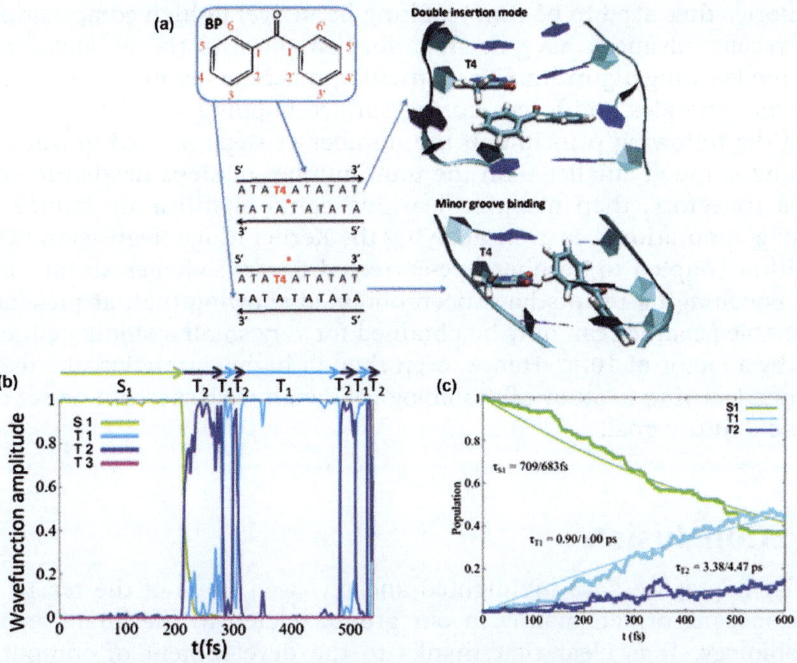

Figure 6.12 (a) Benzophenone binding modes with DNA: double insertion and minor groove binding modes. (b) Wavefunction amplitude as a function of time, as extracted from a non-adiabatic trajectory of isolated benzophenone, starting in S_1. After S_1/T_2 intersystem crossing, several crossings between T_2 and T_1 demonstrate that an initial equilibrium between these two states is set. (c) Kinetic model resulting from all surface hopping trajectories: at 600 fs the S_1 decay permits T_1 and, to a lower but non-negligible extent, T_2 population. The corresponding lifetimes are shown, calculated by a parallel (first number) and a serial (second number) kinetic model. The serial model implies formation of T_2 followed by T_1, as experimentally believed, while the parallel model implies formation of both T_2 and T_1 from S_1, as observed using non-adiabatic MD.
Part (a) is reproduced from ref. 152 with permission from American Chemical Society, Copyright 2013. Parts (b) and (c) are reproduced from ref. 151 with permission from American Chemical Society, Copyright 2016.

The aforementioned methods can be complemented by several other techniques aimed at modelling non-adiabatic molecular dynamics events, such as symmetrical quasi-classical windowing,[153] quantum-classical Liouville approaches,[154] linearized nonadiabatic dynamics,[155] exact-factorization-based mixed quantum/classical algorithms,[156] and Bohmian dynamics.[157] Nevertheless, their application to photobiological systems has, at present, not been attempted or is very limited.

As the main drawbacks of semiclassical non-adiabatic molecular dynamics are the use of an accurate level of theory and the number of trajectories that need to be run, resulting in an overall high computational cost, recent advances have been performed towards the application of machine learning algorithms. In particular, machine learning can be used to predict energies and forces during surface hopping simulations by applying the following principle: if the number of steps needed to train the machine is much smaller than the total number of steps needed to compute a trajectory, then machine learning could significantly reduce the overall computational cost. By applying the Kernel Ridge Regression (KRR) algorithm coupled to decoherence-corrected fewest switches surface hopping, encouraging results have been obtained showing that, at present, a reasonable accuracy can only be obtained for very small systems, reducing costs by a factor of 10.[158] Hence, even though highly attractive, the use of machine learning to study photobiological dynamics issues can be set as a desirable future goal.

6.6 Conclusion

In this chapter we have highlighted and reviewed some of the recent applications, performed mainly in our group, related to the broad area of photobiology. It is clear that thanks to the development of computing power, as well as the development of computational techniques and protocols, the field is ripe to achieve a proper description of complex light-induced phenomena in a complex environment. As a consequence, *in silico* applications have started to emerge as a crucial branch of photobiology that are able to give a complementary vision to, or to complete, experimental interpretations, while in many cases a predictive power can also be observed.

The possibilities offered by molecular modelling and simulation span a wide range of domains and allow the reproduction and interpretation of steady state spectra, either in the case of linear optical properties or for more complex problems such as circular dichroism or two-photon absorptions. More importantly it also allows researchers to clearly map the spectroscopic properties with the structural and dynamic properties sampled using MD simulations, hence offering an invaluable and complementary tool to analytical and experimental chemistry techniques.

The exploration of the complex PES for the ground and excited states, either performed statically or dynamically, can even allow a much deeper comprehension of photobiology, and in particular unravel, rationalize, and preview the balance between the different competitive relaxation pathways that occur following light-absorption and hence shed light on the elementary photochemical and photophysical processes taking place in living organisms. It has to be noted that the inclusion of the effects of the surroundings, for instance *via* the QM/MM approaches, permits researchers to go from the use of simple model systems, to the simulation of the processes in conditions that are closer and closer to those of the actual experiment.

Although some more specialised techniques, such as non-adiabatic molecular dynamics simulations, are in many instances still far from being black-boxes, it is also remarkable how the calculations of linear optical properties, also including vibrational effects, can nowadays be performed almost routinely and do not in general require a deep knowledge of the underlying numerical and mathematical algorithms. Hence, they can easily be performed by non-specialized groups, once again offering a valuable addition to the experimental results. Furthermore, it has to be expected that this tendency will also be strengthened in the following years, thanks to the use of specific machine learning algorithms that will strongly facilitate the calculations of such properties.

Hence, one can safely affirm, as also proved by the applications presented in this chapter, that the molecular modelling and simulation of excited states and of the interaction between radiation and matter, are ripe enough to allow the actual emergence of *in silico* photobiology. The latter, also thanks to the atomistic and electronic resolution that it can provide, is bound to bring unprecedented answers to fundamental questions in the domain of chemistry and biology and find widespread application in overcoming the traditional borders of computational chemistry.

Acknowledgements

The authors wish to thank Université de Lorraine and CNRS for support. A.F.-M. acknowledges the Generalitat Valenciana and the European Social Fund (contract APOSTD/2019/149 and project GV/2020/226) for the financial support. M.M. acknowledges Universidad de Alcalá for a postdoctoral research grant. The authors also thank the "Phi Science" association for providing a convenient working place in which many discussions related to the themes dealt with in this chapter took place.

References

1. N. Bohr, *Nature*, 1933, **131**, 457–459.
2. H. Hashimoto, C. Uragami and R. J. Cogdell, *Subcell. Biochem.*, 2016, **79**, 111–139.
3. P. S. Nobel, *Physicochemical and Environmental Plant Physiology*, Elsevier, 2009, pp. 228–275.
4. J. C. Dunlap and J. J. Loros, *Cell Res.*, 2016, **26**, 759–760.
5. K. N. Paul, T. B. Saafir and G. Tosini, *Rev. Endocr. Metab. Disord.*, 2009, **10**, 271–278.
6. G. Tosini, I. Ferguson and K. Tsubota, *Mol. Vis.*, 2016, **22**, 61–72.
7. G. Wald, L. Peteanu, R. Mathies and C. Shank, *Science*, 1968, **162**, 230–239.
8. W. Wang, J. H. Geiger and B. Borhan, *BioEssays*, 2014, **36**, 65–74.

9. R. Schoenlein, L. Peteanu, R. Mathies and C. Shank, *Science*, 1991, **254**, 412–415.
10. J. Cadet, E. Sage and T. Douki, *Mutat. Res., Fundam. Mol. Mech. Mutagen.*, 2005, **571**, 3–17.
11. E. Sage, P.-M. Girard and S. Francesconi, *Photochem. Photobiol. Sci.*, 2012, **11**, 74–80.
12. S. Gandini, P. Autier and M. Boniol, *Prog. Biophys. Mol. Biol.*, 2011, **107**, 362–366.
13. J. Cadet and T. Douki, *Photochem. Photobiol. Sci.*, 2018, **17**, 1816–1841.
14. J. A. Marteijn, H. Lans, W. Vermeulen and J. H. J. Hoeijmakers, *Nat. Rev. Mol. Cell Biol.*, 2014, **15**, 465–481.
15. G. E. Zentner and S. Henikoff, *Nat. Struct. Mol. Biol.*, 2013, **20**, 259–266.
16. E. Factors, *Skin Stress Response Pathways*, Springer, 2016.
17. J. Cadet, T. Douki and J. L. Ravanat, *Photochem. Photobiol.*, 2015, **91**, 140–155.
18. Y. Rodriguez and M. J. Smerdon, *J. Biol. Chem.*, 2013, **288**, 13863–13875.
19. R. H. D. Lyngdoh and H. F. Schaefer, *Acc. Chem. Res.*, 2009, **42**, 563–572.
20. R. González-Luque, M. Garavelli, F. Bernardi, M. Merchán, M. A. Robb and M. Olivucci, *Proc. Natl. Acad. Sci. U. S. A.*, 2000, **97**, 9379–9384.
21. P. O'Neill and P. Wardman, *Int. J. Radiat. Biol.*, 2009, **85**, 9–25.
22. K. C. Smith, *The Science of Photobiology*, Springer US, New York, USA, 1989.
23. N. E. Henriksen and V. Engel, *Int. Rev. Phys. Chem.*, 2001, **20**, 93–126.
24. A. Ettinger and T. Wittmann, *Methods Cell Biol.*, 2014, **123**, 77–94.
25. B. Huang, M. Bates and X. Zhuang, *Annu. Rev. Biochem.*, 2009, **78**, 993–1016.
26. J. Biteen and K. A. Willets, *Chem. Rev.*, 2017, **117**, 7241–7243.
27. N. Li, R. Zhao, Y. Sun, Z. Ye, K. He and X. Fang, *Natl. Sci. Rev.*, 2017, **4**, 739–760.
28. B. A. West and A. M. Moran, *J. Phys. Chem. Lett.*, 2012, **3**, 2575–2581.
29. A. Nenov, I. Rivalta, G. Cerullo, S. Mukamel and M. Garavelli, *J. Phys. Chem. Lett.*, 2014, **5**, 767–771.
30. T. Stoll, F. Branchi, J. Réhault, F. Scotognella, F. Tassone, I. Kriegel and G. Cerullo, *J. Phys. Chem. Lett.*, 2017, **8**, 2285–2290.
31. D. Polli, L. Lüer and G. Cerullo, *Rev. Sci. Instrum.*, 2007, **78**, 103108.
32. M. Maiuri, J. Réhault, A. M. Carey, K. Hacking, M. Garavelli, L. Lüer, D. Polli, R. J. Cogdell and G. Cerullo, *J. Chem. Phys.*, 2015, **142**, 212433.
33. E. Dumont and A. Monari, *J. Phys. Chem. B*, 2015, **119**, 410–419.
34. K. Hirakawa, T. Hirano, Y. Nishimura, T. Arai and Y. Nosaka, *J. Phys. Chem. B*, 2012, **116**, 3037–3044.
35. J. J. Nogueira, M. Oppel and L. González, *Angew. Chem., Int. Ed.*, 2015, **54**, 4375–4378.
36. M. C. Cuquerella, V. Lhiaubet-Vallet, J. Cadet and M. A. Miranda, *Acc. Chem. Res.*, 2012, **45**, 1558–1570.
37. A. M. Blanco-Rodríguez, M. Towrie, J. Sýkora, S. Záliš and A. Vlček, *Inorg. Chem.*, 2011, **50**, 6122–6134.

38. S. P. Balashov and J. K. Lanyi, *Cell. Mol. Life Sci.*, 2007, **64**, 2323–2328.
39. J. Zhao, W. Wu, J. Sun and S. Guo, *Chem. Soc. Rev.*, 2013, **42**, 5323.
40. L. M. Frutos, T. Andruniow, F. Santoro, N. Ferre and M. Olivucci, *Proc. Natl. Acad. Sci.*, 2007, **104**, 7764–7769.
41. G. Tomasello, O. G. Gloria, P. Altoè, M. Stenta, S. A. Luis, M. Merchán, G. Orlandi, A. Bottoni and M. Garavelli, *J. Am. Chem. Soc.*, 2009, **131**, 5172–5186.
42. M. Marazzi, M. Wibowo, H. Gattuso, E. Dumont, D. Roca-Sanjuán and A. Monari, *Phys. Chem. Chem. Phys.*, 2016, **18**, 7829–7836.
43. A. Francés-Monerris, H. Gattuso, D. Roca-Sanjuán, I. Tuñón, M. Marazzi, E. Dumont and A. Monari, *Chem. Sci.*, 2018, **9**, 7902–7911.
44. M. Navarrete-Miguel, J. Segarra-Martí, A. Francés-Monerris, A. Giussani, P. Farahani, B.-W. Ding, A. Monari, Y.-J. Liu and D. Roca-Sanjuán, *Photochemistry*, Royal Society of Chemistry, vol. 46, 2019, pp. 28–77.
45. X. Assfeld, A. Monari, M. Marazzi and H. Gattuso, *Front. Chem.*, 2018, **6**, 86.
46. A. Hospital, J. R. Goñi, M. Orozco and J. L. Gelpí, *Adv. Appl. Bioinf. Chem.*, 2015, **8**, 37–47.
47. J. R. Perilla, B. C. Goh, C. K. Cassidy, B. Liu, R. C. Bernardi, T. Rudack, H. Yu, Z. Wu and K. Schulten, *Curr. Opin. Struct. Biol.*, 2015, **31**, 64–74.
48. R. Crespo-Otero and M. Barbatti, *Chem. Rev.*, 2018, **118**, 7026–7068.
49. A. Monari, J. L. Rivail and X. Assfeld, *Acc. Chem. Res.*, 2013, **46**, 596–603.
50. C. Ullrich, *Time-dependent Density-functional Theory: Concepts and Applications*, Oxford University Press, 2012.
51. B. O. Roos, R. Lindh, P. Malmqvist, V. Veryazov and P. O. Widmark, *Multiconfigurational Quantum Chemistry*, John Wiley & Sons, Inc., Hoboken, NJ, USA, 2016.
52. A. R. Leach, *Molecular Modelling: Principles and Applications*, Longman, 2001.
53. P. D. Dans, I. Ivani, A. Hospital, G. Portella, C. González and M. Orozco, *Nucleic Acids Res.*, 2017, **45**, 4217–4230.
54. J. Wang, R. M. Wolf, J. W. Caldwell, P. A. Kollman and D. A. Case, *J. Comput. Chem.*, 2004, **25**, 1157–1174.
55. A. Warshel and M. Levitt, *J. Mol. Biol.*, 1976, **103**, 227–249.
56. X. Assfeld and J. L. Rivail, *Chem. Phys. Lett.*, 1996, **263**, 100–106.
57. D. Roca-Sanjuán, A. Francés-Monerris, I. F. Galván, P. Farahani, R. Lindh and Y.-J. Liu, *Photochemistry*, The Royal Society of Chemistry, vol. 44, 2017, pp. 16–60.
58. T. Etienne, T. Very, E. A. Perpète, A. Monari and X. Assfeld, *J. Phys. Chem. B*, 2013, **117**, 4973–4980.
59. T. Etienne, H. Gattuso, A. Monari and X. Assfeld, *Comput. Theor. Chem*, 2014, 1040–1041.
60. M. Marazzi, H. Gattuso and A. Monari, *Theor. Chem. Acc.*, 2016, **135**, 1–11.
61. O. Sengul, M. Marazzi, A. Monari and S. Catak, *J. Phys. Chem. C*, 2018, **122**, 16315–16324.

62. O. Sengul, E. B. Boydas, M. Pastore, W. Sharmouk, P. C. Gros, S. Catak and A. Monari, *Theor. Chem. Acc.*, 2017, **136**, 67.
63. H. T. Turan, Y. Y. Eken, M. Marazzi, M. Pastore, V. Aviyente and A. Monari, *J. Phys. Chem. C*, 2016, **120**, 17916–17926.
64. S. Mai, H. Gattuso, A. Monari and L. González, *Front. Chem.*, 2018, **6**, 495.
65. M. Kasha, *Discuss. Faraday Soc.*, 1950, **9**, 14.
66. R. Gonzalez-Luque, T. Climent, I. Gonzalez-Ramirez, M. Merchan and L. Serrano-Andres, *J. Chem. Theory Comput.*, 2010, **6**, 2103–2114.
67. H. C. Longuet-Higgins, *J. Chem. Phys.*, 1950, **18**, 265–274.
68. C. B. Duke, *Int. J. Quantum Chem.*, 1979, **16**, 267–281.
69. P. R. Andrews, G. D. Smith and I. G. Young, *Biochemistry*, 1973, **12**, 3492–3498.
70. G. Herzberg and H. C. Longuet-Higgins, *Discuss. Faraday Soc.*, 1963, **35**, 77–82.
71. A. Francés-Monerris, M. Merchán and D. Roca-Sanjuán, *J. Chem Phys.*, 2013, **139**, 71101.
72. A. Francés-Monerris, M. Merchán and D. Roca-Sanjuán, *J. Phys. Chem. B*, 2014, **118**, 2932–2939.
73. E. Hayon and M. Simic, *J. Am. Chem. Soc.*, 1973, **95**, 1029–1035.
74. J. Aranda, A. Francés-Monerris, I. Tuñón and D. Roca-Sanjuán, *J. Chem. Theory Comput.*, 2017, **13**, 5089–5096.
75. C. von Sonntag, *Free-Radical-Induced DNA Damage and Its Repair: A Chemical Perspective*, Springer-Verlag, Berlin, 1st edn, 2006.
76. S. V Jovanovic and M. G. Simic, *J. Am. Chem. Soc.*, 1986, **108**, 5968–5972.
77. A. Francés-Monerris, M. Merchán and D. Roca-Sanjuán, *J. Org. Chem.*, 2017, **82**, 276–288.
78. J. Tomasi, B. Mennucci and R. Cammi, *Chem. Rev.*, 2005, **105**, 2999–3093.
79. A. Klamt and G. Schüürmann, *J. Chem. Soc., Perkin Trans. 2*, 1993, 799–805.
80. J. Catalán, J. L. G. de Paz and C. Reichardt, *J. Phys. Chem. A*, 2010, **114**, 6226–6234.
81. T. Etienne, C. Michaux, A. Monari, X. Assfeld and E. A. Perpète, *Dyes Pigm.*, 2014, **100**, 24–31.
82. R. L. Martin, *J. Chem. Phys.*, 2003, **118**, 4775–4777.
83. T. Etienne, X. Assfeld and A. Monari, *J. Chem. Theory Comput.*, 2014, **10**, 3896–3905.
84. A. Francés-Monerris, K. Magra, M. Darari, C. Cebrián, M. Beley, E. Domenichini, S. Haacke, M. Pastore, X. Assfeld, P. C. Gros and A. Monari, *Inorg. Chem.*, 2018, **57**, 10431–10441.
85. T. Duchanois, L. Liu, M. Pastore, A. Monari, C. Cebrián, Y. Trolez, M. Darari, K. Magra, A. Francés-Monerris, E. Domenichini, M. Beley, X. Assfeld, S. Haacke and P. Gros, *Inorganics*, 2018, **6**, 63.
86. O. S. Wenger, *J. Am. Chem. Soc.*, 2018, **140**, 13522–13533.
87. S. Sitkiewicz, D. Rivero, J. M. Oliva, A. Saiz-Lopez and D. Roca-Sanjuán, *Phys. Chem. Chem. Phys.*, 2019, **21**, 455–467.

88. M. Fumanal, S. Vela, H. Gattuso, A. Monari and C. Daniel, *Chem. – Eur. J.*, 2018, **24**, 14425–14435.
89. Y.-C. Zheng, M.-L. Zheng, K. Li, S. Chen, Z.-S. Zhao, X.-S. Wang and X.-M. Duan, *RSC Adv.*, 2015, **5**, 770–774.
90. M. Pawlicki, H. A. Collins, R. G. Denning and H. L. Anderson, *Angew. Chem., Int. Ed.*, 2009, **48**, 3244–3266.
91. T. Dai, Y. Y. Huang and M. R. Hamblin, *Photodiagnosis Photodyn. Ther.*, 2009, **6**, 170–188.
92. P. Agostinis, K. Berg, K. A. Cengel, T. H. Foster, A. W. Girotti, S. O. Gollnick, S. M. Hahn, M. R. Hamblin, A. Juzeniene, D. Kessel, M. Korbelik, J. Moan, P. Mroz, D. Nowis, J. Piette, B. C. Wilson and J. Golab, *Canc. J. Clin.*, 2011, **61**, 250–281.
93. H. Gattuso, E. Dumont, M. Marazzi and A. Monari, *Phys. Chem. Chem. Phys.*, 2016, **18**, 18598–18606.
94. H. Gattuso, T. Duchanois, V. Besancenot, C. Barbieux, X. Assfeld, P. Becuwe, P. C. Gros, S. Grandemange and A. Monari, *Front. Chem.*, 2015, **3**, 67.
95. V. D. Suryawanshi, L. S. Walekar, A. H. Gore, P. V. Anbhule and G. B. Kolekar, *J. Pharm. Anal.*, 2016, **6**, 56–63.
96. H. Gattuso, V. Besancenot, S. Grandemange, M. Marazzi and A. Monari, *Sci. Rep*, 2016, **6**, 28480.
97. A. Penninks, K. Baert, S. Levorato and M. Binaglia, *EFSA J.*, 2017, **15**, 4920.
98. H. Lischka, D. Nachtigallová, A. J. A. Aquino, P. G. Szalay, F. Plasser, F. B. C. Machado and M. Barbatti, *Chem. Rev.*, 2018, **118**, 7293–7361.
99. A. Francés-Monerris, C. Hognon, M. A. Miranda, V. Lhiaubet-Vallet and A. Monari, *Phys. Chem. Chem. Phys.*, 2018, **20**, 25666–25675.
100. I. Aparici-Espert, G. Garcia-Lainez, I. Andreu, M. A. Miranda and V. Lhiaubet-Vallet, *ACS Chem. Biol.*, 2018, **13**, 542–547.
101. I. Conti and M. Garavelli, *J. Phys. Chem. Lett.*, 2018, **9**, 2373–2379.
102. A. A. Beckstead, Y. Zhang, M. S. de Vries and B. Kohler, *Phys. Chem. Chem. Phys.*, 2016, **18**, 24228–24238.
103. C. E. Crespo-Hernandez, B. Cohen, P. M. Hare and B. Kohler, *Chem. Rev.*, 2004, **104**, 1977–2019.
104. R. Improta, F. Santoro and L. Blancafort, *Chem. Rev.*, 2016, **116**, 3540–3593.
105. A. Giussani, J. Segarra-Martí, D. Roca-Sanjuán and M. Merchán, *Top. Curr. Chem*, 2015, **355**, 57–98.
106. D. Markovitsi, F. Talbot, T. Gustavsson, D. Onidas, E. Lazzarotto and S. Marguet, *Nature*, 2006, **441**, E7.
107. C. E. Crespo-Hernandez, B. Cohen and B. Kohler, *Nature*, 2005, **436**, 1141–1144.
108. J. Chen, A. K. Thazhathveetil, F. D. Lewis and B. Kohler, *J. Am. Chem. Soc.*, 2013, **135**, 10290–10293.
109. P. M. Hare, C. E. Crespo-Hernandez and B. Kohler, *Proc. Natl. Acad. Sci. U. S. A.*, 2007, **104**, 435–440.

110. D. Marx and J. Hutter, *Ab initio Molecular Dynamics: Basic Theory and Advanced Methods*, Cambridge University Press, Cambridge, 2009.
111. A. Abo-Riziq, L. Grace, E. Nir, M. Kabelac, P. Hobza and M. S. de Vries, *Proc. Natl. Acad. Sci. U. S. A.*, 2005, **102**, 20–23.
112. A. L. Sobolewski and W. Domcke, *Phys. Chem. Chem. Phys.*, 2004, **6**, 2763–2771.
113. A. L. Sobolewski, W. Domcke and C. Hattig, *Proc. Natl. Acad. Sci.*, 2005, **102**, 17903–17906.
114. V. Sauri, J. P. Gobbo, J. J. Serrano-Pérez, M. Lundberg, P. B. Coto, L. Serrano-Andrés, A. C. Borin, R. Lindh, M. Merchán and D. Roca-Sanjuán, *J. Chem. Theory Comput.*, 2013, **9**, 481–496.
115. K. Rottger, H. J. B. Marroux, M. P. Grubb, P. M. Coulter, H. Bohnke, A. S. Henderson, M. C. Galan, F. Temps, A. J. Orr-Ewing and G. M. Roberts, *Angew. Chem., Int. Ed.*, 2015, **54**, 14719–14722.
116. Y. Zhang, X.-B. Li, A. M. Fleming, J. Dood, A. A. Beckstead, A. M. Orendt, C. J. Burrows and B. Kohler, *J. Am. Chem. Soc.*, 2016, **138**, 7395–7401.
117. Y. Zhang, K. De La Harpe, A. A. Beckstead, R. Improta and B. Kohler, *J. Am. Chem. Soc.*, 2015, **137**, 7059–7062.
118. A. Frances-Monerris, J. Segarra-Marti, M. Merchan and D. Roca-Sanjuan, *Theor. Chem. Acc.*, 2016, **135**, 31.
119. G. Groenhof, L. V Schäfer, M. Boggio-Pasqua, M. Goette, H. Grubmüller and M. A. Robb, *J. Am. Chem. Soc.*, 2007, **129**, 6812–6819.
120. M. H. Beck, A. Jäckle, G. A. Worth and H.-D. D. Meyer, *Phys. Rep.*, 2000, **324**, 1–105.
121. M. Fumanal, E. Gindensperger and C. Daniel, *Phys. Chem. Chem. Phys.*, 2018, **20**, 1134–1141.
122. S. Reiter, D. Keefer and R. De Vivie-Riedle, *J. Am. Chem. Soc.*, 2018, **140**, 8714–8720.
123. D. Keefer, S. Thallmair, S. Matsika and R. De Vivie-Riedle, *J. Am. Chem. Soc.*, 2017, **139**, 5061–5066.
124. B. F. E. Curchod and T. J. Martínez, *Chem. Rev.*, 2018, **118**, 3305–3336.
125. E. Tapavicza, G. D. Bellchambers, J. C. Vincent and F. Furche, *Phys. Chem. Chem. Phys.*, 2013, **15**, 18336–18348.
126. Y. Dou, B. R. Torralva and R. E. Allen, *J. Mod. Opt*, 2003, **50–15**, 2615–2643.
127. W. Wu, S. Yuan, J. She, Y. Dou and R. E. Allen, *Int. J. Photoenergy*, 2015, **2015**, 1–6.
128. P. B. Woiczikowski, T. Steinbrecher, T. Kubař and M. Elstner, *J. Phys. Chem. B*, 2011, **115**, 9846–9863.
129. B. O. Roos, *Ab Initio Methods in Quantum Chemistry - II*, John Wiley & Sons, Ltd, vol. 69, 2007, pp. 399–445.
130. K. Andersson, P.-A. Malmqvist, B. O. Roos, A. J. Sadlej and K. Wolinski, *J. Phys. Chem.*, 1990, **94**, 5483–5488.
131. E. Tapavicza, I. Tavernelli and U. Rothlisberger, *Phys. Rev. Lett.*, 2007, **98**, 023001.
132. N. L. Doltsinis and D. Marx, *Phys. Rev. Lett.*, 2002, **88**, 4.

133. C. F. Craig, W. R. Duncan and O. V. Prezhdo, *Phys. Rev. Lett.*, 2005, **95**, 163001.
134. A. Humeniuk and R. Mitrić, *Comput. Phys. Commun.*, 2017, **221**, 174–202.
135. A. Kazaryan, Z. Lan, L. V. Schäfer, W. Thiel and M. Filatov, *J. Chem. Theory Comput.*, 2011, **7**, 2189–2199.
136. J. C. Tully, *J. Chem. Phys.*, 1990, **93**, 1061–1071.
137. J. E. Subotnik, A. Jain, B. Landry, A. Petit, W. Ouyang and N. Bellonzi, *Annu. Rev. Phys. Chem.*, 2016, **67**, 387–417.
138. D. Polli, P. Altoè, O. Weingart, K. M. Spillane, C. Manzoni, D. Brida, G. Tomasello, G. Orlandi, P. Kukura, R. A. Mathies, M. Garavelli and G. Cerullo, *Nature*, 2010, **467**, 440–443.
139. J. Sasaki and J. L. Spudich, *Photochem. Photobiol.*, 2008, **84**, 863–868.
140. C. H. Slamovits, N. Okamoto, L. Burri, E. R. James and P. J. Keeling, *Nat. Commun.*, 2011, **2**, 183.
141. G. Nagel, D. Ollig, M. Fuhrmann, S. Kateriya, A. M. Musti, E. Bamberg and P. Hegemann, *Science*, 2002, **296**, 2395–2398.
142. K. Deisseroth, *Nat. Methods*, 2011, **8**, 26–29.
143. Y. Guo, F. E. Beyle, B. M. Bold, H. C. Watanabe, A. Koslowski, W. Thiel, P. Hegemann, M. Marazzi and M. Elstner, *Chem. Sci.*, 2016, **7**, 3879–3891.
144. H. E. Kato, F. Zhang, O. Yizhar, C. Ramakrishnan, T. Nishizawa, K. Hirata, J. Ito, Y. Aita, T. Tsukazaki, S. Hayashi, P. Hegemann, A. D. Maturana, R. Ishitani, K. Deisseroth and O. Nureki, *Nature*, 2012, **482**, 369–374.
145. O. Volkov, K. Kovalev, V. Polovinkin, V. Borshchevskiy, C. Bamann, R. Astashkin, E. Marin, A. Popov, T. Balandin, D. Willbold, G. Büldt, E. Bamberg and V. Gordeliy, *Science*, 2017, **358**, eaan8862.
146. Y. Hontani, M. Marazzi, K. Stehfest, T. Mathes, I. H. M. van Stokkum, M. Elstner, P. Hegemann and J. T. M. Kennis, *Sci. Rep.*, 2017, **7**, 7217.
147. I. Dokukina, A. Nenov, C. M. Marian, M. Garavelli and O. Weingart, *ChemPhotoChem*, 2019, **3**, 107–116.
148. M. Barbatti, A. J. A. Aquino, J. J. Szymczak, D. Nachtigallová, P. Hobza and H. Lischka, *Proc. Natl. Acad. Sci.*, 2010, **107**, 21453–21458.
149. M. Richter, P. Marquetand, J. González-Vázquez, I. Sola and L. González, *J. Chem. Theory Comput.*, 2011, **7**, 1253–1258.
150. S. Mai, P. Marquetand and L. González, *Wiley Interdiscip. Rev. Comput. Mol. Sci.*, 2018, **8**, e1370.
151. M. Marazzi, S. Mai, D. Roca-Sanjuán, M. G. Delcey, R. Lindh, L. González and A. Monari, *J. Phys. Chem. Lett.*, 2016, **7**, 622–626.
152. E. Dumont and A. Monari, *J. Phys. Chem. Lett.*, 2013, **4**, 4119–4124.
153. S. J. Cotton and W. H. Miller, *J. Phys. Chem. A*, 2013, **117**, 7190–7194.
154. R. Kapral, *Annu. Rev. Phys. Chem.*, 2006, **57**, 129–157.
155. S. Bonella and D. F. Coker, *J. Chem. Phys.*, 2005, **122**, 194102.
156. F. Agostini, S. K. Min, A. Abedi and E. K. U. Gross, *J. Chem. Theory Comput.*, 2016, **12**, 2127–2143.

157. B. F. E. Curchod and I. Tavernelli, *J. Chem. Phys.*, 2013, **138**, 184112.
158. P. O. Dral, M. Barbatti and W. Thiel, *J. Phys. Chem. Lett.*, 2018, **9**, 5660–5663.
159. R. O. Dror, R. M. Dirks, J. P. Grossman, H. Xu and D. E. Shaw, *Annu. Rev. Biophys.*, 2012, **41**, 429–452.
160. P. Norman and M. Linares, *Chirality*, 2014, **26**, 483–489.
161. D. M. Florant, M. N. Pedersen, J. Rubio-Magnieto, M. Surin, M. Linares and P. Norman, *J. Phys. Chem. Lett.*, 2015, **6**, 355–359.
162. P. Norman, J. Parello, P. L. Polavarapu and M. Linares, *Phys. Chem. Chem. Phys.*, 2015, **17**, 21866–21879.
163. N. Holmgaard List, J. Knoops, J. Rubio-Magnieto, J. Idé, D. Beljonne, P. Norman, M. Surin and M. Linares, *J. Am. Chem. Soc.*, 2017, **139**, 14947–14953.
164. E. Bignon, H. Gattuso, C. Morell, E. Dumont and A. Monari, *Chem. – Eur. J.*, 2015, **21**, 11509–11516.
165. D. Padula, S. Jurinovich, L. Di Bari and B. Mennucci, *Chem. – Eur. J.*, 2016, **22**, 17011–17019.
166. D. Loco, S. Jurinovich, L. Di Bari and B. Mennucci, *Phys. Chem. Chem. Phys.*, 2016, **18**, 866–877.
167. S. Jurinovich, G. Pescitelli, L. Di Bari and B. Mennucci, *Phys. Chem. Chem. Phys.*, 2014, **16**, 16407–16418.
168. H. Gattuso, X. Assfeld and A. Monari, *Theor. Chem. Acc.*, 2015, **134**, 36.
169. H. Gattuso, A. Spinello, A. Terenzi, X. Assfeld, G. Barone and A. Monari, *J. Phys. Chem. B*, 2016, **120**, 3113–3121.
170. H. Gattuso, C. García-Iriepa, D. Sampedro, A. Monari and M. Marazzi, *J. Chem. Theory Comput.*, 2017, **13**, 3290–3296.

CHAPTER 7

Bridging the Gap Between Atomistic Molecular Dynamics Simulations and Wet-lab Experimental Techniques: Applications to Membrane Proteins

LUCIE DELEMOTTE

Department of Applied Physics, Science for Life Laboratory, KTH Royal Institute of Technology, Stockholm, Sweden
Email: lucie.delemotte@scilifelab.se

7.1 Classical Molecular Dynamics Simulations

Molecular modelling and simulations encompass all the computational techniques used to reproduce and predict the structure and behavior of molecular systems, based on their physical properties.[1] All levels of description are models that carry their approximations and their associated advantages and drawbacks (Figure 7.1). Quantum chemistry approaches, in which electrons are considered explicitly, can provide an accurate description of the physical properties of molecular systems, but the computational cost associated with their use often prohibits sampling of the times scales that are of interest to the practitioner, and estimation of the properties and the level of

Theoretical and Computational Chemistry Series No. 20
Computational Techniques for Analytical Chemistry and Bioanalysis
Edited by Philippe B. Wilson and Martin Grootveld

Published by the Royal Society of Chemistry, www.rsc.org

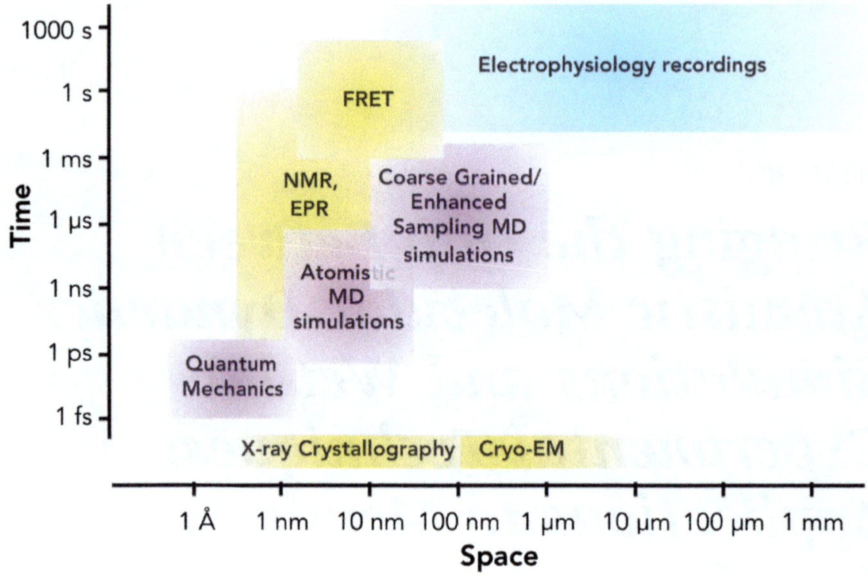

Figure 7.1 Space and time scales accessed by selected methods in biomolecular research: structural/spectroscopic (yellow), electrophysiology (blue), and molecular simulations (purple) approaches.

uncertainty associated with the measurement may not be possible at that level of detail. Classical atomistic simulations consider individual atoms as a single interaction center that has a partial charge and mass. Such methods lose the electronic degree of freedom, but allow sampling of much larger space and time scales, up to the microsecond range. To access time scales beyond this, a further compromise is needed: on the one hand, coarse grained simulations consider a single interaction site for several heavy atoms, thus losing atomistic details; on the other, enhanced sampling atomistic molecular dynamics (MD) simulations use various strategies to cross large free energy barriers and increase sampling of phase space. With each strategy, some information is necessarily lost, but the protocols can be advantageously combined for the needs of the user. This section covers classical atomistic and coarse grained simulations, and briefly reviews enhanced sampling techniques of use to characterize experimentally measurable quantities.

7.1.1 MD Simulation Algorithm

Molecular dynamics simulations aim to reproduce macroscopic behaviors by numerically integrating the classical equations of motion of a microscopic system consisting of a large number of particles (up to a few millions). The output of a MD simulation is called a trajectory, and consists of particle coordinates and velocities recorded along time. Macroscopic properties are then expressed as functions of these.[1]

7.1.1.1 Propagating the Equations of Motion

The classical equations of motion of Newton are used to generate a set of atomic positions and velocities along time: positions, velocities or momenta, and forces at a given instant in time, t, are used to predict the positions and momenta at a later time, $t+\delta t$.

$$\nabla U(x(t)) = f(x(t)) = m \cdot a(t) = m \cdot \ddot{x}(t) = m \cdot \frac{\mathrm{d}^2 x}{\mathrm{d}t^2}$$

The algorithm is deterministic, the forces that guide the dynamics ($f(x(t))$) are derived from a complex potential energy function ($U(x)$) that depends on all positions of the atoms in the system, x.

To numerically integrate these equations, several algorithms can be used: the positions and momenta at time $t+\delta t$ indeed depend on the positions, on the first derivative of the positions (velocities) and on the second derivative of the positions (acceleration) at time t, and algorithms based on Taylor decomposition in the vicinity of time point t are used to numerically allow the integration.

The Verlet algorithm is the most common.[2] It is based on combining the expressions for the expansion of the positions in the vicinity of time t:

$$x(t+\delta t) = x(t) + \delta t v(t) + \frac{1}{2}\delta t^2 a(t) + \ldots$$

$$x(t-\delta t) = x(t) - \delta t v(t) + \frac{1}{2}\delta t^2 a(t) + \ldots$$

By adding these two equations one obtains the positions at time $t+\delta t$ from the positions at time t and at time $t-\delta t$:

$$x(t+\delta t) = 2x(t) - x(t-\delta t) + \delta t^2 a(t)$$

The velocities do not appear explicitly, but may be computed. In the simplest version:

$$v(t) = \frac{[x(t-\delta t) + x(t+\delta t)]}{2\delta t}$$

Other algorithms that derive from the Verlet algorithm have been developed, in which the velocities are taken into account explicitly (leap-frog[3] or velocity Verlet[4]) or considering higher-order expansion terms in the Taylor series. The choice of the algorithm is nowadays usually not considered crucial, as even the simplest are shown to yield stable simulations, provided an appropriate choice of simulation parameters is used.[1] An important aspect of the algorithm is that the Taylor expansion is only valid in the near vicinity of t. The time step, δt, should thus be small enough for the approximation to hold. In practice, it should be chosen to be smaller than the fastest oscillations in the system: in atomistic simulations, it is usually femtoseconds (10^{-15} s) at most; in coarse grained MD simulations, it can be up to two orders of magnitude larger (see Section 7.1.2).

7.1.1.2 Periodic Boundary Conditions

In most use-cases, the number of atoms (N) in a MD simulation system is of the order of 10^4–10^6, yielding a simulation box of dimensions ranging from a few Ångström (Å) to a few nanometers (nm) at most. The length scales of interest for biological processes are generally much larger, which makes it important to pay attention to the treatment of the boundaries of the system. It is thus customary to use full 3D periodic boundary conditions (PBCs), in which each cell is surrounded on all sides by periodic images of itself.[5] In the case of membrane systems, the simulated system considering the replicas thus looks more like a multilamellar vesicle than a single 2-dimensional membrane. When a particle leaves the simulation box from one edge, it then re-enters the box from the opposing edge. To calculate interactions, the minimum image convention is used: the closest neighbors are used even if they are in different simulation boxes.

7.1.1.3 Simulation Ensembles

In its most simple form, molecular dynamics simulations allow sampling of the time evolution of a system with a constant number of particles (N) and using a simulation box of constant dimensions, for example volume (V). Using the equations of Newton to propagate the positions and velocities of the system allows conservation of the total energy of the system (E).[1] In this case, the temperature of the system (T) will vary according to the kinetic energy of the ensemble of particles, as well as the pressure acting on the simulation box (P). Often MD simulations are aimed at reproducing the time evolution of systems that are subject to a constant temperature and pressure, meaning that the algorithm needs to be modified.

7.1.1.3.1 Thermostatting. A number of schemes have been proposed to carry out isothermal MD simulations.

The simplest strategy consists of periodically rescaling the velocities by a factor $\sqrt{T_0/T(t)}$, in which $T(t)$ denotes the instantaneous kinetic temperature[6] and T_0 the target temperature:

$$T(t) = \sum_{i=1}^{N} \frac{m_i v_i^2(t)}{k_B N_f}$$

for a system of N particles of masses m_i and velocities v_i, and in which N_f is the number of degrees of freedom of the system, for example $N_f = 3N - 3$ for a fixed total momentum. The reference temperature, however, does not appear explicitly in the equations that are integrated, leading to large fluctuations in the value of T_0.

The weak coupling reported by Berendsen[7] consists of letting the instantaneous kinetic temperature $T(t)$ "relax" towards the reference temperature, T_0, following:

$\frac{dT}{dt} = \frac{T(t) - T_0}{\tau_T}$, in which τ_T is the relaxation time associated with the fluctuations of the temperature. In this scheme, the velocities are rescaled by a factor λ:

$$\lambda = \sqrt{1 + \frac{\delta t}{\tau_T}\left(\frac{T_0}{T(t)} - 1\right)}$$

This scheme thus consists of coupling the simulation to a "heat reservoir" by means of a first-order process and has the advantage of avoiding large oscillating responses to temperature changes.

Third, resorting to Langevin dynamics involves controlling the temperature by adding damping and random forces to the simulation system.[5] This is a way to control the temperature by simulating the influence of the external medium on the system *via* $m_i a_i(t) = F_i(t) + f_i(t) - m_i \beta_i \nu_i(t)$, in which m_i and a_i are the mass and the acceleration of atom i, respectively. F_i is the sum of forces exerted by all other atoms on atom i, β_i is a friction term and v_i the velocity of the atom i at time t. f_i is a random force depending on the temperature and whose components are normally distributed according to: $\sqrt{\frac{2 m_i \beta_i k_B T_0}{\delta t}}$.

A final approach consists of augmenting the simulation system by adding an additional degree of freedom, s, to which the system is coupled. Particle s is then the thermostat of the system, and its dynamics are governed by equations of motion, so that it itself has a potential and kinetic energy term associated with it. This approach is known as the Nose–Hoover approach.[8,9]

7.1.1.3.2 Barostatting. Similar to the control of temperature, the control of pressure can also be achieved in several ways.

The pressure can be controlled by an extension of the weak coupling algorithm.[7] The equations of motion are modified in response to the relaxation of the instantaneous pressure, $P(t)$, towards its reference value, P_0, according to $\frac{dP}{dt} = \frac{P(t) - P_0}{\tau_P}$, in which τ_P is the relaxation time associated to the fluctuations of the pressure. In this scheme, the velocities are rescaled by a factor κ:

$$\kappa = \left[1 - \beta \delta t\left(\frac{P_0 - P(t)}{\tau_P}\right)\right]^{\frac{1}{3}}$$

In which β denotes the isothermal compressibility.

Other strategies are used to dampen the oscillations in the pressure using a Langevin approach or coupling the system to an external variable, V.[1]

7.1.2 Force Fields and the Potential Energy Function

7.1.2.1 Classical All-atom Models

The propagation of the equations of motion requires knowledge of the forces F_i acting on all particles, i, of the system. The forces are derived from the potential $U(r)$, which depends on the positions of all the atoms in the system. $U(r)$ is also known as the force field.[1,5,10]

The classical force fields that are most commonly used in biomolecular applications usually consist of bonded and non-bonded interactions. The bonded interactions are a sum of energies owing to the intramolecular harmonic bond stretching (E_{bond}), angle bending (E_{angle}), and dihedral angle deformation (E_{torsion}). Non-bonded interactions come from intermolecular electrostatic (E_{el}) and van der Waals terms (E_{vdw}).

The common mathematical form of the force field is:

$$U = E_{\text{bond}} + E_{\text{angle}} + E_{\text{torsion}} + E_{\text{el}} + E_{\text{vdw}}$$

In which the bonded terms are expressed as:

$$E_{\text{bond}} = \sum_{i,j} k_{ij}^{b}\left(r_{ij} - r_{ij}^{0}\right)^2$$

$$E_{\text{angle}} = \sum_{i,j,k} k_{ijk}^{a}\left(\alpha_{ijk} - \alpha_{ijk}^{0}\right)^2$$

$$E_{\text{torsion}} = \sum_{i,j,k,l} k_{ijkl}^{t}\left(1 + \cos(n\omega_{ijkl} - \omega_{ijkl}^{0})\right)$$

In which r_{ij} is the distance between atoms i and j, α_{ijk} is the angle between atoms i,j and k, ω_{ijkl} is the dihedral angle between i,j, k and l, and k_{ij}^{b}, k_{ijk}^{a} and k_{ijkl}^{t} on one hand and r_{ij}^{0}, α_{ijk}^{0}, and ω_{ijkl}^{0} on the other are the force constants and equilibrium values for the bond stretch, angle bend, and dihedral torsion deformations.

A coulombic potential describes the interaction between the partial charges q_i assigned to atoms:

$$E_{\text{el}} = \sum_{i,j} \frac{q_i q_j}{4\pi\varepsilon_0 r_{ij}}$$

In which q_i is the partial charge on atom i and ε_0 the vacuum permittivity, and a Van der Waals potential describes short-range repulsion and long-range attraction between the pairs of atoms:

$$E_{vdw} = \sum_{i,j} \varepsilon_{ij}\left[\left(\frac{\sigma_{ij}}{r_{ij}}\right)^{12} - 2\left(\frac{\sigma_{ij}}{r_{ij}}\right)^{6}\right]$$

In which ε_{ij} and σ_{ij} are Lennard–Jones parameters for the van der Waals interactions.

The parameters of the models have been derived and refined over the years from comparison to quantum calculations and/or reproduction of experimentally measured quantities. The most commonly used force fields in biomolecular simulations are GROMOS,[11] CHARMM[12] and AMBER.[13]

7.1.2.2 Polarizable Models

The type of classical force fields described above consider atoms as particles with a partial charge, but no higher order terms of the electric potential expansion. That is, the dipolar, quadrupolar and higher order terms are ignored in such force fields. Molecular polarizability can thus arise from the orientation of molecules or fragments, but no electronic polarizability is taken into account. For an increased accuracy in the treatment of electrostatic properties, electronic polarization can be accounted for.[14] Two types of models exist: one considers the charge redistribution within each atom (*via* an induced dipole[15] or a charge-on-spring model, also called the Drude model);[16] the other allows charge flow between atoms (fluctuating charge (FQ) model).[17]

In these models the total electrostatic energy comes from the sum of two terms:

$$E_{\text{el}} = E_{\text{self}} + E_{\text{coulomb}}$$

In which the self-energy depends on the type of polarizable model considered:

$$E_{\text{self}}^{\text{Ind}} = \sum_i \frac{1}{2} \alpha_i^{-1} \mu_i^2$$

$$E_{\text{self}}^{\text{Drude}} = \sum_i \frac{1}{2} k_i^D d_i^2$$

$$E_{\text{self}}^{\text{FQ}} = \sum_i (\chi_i q_i + \eta_i q_i^2)$$

In which α_i and μ_i are the atomic polarizability and induced dipole of atom i, k_i^D and d_i are the force constant and displacement of the Drude particle, and χ_i, η_i and q_i are the electronegativity, chemical hardness, and partial charge of atom i.

AMBER,[18,19] AMOEBA,[15] CHARMM Drude,[16] CHARMM-FQ,[17,20] SIBFA,[21] and ABEEM$\sigma\pi$[22] are all examples of polarizable force fields that have been developed for biological systems.

7.1.2.3 Coarse-grained Models

Despite the progress made in hardware and software, the timescales of interest in biology are often beyond the scope of atomistic simulations.

In such cases, the description of the system can be coarse-grained. Atomistic details are thereby lost but the decreased resolution allows us to use a large time-step and the reduced number of particles leads to a decreased computational cost when calculating the interaction potential.[23] The overall gain is two to three orders of magnitude in the time scales attainable using such models. Coarse-grained force fields range from a supra-coarse grain (CG) level of resolution to near-atomistic models. The most commonly used models distinguish chemical groups and consider on average three to six non-hydrogen atoms as part of a single interaction site. The functional form remains the same as in atomistic models, which allows researchers to run the simulations using the same code. The parametrization of the force field is often aimed at reproducing properties observed at the atomistic level of resolution and/or at reproducing thermodynamic observables. In the most commonly used force fields, an effort is made to keep the parameters transferable from one chemical group to the other in order to ensure modularity of the model.

Martini,[24,25] Shinoda/Devane/Klein (SDK),[26] ELBA,[27] and SIRAH[28] are all examples of common coarse-grained force fields in use for the study of biomolecular systems. The main area of applicability of such coarse-grained models has been the study of biomembranes, and of the interaction of lipids with membrane proteins.

7.1.2.4 Treatment of Long-range Non-bonded Interactions

Van der Waals interactions decay rapidly and are typically turned off beyond a certain cutoff distance (generally 10–15 Å). Electrostatic interactions described by a coulomb potential decay much more slowly, which makes it necessary to consider their contribution beyond this cutoff distance.[1] Ewald summation (and its modern computational equivalent Particle Mesh Ewald – PME) is commonly used: the summation of interaction energies can be carried out in Fourier space, a technique that has the advantage of converging rapidly.[29] In fact, it is maximally efficient to decompose the interaction potential into a short-range component summed in real space and a long-range component summed in Fourier space:

$$E_{\mathrm{el}}(r) = E_{\mathrm{sr}}(r) + E_{\mathrm{lr}}(r)$$

As both summations converge quickly in their respective spaces, they may be truncated with little loss of accuracy and a significant improvement in computational efficiency.

7.1.3 Enhanced Sampling Simulations Schemes

Inferring statistically converged probability distributions of degrees of freedom of interest requires repeated sampling of the latter. The quantities of interest in biological systems often equilibrate over milliseconds to minutes, whereas the time step needed to integrate the equations of motion is of the order of femtoseconds. Twelve order of magnitudes thus separate these two time scales making it challenging to exhaustively sample biological phenomena.[30]

One approach to reach these time scales was spearheaded by David E Shaw: the design and construction of a special-purpose machine, Anton (and subsequently Anton2), has allowed him to reach unprecedented time scales using brute-force, all-atom classical MD simulations.[31] Although most of the Anton and Anton2 machines are accessible only to D.E. Shaw Research scientists, one machine was donated to the Pittsburgh Computer Center and can be used by US-based researchers.

Enhanced sampling refers to a class of techniques that attempts to increase the sampling of the configurational space.[30] The aim of enhanced sampling can be one of three kinds: one can simply attempt to discover new configurations that were not resolved previously (increase exploration of the configurational space). In an additional level of complexity, the aim can be to obtain a converged estimate of probability distributions, or of free energies, which requires computing the probability of a state of interest relative to another, or relative to the whole ensemble of possible configurations. Finally, the aim could be to characterize the rates of interconversion between states, which requires observing transitions between states enough times to estimate the average time it takes to interconvert between states.

7.1.3.1 Constraints

Certain degrees of freedom can be frozen throughout the simulation through the use of holonomic constraints. If the fastest degrees of freedom are frozen (the bonds involving hydrogen atoms, typically), the time step that can be used can be increased and the total simulation time obtained for an equivalent computer investment, increased. Freezing certain degrees of freedom is equivalent to solving constrained equations of motion:[32]

In the case of a fixed bond length, for example, the constraint is written:

$$\Theta_{ij}(t) = |r_j(t) - r_i(t)|^2 - d_{ij}^2$$

In which d_{ij} is the equilibrium length of the bond. Atom i then feels not only the force F_i owing to the force field, but also a constraint force, g_i, that now appears in the equations of motion that guide the dynamics of particle i:

$$m_i \frac{d^2 r_i(t)}{dt^2} = F_i + g_i$$

This constraint force is defined by $g_i = -\sum_j \lambda_{ij}(t)\nabla_i\Theta_{ij}(t) = -2\lambda_{ij}(t)r_{ij}(t)$

In which $\lambda_{ij}(t)$ is the Lagrange multiplier associated to the constraint enforced along the bond connecting atoms i and j.

These constrained equations of motion are solved in most cases following an iterative scheme, solving one by one the equations of a linear system until each holonomic constraint is satisfied.

7.1.3.2 Strategies to Enhance the Sampling

Several other strategies have been proposed to modify the simulation protocol to allow researchers to both explore the configurational space, and to obtain converged estimates of probability distributions.[33]

A class of widely used methods modifies the potential that controls the interactions between particles. In so-called importance sampling, the quantities are estimated for the modified potential, and reweighted to estimate their distribution considering the original potential. Metadynamics,[34] the accelerated weighted histogram (AWH)[35] and adaptive biasing force (ABF)[36] fall into this category. Similarly, the equations that govern the dynamics can be modified, by softening the degrees of freedom that trap the system in local minima. Accelerated MD,[37] in which the dihedral potentials are modified to be softer, follows this approach.

Non-equilibrium methods act by adding a force on selected degrees of freedom, but contrary to importance sampling, do not wait for the system to equilibrate. The original probability distribution can be estimated using Jarzynsky's equality[38] or Crooks theorem.[39] Steered MD (SMD) and targeted MD (TMD) are examples of these.[40,41]

Alchemical methods calculate free energy differences between states by joining the two states *via* alchemical perturbation, in which non-physical, partially decoupled, states are simulated. The free energy difference can then be estimated using free energy perturbation (FEP)[42] or thermodynamic integration (TI).[32] This transformation may require a step-wise approach to ensure thermodynamic overlap between states.

A similar stratification strategy may be used in physical space: the system is restrained to a specific part of conformational space (typically using a restraint potential along a desired degree of freedom) in a step-wise manner. Here too, if adjoining "windows" overlap, the free energy difference can be estimated. Umbrella sampling (US)[43] or boxed MD[44] are examples of those, and the weighted histogram analysis method (WHAM)[45] or multiple state-Bennet Acceptance Ratio (mBAR)[46] estimators are used to recover the free energy profile.

As higher temperatures promote the exploration of regions of the configurational space of lower probability, a class of methods exploits fluctuations induced by higher temperatures. Simulated annealing, in which the simulations are run at decreasing temperatures, until reaching the temperature of interest, fall into this category.[47] Adiabatic decoupling schemes are also based on this idea: a degree of freedom of interest is simulated at a high temperature to promote exploration of the space. It is also given a high mass to decouple it from the rest of the system and to avoid heat transfer. Adiabatic free energy decoupling (AFED)[48] and temperature-accelerated molecular dynamics (TAMD)[49] are example implementations of this principle.

Another simple strategy to enhance the sampling is to simply run multiple parallel simulations. Such an approach can be combined with the previous one: in temperature-based replica exchange,[50] simulations are run at different temperatures that overlap in energy space; periodically, exchanges between coordinates (or, conversely, temperatures) are attempted, and accepted with a Metropolis criterion. The replica at the lower temperature is exploited to recover the probability distribution. Replica-exchange can also be done between replicas that are governed by a different set of equations (Hamiltonian-replica exchange MD, H-REMD).[51]

Another approach that also hinges on the use of multiple replicas is to initiate simulations for different regions of the space. More interestingly, simulations can then be stopped when they sample a well-known region of the space, while those that discover "interesting" regions are seeded from. This class of method is called adaptive seeding.[52] This strategy is particularly useful when attempting to find pathways linking known states of the phase space. Transition path sampling,[53] transition interface sampling,[54] forward flux sampling,[55] adaptive multilevel splitting[56] are all versions of this scheme. Other pathway finding methods such as the string method and its derivatives can also be mentioned as popular enhanced sampling techniques.[49,57,58]

All the schemes cited in this paragraph require fine tuning of a set of parameters, a task that is far from trivial.

Most of these schemes have been proposed in an adaptive version, in which the enhanced sampling simulation parameters are modified on-the-fly. This generally allows researchers to take advantage of the principle that allows them to enhance the sampling while considering the particularities of the system of interest. Choosing an adaptive scheme and its parameters generally also comes with its own challenges and it appears important to understand the scheme to be able to wisely do so.

A class of analysis that has emerged in recent years and become extremely popular is Markov State Modeling (MSM).[59,60] It is thanks to this approach that it has become possible to combine independent simulations, whether they were obtained in an enhanced sampling way or not. Markov state models hinge on dividing phase space into small bins, and counting the transition between bins. The transition count matrix is then diagonalized and the eigenvalues sorted along decreasing values. The associated eigenvectors represent the stationary probability distribution and the slow decaying collective degrees of freedom. Clustering can be performed in this space to uncover macrostates (metastable states) and characterize the time scales associated with transitions between macrostates. This approach has been shown to be powerful, and significant effort has been placed on defining good practices and providing code to conduct this type of analysis.

As a final note, it is worth mentioning that the most novel and possibly successful approaches combine different strategies: for example, adiabatic decoupling can be combined with importance sampling in a scheme in which the hot degree of freedom is accelerated by a metadynamics-type of potential, in a scheme called unified free energy dynamics (UFED).[61] A difficulty in this area of research is to compare these schemes and choose one for a new system of interest. Most schemes are tested on the sampling of the configurational space spanned by alanine-dipeptide, and are shown to perform well in that case. An effort is now needed to define good benchmark systems and to systematically test these schemes.[62] We note that this is increasingly made possible by the increased practice to share the code associated with the methodological development.

7.2 Interfacing MD Simulations and Experimental Results

The raw output of an MD simulation is a list of positions and velocities of particles along time. Any experimental observable that can be expressed as a function of these, can be computed along time, an approach often referred to as "forward modelling". The probability distribution of a certain observable, or other statistical properties is then often of interest. However, MD simulations suffer from approximations owing to imprecisions in the physical model (force field) and the limited sampling afforded by the bounded computational resources. Experimental data can be leveraged to mitigate these imprecisions: sparse experimental data, coming, for example, from spectroscopies, is usually not sufficient to derive a high-resolution conformational ensemble, but the combination with MD simulation can be powerful. We list here different approaches that combine MD simulations and the incorporation of experimental data to compute high-resolution conformational ensembles and propose mechanistic insights. This type of approach can be referred to as integrative modeling and mainly takes two routes: in the first the conformational ensemble obtained from non-modified MD simulations is reweighted to match the experimental observables, while in the second, the equations of motion or the sampling scheme are modified to let the experimental data drive the simulation.[63]

7.2.1 Forward Modelling

7.2.1.1 Expectation Values

Given a probability density $P(y)$ of a continuous variable y, we are often interested in computing its ensemble average, y, and the uncertainty of the measurement associated with the variable, often estimated assuming a gaussian distribution of the observable, as the standard deviation, or the square root of the variance, $\sqrt{\sigma_y^2}$.

$$\langle y \rangle = \int \mathrm{d}y P(y) y$$

and

$$\sigma_y{}^2 = \int \mathrm{d}y P(y)(y - \langle y \rangle)^2$$

As MD simulations output time series, the ensemble average is calculated as the expectation value, in which the integral is computed over time. The ensemble average will only be correctly estimated if the ensemble sampled by the MD simulation represents the experimental ensemble well, that is if sampling is extensive enough.[64]

7.2.1.2 *Probability Distributions and Free Energies*

Upon thorough sampling of the configurational space, the probability distribution of a given degree of freedom Q can be estimated *via*:

$$P_\mathrm{Q} = \frac{\int_\mathrm{Q} \mathrm{d}x e^{-\frac{U(r)}{k_\mathrm{B}T}}}{Z}$$

In which k_B is the Boltzmann constant, T the temperature of the system, $U(r)$ is the internal energy of a configuration described by its atomic coordinates r, and Z is the partition function that lists all the possible configurations that the system can assume:

$Z = \int_\Omega \mathrm{d}x e^{-\frac{U(r)}{k_\mathrm{B}T}}$, in which Ω represents the entire configurational space. Computing Z requires exhaustive sampling of the configurational space, a situation that is never encountered in practice.

The free energy of the configuration is related to the probability distribution *via* the Boltzmann relationship: $\Delta G_\mathrm{Q} \propto -k_\mathrm{B}T\log P_\mathrm{Q}$, meaning that the free energy difference between two states A and B is related to the relative probability of the system to be found in a given state A (P_A) and in a state B (P_B) via:

$$\Delta G_\mathrm{AB} = -k_\mathrm{B}T\log\frac{P_\mathrm{B}}{P_\mathrm{A}} = -k_\mathrm{B}T\log\frac{\int_\mathrm{B}\mathrm{d}x e^{-\frac{U(r)}{k_\mathrm{B}T}}}{\int_\mathrm{A}\mathrm{d}x e^{-\frac{U(r)}{k_\mathrm{B}T}}}$$

Estimation of these differences requires a significant overlap between the two configurational ensembles, meaning that enhanced sampling is often needed.[30,65]

7.2.2 Reweighting Schemes

Often experimental data is sparse and the problem of determining a unique conformational ensemble matching the data is severely underdetermined. The maximum (relative) entropy principle seeks the smallest possible perturbation to the existing conformational ensemble (prior distribution, derived from MD simulations for example) that satisfies constraints derived from experiments.[66] The entropy, given a prior distribution $P_0(q)$, expressed as $S(P\,|\,P_0) = -\int dqP(r)\ln\frac{P(r)}{P_0(r)}$ is maximized subject to constraints that reflect the compliance of the data with the experimental observables:

$\int dqs_i(r)P(r) = \langle s_i(r)\rangle = s_i^{exp}$, in which s_i^{exp} is the experimental observables and $s_i(r)$ is the observable calculated from the MD simulation trajectory by forward modelling.

The entire probability distribution is also normalized, such that $\int dqP(r) = 1$.

In the reweighting approaches, this maximization is performed by finding the weights of the ensemble configurations that maximize the agreement with experimental data

$$\langle s_i^{calc}\rangle = \sum_{j=1}^{n} w_j s_i(r_j)$$

The weights are obtained by:

$$w_j = \frac{w_j^0 \exp\left[-\sum_i^m \lambda_i s_i(r_j)\right]}{\sum_{j=1}^{n} w_j^0 \exp\left[-\sum_i^m \lambda_i s_i(r_j)\right]}$$

In which λ_i are the Lagrange multipliers determined by finding the stationary points of the Lagrange function:

$$\mathscr{L} = S(P|P_0) - \sum_{i=1}^{M} \lambda_i\left(\int dqs_i(r)P(r) - s_i^{exp}\right) - \mu\left(\int dqP(r) - 1\right)$$

Reweighting can also be performed by selecting only a small number of components in the prior ensemble and weighting them equally.

The SAXS ensemble-refinement method,[67] the convex optimization for the ensemble reweighting method,[68] and the ENSEMBLE method[69] are all reweighting schemes. The maximum parsimony approach, on the other hand, seeks the smallest number of components that allow one to fit the data: among which are the ensemble optimization method (EOM),[70] the selection tool for ensemble representations of intrinsically disordered states (ASTEROIDS),[71] the sparse ensemble selection (SES) method,[72] the sample and select method,[73] the maximum occurrence method,[74] the minimal ensemble search method,[75] and the basis-set supported SAXS reconstruction (BSS-SAXS) method.[76]

In these approaches, the inaccuracy of the experimental measurement is not explicitly taken into account. Bayesian approaches, on the other hand, weigh all the ingredients of the model (the experimental measurement, the forward modeling procedure) according to their certainty. These approaches cover the Bayesian ensemble refinement method,[77] the Bayesian ensemble SAXS (BE-SAXS) method,[78] the experimental inferential structure determination (EISD) method,[79] the Bayesian energy landscape tilting method,[80] the integrated Bayesian approach,[81] the method designed by Sethi *et al.*,[82] the reference ratio method,[83] the Bayesian inference of conformational populations (BICePS) method,[84] the Bayesian inference of EM method (BioEM),[85] the Bayesian weighting (BW) method[86] and the Bayesian/maximum entropy method (BME).[87]

The reweighting schemes assume that the prior distribution largely overlaps with the final, reweighted distribution. If this is not the case, an approach in which the simulation is explicitly driven to match the simulation data is better suited, as described below.[63]

7.2.2.1 Guiding the Sampling Using Experimental Data

The most straightforward approach to exploit experimental observables to drive molecular dynamics simulations, whether to correct for inaccuracies in the force field, or to discover states that were not yet modeled, is to add external forces to the forces derived from the force field in the equations of motion:

$$\nabla U^{ff}(x(t)) + \nabla U^{\exp}(x(t)) = f^{ff}(x(t)) + f^{\exp}(x(t)) = m \cdot \frac{\mathrm{d}^2 x}{\mathrm{d}t^2}$$

In which *ff* represents the potential/forces arising from the force field and exp represents the ones originating from the experimental data.

For example, molecular dynamics flexible fitting (MDFF) aims to flexibly fit atomic structures into electron density maps. The external forces are derived directly from the density map.[88,89] This approach, however, does not work well when the density is of a high resolution, as parts of the system may get trapped in rugged density regions, the external force confining particles behind high free energy barriers. Cascade MDFF and resolution exchange MDFF aim to alleviate these drawbacks.[90]

Such pitfalls can be avoided by resorting to thermodynamic sampling. In correlation-driven molecular dynamics (CDMD),[91] for example, the agreement between the electron density map computed from the available structural ensemble and the experimental electron density map is evaluated by computing their correlation coefficient. The potential derived therefrom is added to the force field, and the weight of the external potential is gradually increased during the simulation to avoid trapping in regions of high-resolution.

Maximum entropy methods may be used here too: the external forces are functions of the calculated experimental observables and their intensities are determined by the Lagrange multipliers described above. New methods take into account the distributions of experimental data by introducing additional restraints, sometimes in the form of metadynamics bias potentials.[92,93] To further improve the convergence of the conformational ensemble one can consider a replica-based approach, in which the restraint acts on the instantaneous average of the observable among the replicas.[94–96]

Similar to the reweighting schemes, Bayesian variants of the "driving schemes" have been suggested. Those allow researchers to account for uncertainties in the data and forward modelling, among these methods are, multi-state Bayesian modelling,[97,98] the Bayesian ensemble refinement method[77] and the metainference method.[99] The two latter methods also involve simulating a set of replicas in which the restraint derives from prior information and from the agreement between the experimental data and the average of the observable over the replicas.

Finally, it has been shown that MD simulations can be used in combination with *ab initio* modeling (building a model from the experimental data directly, using Rosetta,[100,101] for example),[102] in an iterative procedure that improves sampling and can yield better agreement with experimental data.[103]

7.3 Case Studies: Applications to Membrane Protein Function

Membrane proteins perform tasks that are crucial to the cell and can be difficult to study by experimental methods owing, in part, to their low abundance in physiological settings and the particular environment they function in. Thus, supplementing wet-lab experiments with computational methods such as molecular dynamics simulations has proven important towards obtaining mechanistic insights.

Membrane proteins enable the communication between the cell and the extracellular medium. The plasma membrane, of which the main ingredient is a lipid bilayer, would mostly isolate the cell from polar, charged and large chemical species, if it wasn't for dedicated membrane proteins that selectively transport molecules or translate the binding of chemical species into secondary signals that have dramatic consequences for the cell. The plasma membrane thus plays an essential role in both shielding the cell's intracellular milieu to enable it to perform its function, while allowing it to respond and act on the extracellular milieu it lives in.

In this third part, we review a few studies of mechanistic insights made possible by a joint MD simulation/experimental approach. We show in particular how the approaches were synergistically exploited, and highlight how the experimental observables can be predicted by MD simulations and tested using wet-lab experiments (first three case studies), how the MD simulation can sometimes help to interpret experimental measurables and thereby generate mechanistic insights (fourth and fifth case study), or else how experimental measurements can be used to drive simulations, thus enhancing the possible insights (last case study).

7.3.1 Testing Predictions from MD Simulations

7.3.1.1 KscA Ion Channel Inactivation: Testing Predictions from MD Simulations Using Electrophysiology Recordings

Ion channels conduct ions across the cell membrane in a passive manner, that is, down their electrochemical gradient, and in a way that is regulated by selected stimuli: ion channels can open and close (a process called gating) in response to changes in the membrane potential, pH, ligand binding, temperature and so forth. They can also be classified according to their selectivity: indeed, at the heart of their function is the ability to discriminate between the ion types they let through: cations *versus* anions, monovalent *versus* divalent, or even more subtle changes such as Na^+ *versus* K^+. Patch clamp electrophysiology is the wet-lab technique of choice that is used to study ion channels in a quantitative manner. Electrodes placed on both sides of the membrane record currents due to ions passing through the channels in a time-resolved way.

Most channels open in response to an incoming stimulus, and start letting ions through, before spontaneously closing when the stimulus is left on over an extended period of time. This phenomenon is called desensitization, or inactivation. The structural and dynamical basis of these phenomena is of great interest, but the time scales over which these processes occur are often long compared to the capabilities of MD simulations.

A spontaneously inactivating channel that has been heavily studied is the pH-gated, K^+-selective channel KcsA. This channel is made of the tetrameric assembly of two transmembrane helices, which form a hydrated central pore, and gating occurs at the intracellular side, in which the transmembrane helices gather into the so-called bundle-crossing (Figure 7.2a). A re-entrant loop containing a short helix forms the selectivity filter, which allows the channel to select K^+ ions over other more abundant species such as Na^+. Crystal structures have captured the channel's selectivity filter in essentially two states: in high potassium concentrations and when the intracellular gate is closed, the selectivity filter delineates five K^+ ion binding sites arranged in a straight line; in a low potassium concentration or when the intracellular gate is wide open, the selectivity filter adopts a more "collapsed" (or "pinched") configuration, in which the central binding sites vanish (Figure 7.2b). MD simulation studies have shown that the former structure can be conductive for K^+ ions, whereas the latter is not. Although pH-dependent gating has been ascribed to intracellular gate conformational changes, inactivation has been proposed to be related to changes in the selectivity filter configuration. As the conformational change involves re-arrangements that are only a few Ångström large, the reason why recovery from inactivation occurs on the milli-second time scale has remained a mystery.

In a 2013 study, the groups of Roux and Perozo joined forces to provide an answer to this question.[104] MD simulations of the KcsA channel in a state

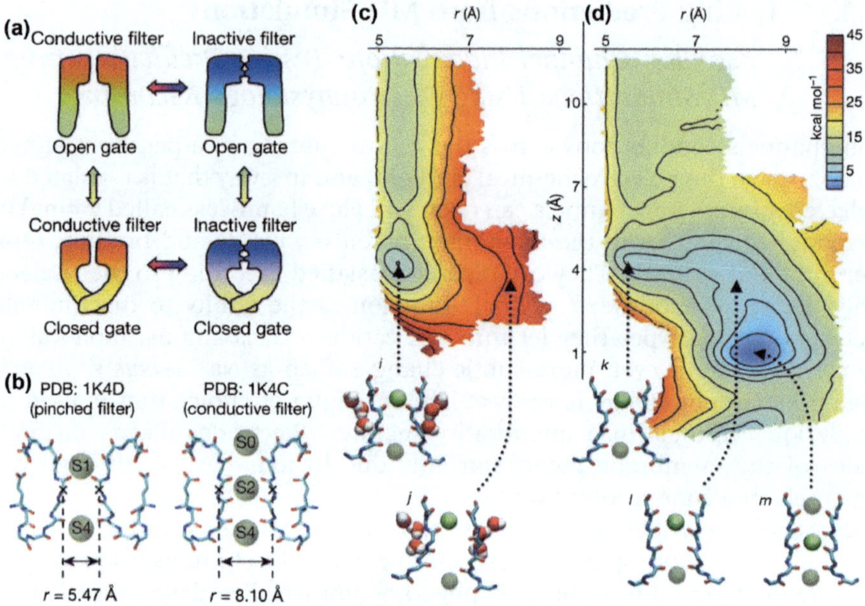

Figure 7.2 The effect of water molecules on the conformational landscape of the selectivity filter in potassium channels. (a) Schematic depiction of the four dominant functional states of the KcsA channel, highlighting the change in the selectivity filter from a conductive to an inactive state, and of the intracellular gate (bundle crossing) from closed to open. (b) The conductive state of the selectivity filter compared to the pinched state is characterized by a small increase in the minimum inter-subunit distance between the backbone atoms of the filter and an increase in the ion occupancy. (c) Free energy landscape calculated with inactivating water molecules present behind the selectivity filter. The pinched filter rests in a free energy minimum with a K^+ in position S1 (snapshot i). The transition from a pinched to a conductive conformation (snapshot j) of the selectivity filter is impeded by a ~25 kcal mol^{-1} free energy barrier relative to the local minimum, resulting in an unstable conformation of the conductive filter. (d) Free energy landscape calculated with the inactivating water molecules absent. The pinched filter with a K^+ ion in position S1 (snapshot k) recovers spontaneously, following the downhill slope of the free energy landscape. The filter recovers to a conductive conformation by moving first to an open conformation (snapshot l) before ions in the filter adopt a conductive configuration (snapshot m). The horizontal reaction coordinate r describes the width of the selectivity filter and is defined as the cross-subunit pinching distance between the Cα atoms of selectivity filter residue Gly 77. The vertical reaction coordinate z is the height of a K^+ ion relative to the center of mass of the selectivity filter.
Adapted from ref. 104 with permission from Springer Nature, 2013.

with a pinched selectivity filter and a closed gate revealed that this state remained stable over the tens of micro-second time scale, despite experimental evidence that equilibration to a state with a conductive selectivity filter and a closed gate should be spontaneous. The MD simulations showed that three water molecules per subunit were spontaneously recruited to the back of the selectivity filter, and their localization there was presumed to be important for the stability of this state and for the long time taken for recovery from the inactivated state (Figure 7.2c). To further confirm this, Roux *et al.* conducted enhanced sampling MD simulations, using an umbrella sampling approach in a replica-exchange scheme (RE-US). Umbrella potentials were placed to constrain the positions of the ions along the channel axis (z), and the diameter of the selectivity filter at the potential pinching site (r), allowing construction of a free energy landscape in this two-dimensional space. These simulations revealed that the presence of the three water molecules bound to the back of the selectivity filter made the conductive state higher in free energy by tens of kcal per mol, thus providing an explanation for the long time scales taken for recovery from inactivation (Figure 7.2c and d). The predictions made from the MD simulation were therefore that reducing the osmotic pressure by depleting the extracellular medium from water should promote recovery from inactivation by promoting release of the three water molecules from the binding site in the back of the selectivity filter. This was tested using electrophysiology, by comparing inactivation rates measured in the absence and in the presence of 2 M external sucrose. Indeed, the rates were found to be higher in the presence of sucrose, confirming that when it was favorable for water to move out of their binding pockets, recovery became faster. This paper is a very elegant illustration of using MD simulations to make a simple prediction that can be tested using electrophysiology, ultimately providing an insight into the structural and dynamical basis of a physiologically important phenomenon.

7.3.1.2 GPCR-mediated Arrestin Activation: Testing Predictions from Large-scale Atomistic Simulations Using Fluorescence Spectroscopy

G-protein coupled receptors (GPCRs) trigger intracellular signal transduction pathways upon ligand binding from the extracellular domain. Ligands range from small molecules such as neurotransmitters, odorant or light activated compounds to peptidic compounds such as specific hormones. The most canonical signaling pathway proceeds *via* G-protein binding, but GPCRs have also been shown to bind arrestin, which often results in G-protein coupled receptor (GPCR) internalization and degradation. Arrestin binding is promoted by GPCR activation and by phosphorylation of its intracellular tail.

Arrestin is made of two domains, the N- and the C-domains, and activation involves twisting of the C-domain with respect of the N-domain by approximately 20 Å (Figure 7.3a and b). To investigate how rhodopsin (a light-sensitive GPCR) activation and phosphorylation contribute to arrestin activation, Dror *et al.* conducted large-scale all-atom simulations of

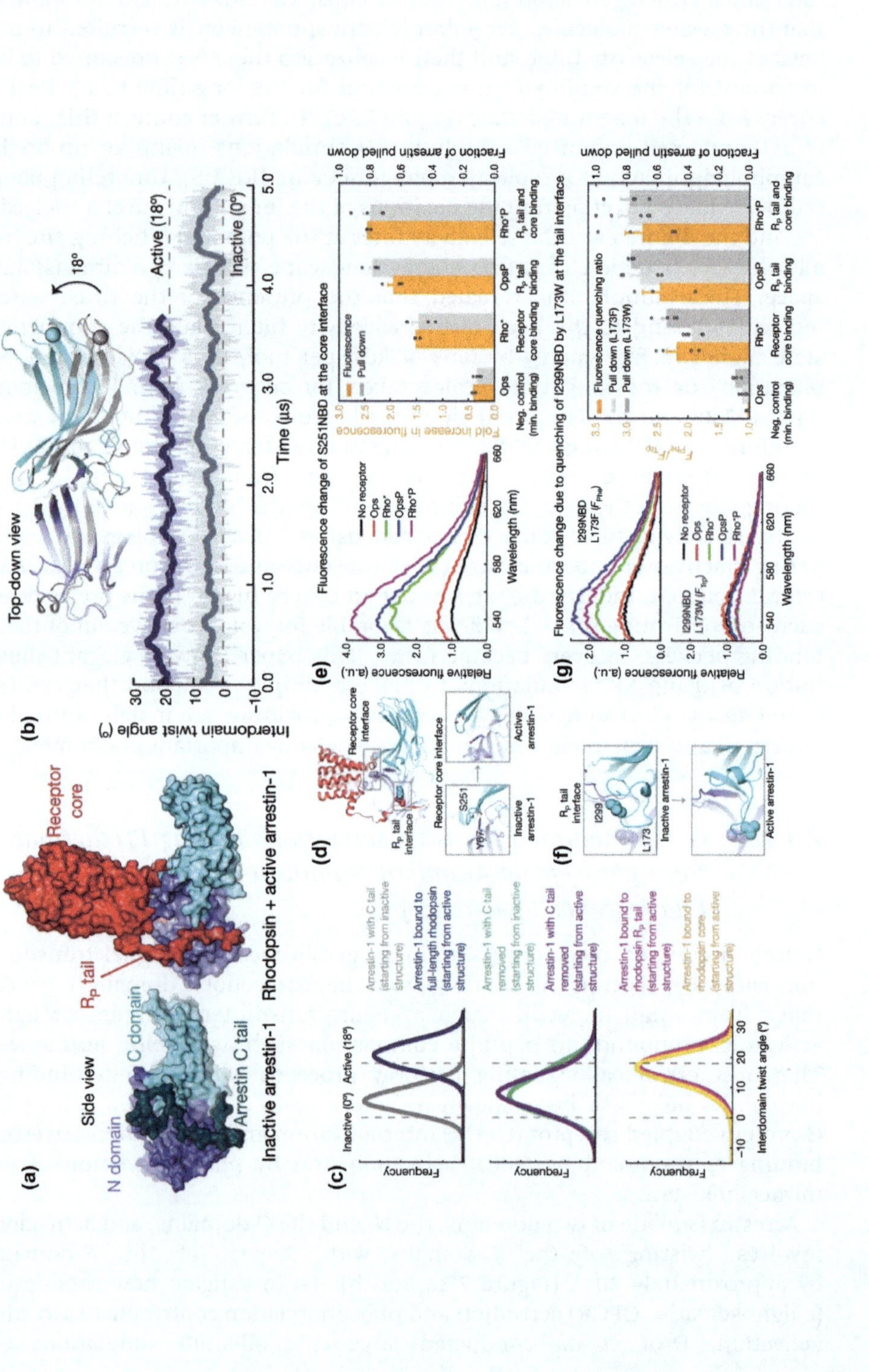
(a)
Side view
N domain
C domain
Arrestin C tail
R_P tail
Receptor core
Inactive arrestin-1
Rhodopsin + active arrestin-1
(b)
Top-down view
18°
Active (18°)
Inactive (0°)
Interdomain twist angle (°)
Time (μs)
(c)
Frequency
Inactive (0°)
Active (18°)
Arrestin-1 with C tail (starting from inactive structure)
Arrestin-1 bound to full-length rhodopsin (starting from active structure)
Arrestin-1 with C tail removed (starting from inactive structure)
Arrestin-1 with C tail removed (starting from active structure)
Arrestin-1 bound to rhodopsin R_P tail (starting from active structure)
Arrestin-1 bound to rhodopsin core (starting from active structure)
Interdomain twist angle (°)
(d)
Receptor core interface
R_P tail interface
S251
Y67
Inactive arrestin-1
Active arrestin-1
(e)
Fluorescence change of S251NBD at the core interface
Relative fluorescence (a.u.)
Wavelength (nm)
No receptor
Ops
Rho*
OpsP
Rho*P
Fold increase in fluorescence
Fraction of arrestin pulled down
Fluorescence
Pull down
Neg. control (min. binding)
Receptor core binding
R_P tail binding
R_P tail and core binding
(f)
R_P tail interface
I299
L173
Inactive arrestin-1
Active arrestin-1
(g)
Fluorescence change due to quenching of I299NBD by L173W at the tail interface
I299NBD L173F (F_{Phe})
I299NBD L173W (F_{Trp})
F_{Phe}/F_{Trp}
Fluorescence quenching ratio
Pull down (L173F)
Pull down (L173W)

arrestin bound to various parts of the receptor, a set of simulations was performed in which the active receptor had its phosphorylated tail removed, and a set of simulations was performed in which the sole phosphorylated tail was present, in the absence of the receptor core.[105] Both conditions yielded structures that were closer to the active structure of arrestin, but less active than when both the receptor core and the phosphorylated tail were both bound (Figure 7.3c). The interactions responsible for stabilizing the active arrestin state from the receptor core were identified as contacts to intracellular

Figure 7.3 Activation of arresting by GPCRs *via* binding to the receptor core and to the phosphorylated tail. (a) Inactive-state (left; PDB 1CF1) and receptor-bound, active-state arrestin-1 (right; PDB 5W0P). (b), Upon activation, the arrestin C domain twists with respect to the N domain. In simulations of arrestin-1 starting from its inactive state, the interdomain twist angle remains close to 0° (grey trace), while in simulations starting from the active state with rhodopsin bound, the twist angle remains close to 20° (blue trace). (c) Distributions of interdomain twist angles under different simulation conditions: grey, arrestin-1 with C tail (starting from the inactive structure); blue, arrestin-1 bound to full-length rhodopsin (starting from active structure); green, arrestin-1 with C tail removed (starting from inactive structure); purple, arrestin-1 with C tail removed (starting from active structure); magenta, arrestin-1 bound to rhodopsin RP tail (starting from active structure); yellow, arrestin-1 bound to rhodopsin core (starting from active structure). Removal of the arrestin C tail leads to an increased range of interdomain twist angles. Binding of either part of the receptor in simulation substantially increases the fraction of time that arrestin spends in active conformations. Binding of the entire receptor has an even stronger effect. (d) Fluorescently labelled arrestin mutants used to monitor conformational changes at the core interface (S251NBD) and interdomain twisting in arrestin mutant S251NBD separates a quenching tyrosine at site 67 from the fluorophore at site 251, resulting in increased fluorescence. (e) Increased fluorescence relative to Ops was observed in the presence of light-activated, non-phosphorylated Rho*, inactivated phosphorylated opsin OpsP and light-activated, phosphorylated Rho*P (orange bars show fold-increase relative to unbound condition). (f) Fluorescently labelled arrestin mutants used to monitor conformational changes at the phosphorylated tail interface (I299NBD/L173F and I299NBD/L173W). Disruption of the polar core is accompanied by a movement of the gate loop, which can be detected by quenching of the NBD fluorophore at site 299 by a tryptophan at site 173. As a control, site 173 is also mutated to non-quenching phenylalanine. The quenching ratio (FPhe/FTrp) is calculated from the spectra of I299NBD/L173F and I299NBD/L173W mutants. (g) Substantial increases in the quenching ratio over Ops were observed in the presence of the receptor core, phosphorylated tail or both, indicating that all three favored active-like conformations of the gate loop (Rho*, Rho*P, *, inactivated non-phosphorylated opsin Ops, OpsP). Centrifugal pull-down analysis for all mutants (grey bars) indicates that fluorescence quenching corresponded to arrestin activation and receptor binding. Adapted from ref. 105 with permission from Springer Nature, Copyright 2018.

loops that stabilized the so-called C-loop from arrestin in an active-like state (Figure 7.3d). The interactions responsible for stabilizing the active arrestin state from the phosphorylated tail, on the other hand, were attributed to a specific contact with a positively charged residue that stabilizes the "gate loop" in an active-like state (Figure 7.3f). The prediction from MD simulations was thus that both the motion of the gate loop and of the C-loop were coupled to the interdomain twisting characteristic of activation.

Dror *et al.* sought to confirm this prediction using fluorescence spectroscopy (Figure 7.3e and g), a label was introduced at the core interface (S251, reporting on the C-loop conformation) and was quenched by the endogenous Y67 in the inactive state (Figure 7.3d), although the fluorescence increased in the active state. The receptor core binding and phosphorylated tail binding were shown to both independently promote arrestin activation of the C-loop (Figure 7.3e). A label introduced at the tail interface (I299) was combined with a mutation of a nearby residue (L173) to tryptophan (W) to promote quenching, or to non-quenching phenylalanine (F), and the quenching ratio was shown to increase as the gate loop moved to the active conformation (Figure 7.3f). Here too, binding of the receptor core and the phosphorylated tail independently promoted transitions to the active state (Figure 7.3g). These observations are consistent with several recent studies, but also provide an explanation for how some receptors that lack a C-terminal phosphorylation can still undergo arrestin-mediated internalization. It also shows that the regions of the GPCRs implicated in activation of arrestin and G-protein differ, even though similar active conformations of GPCRs are usually able to bind both arrestin and G-proteins. This study again provides a clear example of how MD simulations can generate simple predictions that can be tested by straightforward experiments.

7.3.1.3 Binding Mode of Allosteric Modulators of the M2 Muscarinic Receptor: Testing Predictions from MD Simulations Using a Radioligand Binding Assay

The GPCR function can also be regulated by allosteric drugs, which bind away from the canonical orthosteric ligand binding site and affect both the binding affinity of the usual ligand, and the activation profile of the receptor. Allosteric modulators can act as positive allosteric modulators, in which case they enhance the effect of the orthosteric ligand, or as negative allosteric modulators, in which case they inhibit the effect of the orthosteric ligand. At the time of the study by Dror *et al.*, no crystal structures of a bound allosteric modulator had been resolved, and the authors used long atomistic molecular dynamics simulations to characterize the binding mode of several allosteric compounds and understand the mechanism of their action.[106]

Through long atomistic MD simulations, Dror *et al.* observed reproducible binding of C_7/3-phth, a negative allosteric modulator, to a pocket in the extracellular vestibule of the M2 receptor (Figure 7.4a). The pose was

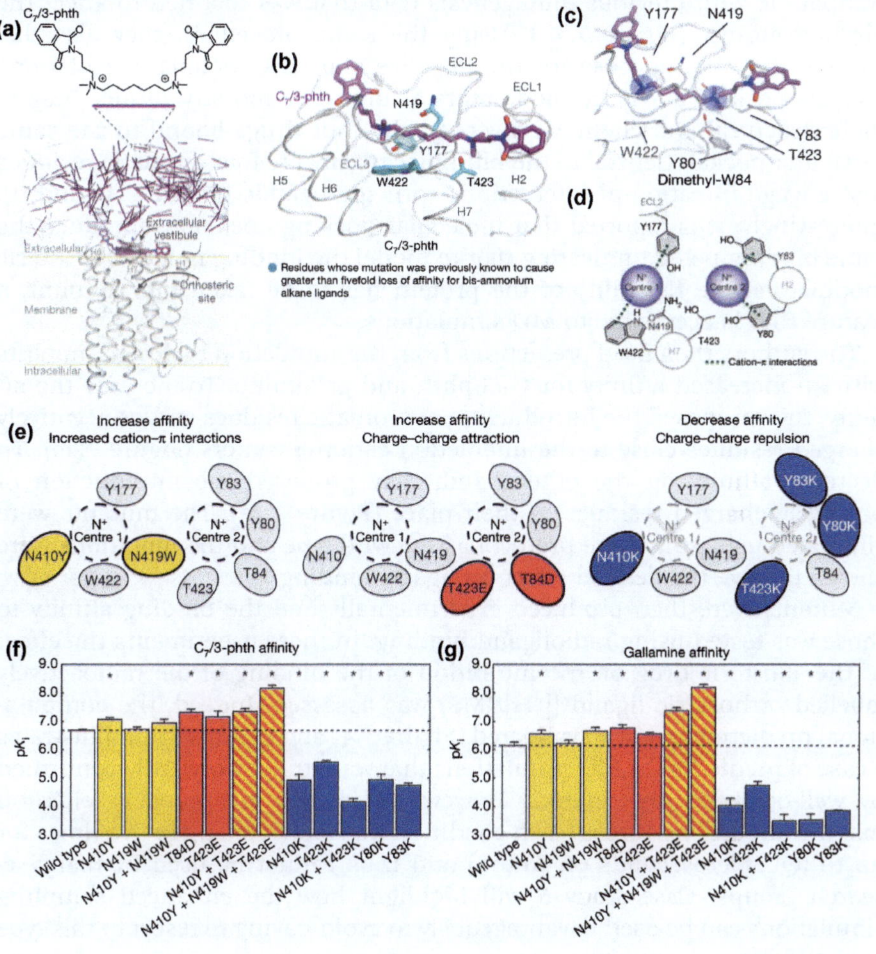

Figure 7.4 Mode of action of allosteric modulators binding to the M2 muscarinic receptor. (a) C7/3-phth diffuses freely before binding stably in the M2 receptor extracellular vestibule. The C7/3-phth position at various times is represented by a stick connecting its two ammonium groups. (Top: C7/3-phth structure.) (b) Typical bound pose of C7/3-phth (purple). Residues known to reduce binding greater than fivefold upon mutation (cyan) all contact C7/3-phth more than 90% of the time after it binds. (c) Typical bound pose for dimethyl-W84. Ammonium groups are highlighted (blue disks). (d) Schematic representation of the ammonium binding centers. (e) Aromatic residues (N410Y, N419W) contribute extra cation–p interactions at center 1, increasing affinity. Anionic residues (T423E, T84D) attract an ammonium group at center 2, increasing affinity. Cationic residues (N410K, N423K, Y80K, Y83K) near centers 1 and 2 repel or compete with cationic modulators, decreasing affinity. (f)–(g) Mutagenesis results; radioligand binding experiments ([3H]NMS displacement), performed on whole cells expressing wild-type and mutant human M2 receptors, confirmed computational predictions. Adapted from ref. 106 with permission from Springer Nature, Copyright 2013.

compatible with previous mutagenesis data that was reported to affect the binding affinity (Figure 7.4b). Using the same procedure, they also discovered the binding pose for the negative allosteric modulator gallamine and the positive allosteric modulators alcuronium and strychnine. Despite their structural and chemical diversity, the four drugs bound to the same vestibular pocket. Central to the binding capacity are four aromatic residues that engage in cation-pi interactions with the ligands (Figure 7.4c and d). Interestingly, it is reported that molecular docking does not reproduce the same binding poses, indicating that to model the binding mode of allosteric modulators, the flexibility of the protein has to be taken into account, a feature that is accessible to MD simulations.

The authors then used predictions from the simulation to design mutants with an increased affinity for C_7/3-phth and gallamine. To increase the affinity, they proposed the introduction of aromatic residues and/or negatively charged residues close to the interaction cation-π centers (Figure 7.4e). To decrease affinity on the other hand, they proposed the introduction of positively charged residues in their place (Figure 7.4e). The mutants were simulated using the same protocol as the wild type and the mutations were shown to have the desired effect on ligand binding.

Mutants were then produced experimentally and the binding affinity to those was tested using radioligand binding: in these experiments the effect of the allosteric drug on the inhibition of the binding of the radioactively labelled orthosteric ligand ([^{3}H]NMS) was assessed. Indeed, the computational predictions were all validated (Figure 7.4f and g). This again illustrates a case of prediction by MD simulations that was straightforwardly confirmed by well-designed experiments. However, this type of approach relies on major computer resources (each condition was replicated up to five times for up to ten microseconds on Anton) and thus cannot be used by many research groups. Case study 5 will highlight how the enhanced sampling simulations can be used advantageously to avoid having to resort to this type of computer resource.

7.3.2 Using MD Simulations to Interpret Experimental Measurements

7.3.2.1 Lipids Interacting with Na^+/H^+ Antiporters: Using MD Simulations to Interpret Mass Spectrometry Measurements

Contrary to ion channels, transporters shuttle substrates across the membrane using an alternating access mechanism: the protein alternates between inward facing and outward facing conformations, without ever displaying a pore open to both sides of the membrane simultaneously. Some transporters transport their substrate in a passive manner, down their electrochemical gradient, whereas others harness energy from an external source to transport their substrate against their electrochemical gradient.

Secondary active transporters, specifically, couple the transport of one substrate along the electrochemical gradient (providing a source of energy) to the transport of another against it (exploiting the energy to conduct an unfavorable process).

Na^+/H^+ antiporters regulate the intracellular pH by harnessing the energy coming from the transportation of Na^+ down their electrochemical gradient. Structurally, this is made possible by the so-called elevator mechanism: these transporters function as homodimers of 12 or 13 TM segment monomers, the static scaffold domain is made by the dimeric assembly of segments (TM(-1),) TM1-2 and TM 7-9, while the core ion-translocation domain is made by segments TM3-5 and TM10-12. The function of these transporters thus hinges on the fact that the scaffolding domain remains relatively static while the core domain translates across the membrane. It is of particular interest to understand the role lipids play in supporting this function.

In their 2017 paper, Landreh *et al.* used mass spectrometry to probe the interaction strength between NapA, a bacterial Na^+/H^+ antiporter that shows tight association between monomers, and lipid molecules.[107] They first found that the transporter is stable in detergent, in the absence of lipids (Figure 7.5a), and that titrating in lipids showed a concentration-dependent increase of lipid protein complexes with no saturation, indicative of non-specific binding of lipids to the protein. Lipids with negatively charged headgroups showed a more stable interaction with the antiporter than zwitterionic ones.

In a next step, Landreh *et al.* conducted collision induced unfolding. In these experiments, they monitored the collision cross section, which reports on the size of the assembly reaching the detector, as a function of the activation energy. Compact states reflect folded proteins while expanded states are indicative of more unfolded ones. More stable assemblies will remain compact at higher activation energies than less stable ones. These experiments showed that lipid-free assemblies require less energy to unfold than assemblies containing a single phosphatidyl-ethanolamine (PE) molecule, and adding a second PE molecule further stabilizes the assembly (Figure 7.5b).

Atomistic MD simulations were then used to provide a rationale for these observations at a molecular level. First, simulations of the NapA dimer were conducted in a PE bilayer, to observe the lipid distribution: interestingly, lipids did not distribute evenly around the protein, and instead tended to cluster at the dimer scaffold domain, and at the hinge region between the scaffold and core domains (Figure 7.5c). The membrane was also compressed around the scaffold domain because the hydrophobic thickness of the protein in this region was smaller than in the rest of the protein (Figure 7.5d). The interaction between the lipid headgroups and the protein was mediated by lipid facing positively charged residues, providing an explanation for the larger binding affinity of the negatively charged lipids than the zwitterionic ones.

A second set of MD simulations was subsequently conducted, this time *in vacuo*, to gain an insight into the behavior of the protein in the mass spectrometry conditions. Simulations in the presence of annular PE lipids were compared at increased temperatures, and NapA remained folded at lower temperatures while it started unfolding at higher temperatures, an observation compatible with the mass spectrometry results (Figure 7.5e). Increasing the temperature also led to the detachment of lipid molecules: the hydrophobic chains detached first, while the polar headgroups remained

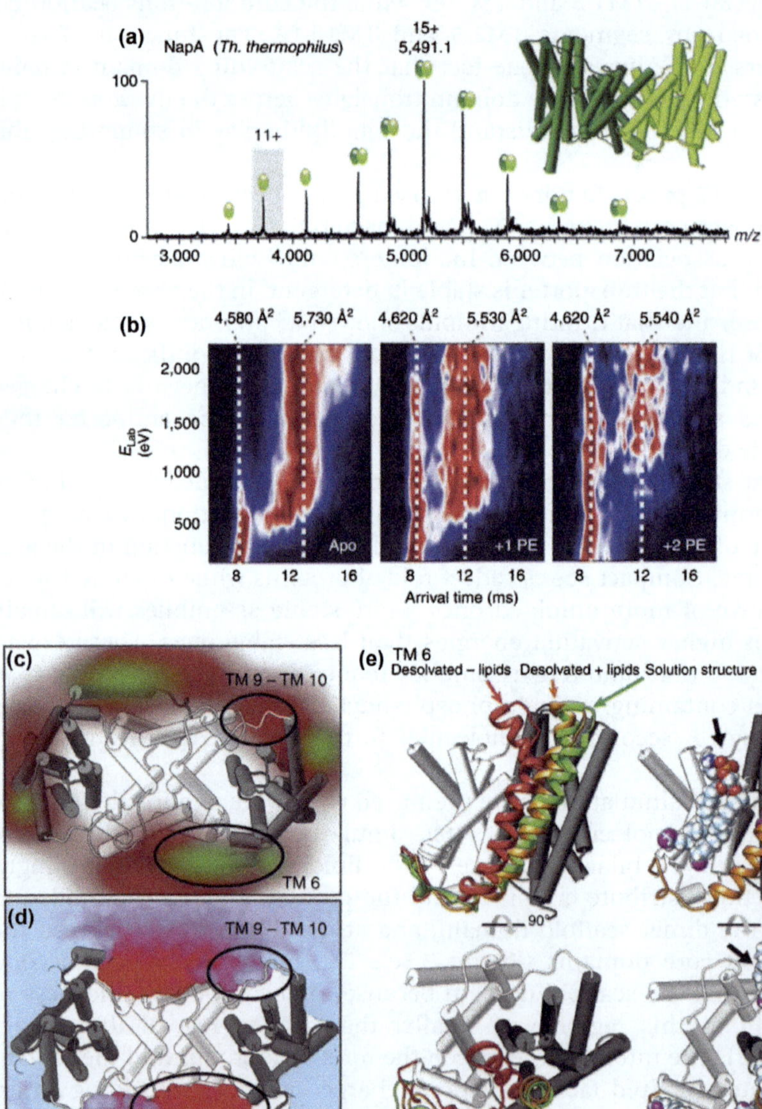

bound for longer *via* hydrogen bonds. These weak interactions are thought to mitigate the effect of collisions on structure destabilization. In this work, mass spectrometry thus generated macroscopic observations on the effect of lipids on protein stability, and MD simulations provided the necessary molecular level counterpart to those observations.

7.3.2.2 Secondary Active Transporter BetP Conformational Change: Interpreting Data from DEER Spectroscopy Using Enhanced Sampling MD Simulations

The sodium-coupled betaine transporter BetP belongs to the family of secondary active transporters. This bacterial transporter is involved in osmotic stress detection and response and acts as a symporter: betaine is transported at the same time as sodium, along the electrochemical gradient of sodium.

Structurally, it belongs to a different family than the Na^+/H^+ exchanger presented above: BetP is a homotrimer and each monomer functions as a betaine transporter. Each monomer consists of 12 helices arranged in a so-called LeuT fold (Figure 7.6a). The monomer consists of three domains, the bundle domain, consisting of TM1-2 and TM6-7, the hash motif formed by helices TM3-4 and TM8-9 and the mobile elements forming the so-called

Figure 7.5 Lipid binding stabilizes the 3D assembly of the sodium proton exchanger NapA. (a) Mass spectrum of NapA following release from detergent micelles, presented as a bar graph, in which each bar represents an ion having a specific mass-to-charge ratio (*m*/*z*). The length of the bar indicates the relative abundance of the ion, with the most intense ion assigned an abundance of 100. The charge of the protein ion is indicated above the important peaks. NapA appears as an intact dimer following release from detergent micelles, in agreement with the crystal structure. (b) Unfolding trajectories (arrival time as a function of the ion kinetic energy, E_{Lab}) of the 12+ charge state of the lipid-free and PE-bound NapA show that the protein–lipid complexes exhibit broader unfolding transitions and that the native-like population is lost at higher collision energies than the lipid-free form, the arrival time for the folded and unfolded protein are marked as white dashed lines. (c) MD simulations of NapA in a PE bilayer show high phosphate densities around the dimer and core domain connections (TM 9–10, red, and TM 6, green). (d) The region around the core–dimer interface exhibits the highest degree of membrane compression (red). (e) Lipid binding stabilizes the domain contacts in the gas phase. In lipid-free NapA, TM 6, which connects core and dimer domains, unfolds rapidly (red). In the lipid-bound system (orange), lipids prevent structural collapse *via* head-group interactions and by intercalating into the interface between the core and dimer domains (black arrows). Core and dimer domains are shown in black and white, respectively.

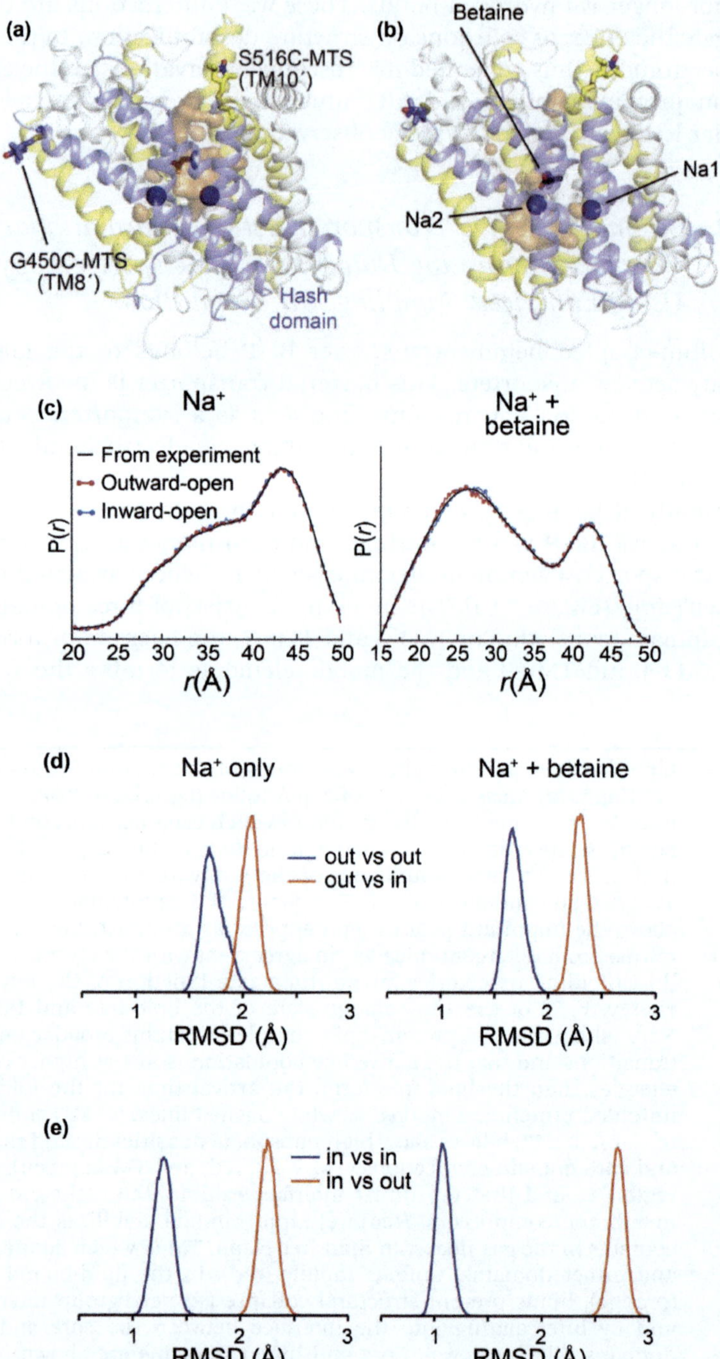
(a)
S516C-MTS
(TM10′)
G450C-MTS
(TM8′)
Hash
domain
(b)
Betaine
Na1
Na2
(c)
Na⁺
Na⁺ +
betaine
From experiment
Outward-open
Inward-open
P(r)
r(Å)
(d)
Na⁺ only
Na⁺ + betaine
out vs out
out vs in
RMSD (Å)
(e)
in vs in
in vs out
RMSD (Å)

thin gates TM5 and TM10. The transition between the outward-facing and the inward-facing conformations *via* occluded states involves many re-arrangements, and capturing structures using structural biology techniques does not allow painting of the whole picture: it can be difficult to assign the functional state of a specific structure, the conditions under which the structures were obtained may have modified the protein away from its physiological state (for membrane proteins, the solubilization in detergents can be particularly disruptive), and structures do not provide information on the ensemble properties of the protein, such as how populated a given conformational state is under specific conditions of ligand binding for example.

Spectroscopy techniques have been valuable in answering these questions, as a complement to structural biology techniques. In their 2019 paper, Leone *et al.* used double electron–electron resonance spectroscopy (DEER, also called pulsed electron–electron double resonance (PELDOR)), a form of electron paramagnetic resonance (EPR) spectroscopy, to probe the conformational ensemble of BetP in the presence and absence of betaine.[108]

Figure 7.6 Using MD simulations to define the betaine-dependent conformational ensemble of BetP compatible with DEER spectroscopy experiments. (a) and (b) Simulation system for spin-labeled monomeric BetP. Snapshots indicate the location of the spin labels with betaine bound in either outward-open (a) or inward-open (b) conformations. The protein is viewed from within the plane of the membrane. The MTS labels and bound betaine are shown as sticks, and sodium ions are shown as blue spheres. Water molecules within 10 Å of the ligand are shown as an orange surface, highlighting the pathways on the extracellular or intracellular sides. Helices 3′, 4′, 8′ (including G450C), and 9′, which comprise the so-called hash domain, are colored blue. Helices 1′, 5′, 6′, and 10′ (including S516C) are colored yellow. (c) Compatibility of simulated and experimental distance distributions. The probability of a distance P(r) is plotted *versus* distance (r). The PELDOR-based distances (black lines), measured in the presence of 500 mM NaCl or 300 mM NaCl plus 5 mM betaine, are compared with distances obtained in 1 μs-long EBMetaD MD simulations, performed for BetP monomers in the presence of two sodium ions or with two sodium ions plus a betaine substrate. The simulations were started with BetP structures of either outward-facing (PDB accession no. 4LLH chain A, red circles) or inward-facing (PDB accession no. 4C7R, blue circles) conformations. (d) and (e) The EBMetaD bias does not change the overall conformation of the protein. The structural similarity (in RMSD) of each simulated ensemble with respect to the two extreme conformations of BetP is shown. EBMetaD simulation trajectories initiated with either outward-facing (d) or inward-facing (e) conformation in the presence of two sodium (left) or two sodium and betaine (right) are compared with either the initial structure (blue) or with the structure of the opposite state (orange).
Adapted from ref. 108, https://doi.org/10.1085/jgp.201812111, under the terms of a CC BY 4.0 license, https://creativecommons.org/licenses/by/4.0/.

DEER spectroscopy involves introducing spin labels, often nitroxide radicals, at two engineered cysteines (in a cysteine-less background) and measuring their interaction to derive the probability distribution of the spin-to-spin distance. In the BetP experiments, the probes were placed at the extracellular end of TM8 and TM10, two positions that move closer to one another by a few Å in the inward-facing conformation (Figure 7.6b).

Leone *et al.* showed that when applying a library of probe orientations to static structures obtained from crystallography, the difference in the distance between the crystal structures is much smaller than the difference in distributions inferred from the DEER measurement. This led them to conduct all-atom MD simulations in which the probes were modeled explicitly, the protein was embedded in its near-native environment, and to use the enhanced sampling method called EBMetaD (see Section 7.2.2) to guide the simulations towards a conformational ensemble that is compatible with the experiment (Figure 7.6c).

Simulations initiated both in the inward facing and in the outward facing conformations were able to converge to the distribution of distances inferred from DEER, both in the presence and absence of betaine (Figure 7.6c). Interestingly, the enhanced sampling ended up modifying the sampling of the distribution of the spectroscopic probe conformations rather than modifying the conformation of the protein towards the opposite state (Figure 7.6d and e). In EBMetaD, it is possible to measure the force that was applied to match the experimental distribution, the smaller the force, the less work was needed to match the experimental distribution, and the more compatible the initial state must have been with the experimental data. Counterintuitively, both in the absence and presence of BetP, the authors found that the inward facing state required less work than the outward facing one. They noted however, that the amount of work was higher in the absence of betaine, indicating that the absence of BetP may lead the protein to assume a mix of both states, or possibly another state not resolved by structural techniques.

This study thus highlighted that although experiments can be difficult to interpret because the signals recorded originate from a mix of factors, MD simulations can provide interpretations. In DEER experiments, the probes can find themselves in a multitude of orientations, and the conformational ensemble of even the protein can deviate substantially from the structure resolved by high resolution structural biology techniques. Explicit modelling using atomistic MD can assist in linking conformational states to the experimental signal. Furthermore, the experimental signal can even be used to enhance the sampling of the simulations towards sampling an ensemble compatible with the latter. The force necessary to match the experimental data can provide information on how compatible the starting structure is with the DEER signal and thus *a posteriori* rationalize how likely the conformational state is to have contributed to the experimental signal.

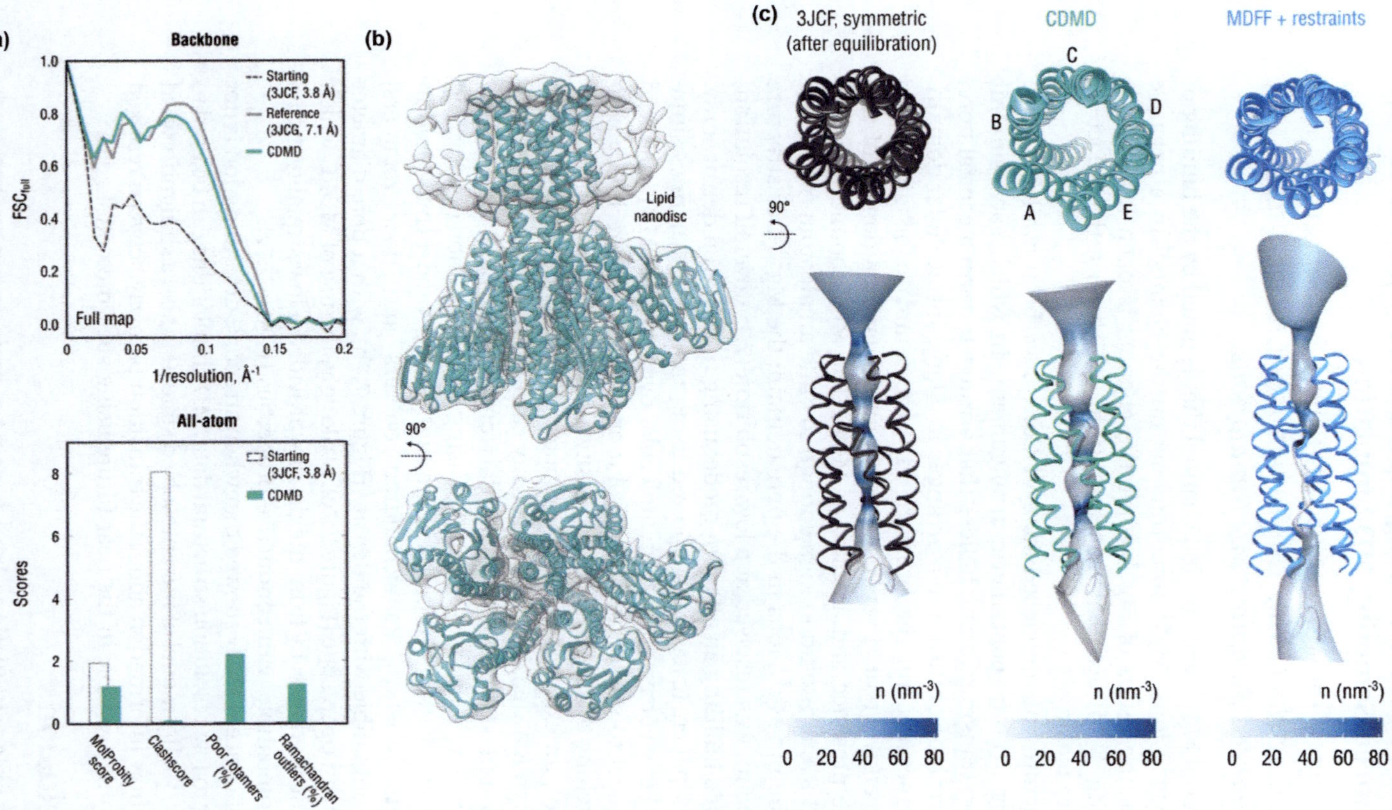

Figure 7.7 Guiding the CorA magnesium transporter from the symmetric closed state into an asymmetric open state using CDMD. (a) Reciprocal-space agreement of the starting (black dashed), the reference (gray) and the model obtained using CDMD model refined with $k = 1.0 \times 10^5$ kJ mol^{-1} (sea green) with the full map (top) and stereochemical quality measure for the three models, assessed using EMRinger and MolProbity (bottom). (b) Overlay of the model obtained using CDMD full-atom model (sea green) with full map. (c) Structure of the gating pore of the closed state model (black), the open state model obtained from CDMD (sea green), and the open state model obtained using MDFF (light blue); only the pore transmembrane helices (residues 281–312) are shown for clarity. Adapted from ref. 91, https://doi.org/10.7554/eLife.43542.001, under the terms of a CC by 4.0 license, https://creativecommons.org/licenses/by/4.0/.

7.3.3 Guiding Processes Using Experimental Data-driven Enhanced Sampling MD Simulations

7.3.3.1 Gating of the Magnesium Transport System CorA: Guiding the Gating Conformational Change Using Enhanced Sampling MD Simulations

CorA is a bacterial Mg^{2+} selective ion channel that is gated by the binding of intracellular Mg^{2+} ions. Mg^{2+} ions thus act on the channel in a negative feedback loop: the influx of Mg^{2+} while the channel is open leads to an increase in Mg^{2+} intracellular concentration, which has the effect of promoting pore closing.

Crystal structures of the closed state of CorA have been available since 2006, revealing a homo-pentameric arrangement, in which each subunit contributes two transmembrane helices that arrange to form a central pore, while the C-terminus of the protein arranges in a large intracellular domain. Open state structures, on the other hand, were found to be difficult to crystallize. Recently, cryoEM was used to obtain the cryoEM density of this channel in the presence and absence of Mg^{2+}. The Mg^{2+} bound structure was solved at 3.8 Å, a resolution high enough to build an all-atom model and pinpoint two bound Mg^{2+} ions in the pore domain; the Mg^{2+}-free structure on the other hand, was resolved at a low resolution (7.1 Å overall resolution) which precluded building an all-atom model using the electron density only.

In their 2019 paper, Igaev *et al.* reported using molecular dynamics simulations under the in correlation-driven molecular dynamics (CDMD) enhanced sampling protocol to guide the high resolution closed structure into a conformation matching the low resolution density obtained in absence of Mg^{2+}.[91] In this protocol, an external potential originating from the cross-correlation between the cryo-EM density and the density calculated using the MD snapshot at hand is applied on top of the potential from the force field. The protocol was adapted to allow successful simulation using a low-resolution density, and was shown to be successful in producing a high quality model (measured in terms of steric clashes, Ramachandran outliers and poor rotamers) (Figure 7.7a) compatible with the density (Figure 7.7b), with a pore diameter consistent with an open state (Figure 7.7c). Strikingly, the use of MDFF, which applies forces derived directly from the density provided a lower quality model that was not compatible with conductive properties (Figure 7.7c).

This proof-of-concept study showed that when structure determination is not possible owing to low-resolution structural data, MD simulations can be used to produce a high-quality model of the unresolved state. It is also a straightforward way to foresee how information about the transition pathway between resolved states will become available in the near future using such methodologies.

7.4 Conclusions

Molecular dynamics simulations have become a mainstream tool to study biomolecular systems and have grown in popularity exponentially in the past

30 years.[109] Although the insights they can provide when taken in conjunction with other experimental techniques are no longer debated, it is still difficult to use them as a black box. This chapter has sought to introduce the principles behind these computational techniques, and has briefly reviewed advanced methodologies to integrate their results with the results from other experimental techniques. After a description of the so-called enhanced sampling strategies, we have explicitly focused on techniques to incorporate experimental data, whether through forward modelling, reweighting of the conformational ensemble or explicit driving of the molecular simulations. To illustrate their applicability, we have then chosen examples of papers in which simulations and experimental work were reported in the same study, an increasingly common trend. We note however, that many more examples of insightful MD simulation studies, either as a follow-up to experimental papers, or that inspired future experimental designs are continuously published.[109] As progress continues to be made in the development of hardware and software to conduct MD simulations, we anticipate that MD simulations will increasingly be used for molecular and mechanistic insights. As this proceeds, the need for educating and sharing simulation and analysis tools grows, and this chapter is a small contribution in that direction.

Acknowledgements

This work was supported by grants from the Gustafsson Foundation and Science for Life Laboratory.

References

1. A. R. Leach, *Molecular Modeling – Principles and Applications*, Pearson Education Limited, 2001.
2. L. Verlet, Computer "experiments" on classical fluids. I. thermodynamical properties of lennard-jones molecules, *Phys. Rev.*, 1967, **159**, 98–103.
3. R. W. Hockney, The potential calculation and some applications, *Methods Comput. Chem.*, 1970, **9**, 136–211.
4. W. C. Swope, H. C. Andersen, P. H. Berens and K. R. Wilson, A computer simulation method for the calculation of equilibrium constants for the formation of physical clusters of molecules: Application to small water clusters, *J. Chem. Phys.*, 1982, **76**, 637–649.
5. M. P. Allen and D. J. Tildesley *Computer Simulation of Liquids*, Clarendon Press, 1987.
6. L. V. Woodcock, Isothermal molecular dynamics calculations for liquid salts, *Chem. Phys. Lett*, 1971, **10**, 257–261.
7. H. J. C. Berendsen, J. P. M. Postma, W. F. van Gunsteren, A. DiNola and J. R. Haak, Molecular dynamics with coupling to an external bath, *J. Chem. Phys.*, 1984, **81**, 3684–3690.

8. S. Nosé, A unified formulation of the constant temperature molecular dynamics methods, *J. Chem. Phys.*, 1984, **81**, 511–519.
9. W. G. Hoover, Canonical dynamics: Equilibrium phase-space distributions, *Phys. Rev. A*, 1985, **31**, 1695.
10. D. Frenkel and B. Smit *Understanding Molecular Simulations: From Algorithms to Applications*, Academic Press, 1996.
11. L. D. Schuler, X. Daura and W. F. van Gunsteren, An improved GROMOS96 force field for aliphatic hydrocarbons in the condensed phase, *J. Comput. Chem*, 2001, **22**, 1205–1218.
12. A. D. MacKerell Jr., *et al.*, All-Atom Empirical Potential for Molecular Modeling and Dynamics Studies of Proteins, *J Phys Chem B*, 1998, **102**, 3586–3616.
13. D. A. Case *et al.*, *AMBER6*, University of California, 1999.
14. Z. Jing, *et al.*, Polarizable Force Fields for Biomolecular Simulations: Recent Advances and Applications, *Annu. Rev. Biophys.*, 2019, **48**, 371–394.
15. J. W. Ponder, *et al.*, Current Status of the AMOEBA Polarizable Force Field, *J. Phys. Chem. B*, 2010, **114**, 2549–2564.
16. J. A. Lemkul, J. Huang, B. Roux and A. D. MacKerell, An Empirical Polarizable Force Field Based on the Classical Drude Oscillator Model: Development History and Recent Applications, *Chem. Rev*, 2016, **116**, 4983–5013.
17. S. Patel and C. L. Brooks, CHARMM fluctuating charge force field for proteins: I parameterization and application to bulk organic liquid simulations, *J. Comput. Chem*, 2004, **25**, 1–16.
18. P. Cieplak, J. Caldwell and P. Kollman, Molecular mechanical models for organic and biological systems going beyond the atom centered two body additive approximation: aqueous solution free energies of methanol and N-methyl acetamide, nucleic acid base, and amide hydrogen bonding and chloroform/water partition coefficients of the nucleic acid bases, *J. Comput. Chem*, 2001, **22**, 1048–1057.
19. J. Wang, *et al.*, Development of Polarizable Models for Molecular Mechanical Calculations. 4. van der Waals Parametrization, *J. Phys. Chem. B*, 2012, **116**, 7088–7101.
20. J. E. Davis and S. Patel, Charge Equilibration Force Fields for Lipid Environments: Applications to Fully Hydrated DPPC Bilayers and DMPC-Embedded Gramicidin A, *J. Phys. Chem. B*, 2009, **113**, 9183–9196.
21. N. Gresh, G. A. Cisneros, T. A. Darden and J.-P. Piquemal, Anisotropic, Polarizable Molecular Mechanics Studies of Inter- and Intramolecular Interactions and Ligand – Macromolecule Complexes. A Bottom-Up Strategy, *J. Chem. Theory Comput*, 2007, **3**, 1960–1986.
22. C. Liu, *et al.*, Development of the ABEEMσπ Polarization Force Field for Base Pairs with Amino Acid Residue Complexes, *J. Chem. Theory Comput*, 2017, **13**, 2098–2111.
23. S. J. Marrink, *et al.*, Computational Modeling of Realistic Cell Membranes, *Chem. Rev*, 2019, **119**, 6184–6226.

24. D. H. de Jong, *et al.*, Improved Parameters for the Martini Coarse-Grained Protein Force Field, *J. Chem. Theory Comput*, 2013, **9**, 687–697.
25. S. J. Marrink, H. J. Risselda, S. Yelfinov, D. P. Tieleman and A. H. de Vries, The MARTINI force field: coarse grained model for biomolecular simulations, *J Phys Chem B*, 2007, **111**, 7812–7824.
26. W. Shinoda, R. DeVane and M. L. Klein, Zwitterionic Lipid Assemblies: Molecular Dynamics Studies of Monolayers, Bilayers, and Vesicles Using a New Coarse Grain Force Field, *J. Phys. Chem. B*, 2010, **114**, 6836–6849.
27. M. Orsi and J. W. Essex, The ELBA Force Field for Coarse-Grain Modeling of Lipid Membranes, *PLOS One*, 2011, **6**, e28637.
28. L. Darré, *et al.*, SIRAH: A Structurally Unbiased Coarse-Grained Force Field for Proteins with Aqueous Solvation and Long-Range Electrostatics, *J. Chem. Theory Comput.*, 2015, **11**, 723–739.
29. A. Y. Toukmaji, Ewald summation techniques in prespective: a survey, *Comput. Phys. Comm.*, 1996, **95**, 73–92.
30. F. Pietrucci, Strategies for the exploration of free energy landscapes: Unity in diversity and challenges ahead, *Rev. Phys.*, 2017, **2**, 32–45.
31. D. Shaw *et al.*, Millisecond-scale molecular dynamics simulations on Anton, in *SC '09: Proceedings of the Conference on High Performance Computing Networking, Storage and Analysis*, ACM, 2009, pp. 1–11.
32. C. Chipot and A. Pohorille, *Free Energy Calculations: Theory and Applications in Chemistry and Biology*, Springer Science & Business Media, 2007.
33. Free Energy Computations. Available at: https://www.worldscientific.com/worldscibooks/10.1142/p579, (Accessed: 5th July 2019).
34. A. Laio and M. Parrinello, Escaping free-energy minima, *Proc. Natl. Acad. Sci.*, 2002, **99**, 12562–12566.
35. V. Lindahl, J. Lidmar and B. Hess, Accelerated weight histogram method for exploring free energy landscapes, *J. Chem. Phys.*, 2014, **141**, 044110.
36. E. Darve, D. Rodríguez-Gómez and A. Pohorille, Adaptive biasing force method for scalar and vector free energy calculations, *J. Chem. Phys.*, 2008, **128**, 144120.
37. D. Hamelberg, J. Mongan and J. A. McCammon, Accelerated molecular dynamics: A promising and efficient simulation method for biomolecules, *J. Chem. Phys.*, 2004, **120**, 11919–11929.
38. C. Jarzynski, Equilibrium free-energy differences from nonequilibrium measurements: A master-equation approach, *Phys. Rev. E*, 1997, **56**, 5018–5035.
39. G. E. Crooks, Nonequilibrium measurements of free energy differences for microscopically reversible markovian systems, *J. Stat. Phys.*, 1998, **90**, 1481–1487.
40. S. Izrailev *et al.*, Steered Molecular Dynamics, in *Computational Molecular Dynamics: Challenges, Methods, Ideas*, Springer, Berlin, Heidelberg, 1999, pp. 39–65, DOI: 10.1007/978-3-642-58360-5_2.

41. J. Schlitter, M. Engels and P. Krüger, Targeted molecular dynamics: A new approach for searching pathways of conformational transitions, *J. Mol. Graph.*, 1994, **12**, 84–89.
42. A. E. Mark Free Energy Perturbation Calculations, in *Encyclopedia of Computational Chemistry*, John Wiley & Sons, Ltd, DOI: 10.1002/0470845015.cfa010, 2002.
43. G. M. Torrie and J. P. Valleau, Nonphysical sampling distributions in Monte Carlo free-energy estimation: Umbrella sampling, *J. Comput. Phys.*, 1977, **23**, 187–199.
44. D. R. Glowacki, E. Paci and D. V. Shalashilin, Boxed Molecular Dynamics: A Simple and General Technique for Accelerating Rare Event Kinetics and Mapping Free Energy in Large Molecular Systems, *J. Phys. Chem. B*, 2009, **113**, 16603–16611.
45. S. Kumar, J. M. Rosenberg, D. Bouzida, R. H. Swendsen and P. A. Kollman, THE weighted histogram analysis method for free-energy calculations on biomolecules. I. The method, *J. Comput. Chem.*, 1992, **13**, 1011–1021.
46. M. R. Shirts and J. D. Chodera, Statistically optimal analysis of samples from multiple equilibrium states, *J. Chem. Phys.*, 2008, **129**, 124105.
47. S. Kirkpatrick, C. D. Gelatt and M. P. Vecchi, Optimization by Simulated Annealing, *Science*, 1983, **220**, 671–680.
48. L. Rosso, P. Mináry, Z. Zhu and M. E. Tuckerman, On the use of the adiabatic molecular dynamics technique in the calculation of free energy profiles, *J. Chem. Phys.*, 2002, **116**, 4389–4402.
49. L. Maragliano, A. Fischer, E. Vanden-Eijnden and G. Ciccotti, String method in collective variables: Minimum free energy paths and isocommittor surfaces, *J. Chem. Phys.*, 2006, **125**, 024106–024115.
50. Y. Sugita and Y. Okamoto, Replica-exchange molecular dynamics method for protein folding, *Chem. Phys. Lett.*, 1999, **314**, 141–151.
51. G. Bussi, Hamiltonian replica-exchange in GROMACS: a flexible implementation, *Mol. Phys.*, 2014, **112**, 379–384.
52. G. R. Bowman, X. Huang and V. S. Pande, Adaptive Seeding: A New Method for Simulating Biologically Relevant Timescales, *Biophys. J.*, 2009, **96**, 575a.
53. P. G. Bolhuis, D. Chandler, C. Dellago and P. L. Geissler, TRANSITION PATH SAMPLING: Throwing Ropes Over Rough Mountain Passes, in the Dark, *Annu. Rev. Phys. Chem.*, 2002, **53**, 291–318.
54. T. S. van Erp, D. Moroni and P. G. Bolhuis, A novel path sampling method for the calculation of rate constants, *J. Chem. Phys.*, 2003, **118**, 7762–7774.
55. R. J. Allen, C. Valeriani and P. R. ten Wolde, Forward flux sampling for rare event simulations, *J. Phys. Condens. Matter*, 2009, **21**, 463102.
56. F. Cérou and A. Guyader, Adaptive Multilevel Splitting for Rare Event Analysis, *Stoch. Anal. Appl.*, 2007, **25**, 417–443.

57. A. C. Pan, D. Sezer and B. Roux, Finding transition pathways using the string method with swarms of trajectories, *J. Phys. Chem. B*, 2008, **112**, 3432–3440.
58. E. Weinan, W. Ren and E. Vanden-Eijnden, String method for the study of rare events, *Phys. Rev. B*, 2002, **66**, 052301.
59. J. D. Chodera and F. Noé, Markov state models of biomolecular conformational dynamics, *Curr. Opin. Struct. Biol.*, 2014, **25**, 135–144.
60. V. S. Pande Understanding Protein Folding Using Markov State Models, in *An Introduction to Markov State Models and Their Application to Long Timescale Molecular Simulation*, ed. G. R. Bowman, V. S. Pande and F. Noé, Springer, Netherlands, 2014, pp. 101–106.
61. M. Chen, M. A. Cuendet and M. E. Tuckerman, Heating and flooding: A unified approach for rapid generation of free energy surfaces, *J. Chem. Phys*, 2012, **137**, 024102.
62. A. Elofsson, *et al.*, Ten simple rules on how to create open access and reproducible molecular simulations of biological systems, *PLOS Comput. Biol.*, 2019, **15**, e1006649.
63. M. Bonomi, G. T. Heller, C. Camilloni and M. Vendruscolo, Principles of protein structural ensemble determination, *Curr. Opin. Struct. Biol.*, 2017, **42**, 106–116.
64. A. Grossfield, *et al.*, Best Practices for Quantification of Uncertainty and Sampling Quality in Molecular Simulations [Article v1.0], *Living J. Comput. Mol. Sci.*, 2018, **1**, 5067.
65. T. J. Harpole and L. Delemotte, Conformational landscapes of membrane proteins delineated by enhanced sampling molecular dynamics simulations, *Biochim. Biophys. Acta*, 2018, **1860**, 909–926.
66. A. Cesari, S. Reißer and G. Bussi, Using the Maximum Entropy Principle to Combine Simulations and Solution Experiments, *Computation*, 2018, **6**, 15.
67. B. Różycki, Y. C. Kim and G. Hummer, SAXS Ensemble Refinement of ESCRT-III CHMP3 Conformational Transitions, *Structure*, 2011, **19**, 109–116.
68. H. T. A. Leung, *et al.*, A Rigorous and Efficient Method To Reweight Very Large Conformational Ensembles Using Average Experimental Data and To Determine Their Relative Information Content, *J. Chem. Theory Comput.*, 2016, **12**, 383–394.
69. W.-Y. Choy and J. D. Forman-Kay, Calculation of ensembles of structures representing the unfolded state of an SH3 domain11Edited by P. E. Wright, *J. Mol. Biol.*, 2001, **308**, 1011–1032.
70. P. Bernadó, E. Mylonas, M. V. Petoukhov, M. Blackledge and D. I. Svergun, Structural Characterization of Flexible Proteins Using Small-Angle X-ray Scattering, *J. Am. Chem. Soc.*, 2007, **129**, 5656–5664.
71. G. Nodet, *et al.*, Quantitative Description of Backbone Conformational Sampling of Unfolded Proteins at Amino Acid Resolution from NMR Residual Dipolar Couplings, *J. Am. Chem. Soc.*, 2009, **131**, 17908–17918.

72. K. Berlin, *et al.*, Recovering a Representative Conformational Ensemble from Underdetermined Macromolecular Structural Data, *J. Am. Chem. Soc.*, 2013, **135**, 16595–16609.
73. Y. Chen, S. L. Campbell and N. V. Dokholyan, Deciphering Protein Dynamics from NMR Data Using Explicit Structure Sampling and Selection, *Biophys. J.*, 2007, **93**, 2300–2306.
74. I. Bertini, *et al.*, Conformational Space of Flexible Biological Macromolecules from Average Data, *J. Am. Chem. Soc.*, 2010, **132**, 13553–13558.
75. M. Pelikan, G. Hura and M. Hammel, Structure and flexibility within proteins as identified through small angle X-ray scattering, *Gen. Physiol. Biophys.*, 2009, **28**, 174–189.
76. S. Yang, L. Blachowicz, L. Makowski and B. Roux, Multidomain assembled states of Hck tyrosine kinase in solution, *Proc. Natl. Acad. Sci.*, 2010, **107**, 15757–15762.
77. G. Hummer and J. Köfinger, Bayesian ensemble refinement by replica simulations and reweighting, *J. Chem. Phys.*, 2015, **143**, 243150.
78. L. D. Antonov, S. Olsson, W. Boomsma and T. Hamelryck, Bayesian inference of protein ensembles from SAXS data, *Phys. Chem. Chem. Phys.*, 2016, **18**, 5832–5838.
79. D. H. Brookes and T. Head-Gordon, Experimental Inferential Structure Determination of Ensembles for Intrinsically Disordered Proteins, *J. Am. Chem. Soc.*, 2016, **138**, 4530–4538.
80. K. A. Beauchamp, V. S. Pande and R. Das, Bayesian Energy Landscape Tilting: Towards Concordant Models of Molecular Ensembles, *Biophys. J.*, 2014, **106**, 1381–1390.
81. X. Xiao, N. Kallenbach and Y. Zhang, Peptide Conformation Analysis Using an Integrated Bayesian Approach, *J. Chem. Theory Comput.*, 2014, **10**, 4152–4159.
82. A. Sethi, D. Anunciado, J. Tian, D. M. Vu and S. Gnanakaran, Deducing conformational variability of intrinsically disordered proteins from infrared spectroscopy with Bayesian statistics, *Chem. Phys.*, 2013, **422**, 143–155.
83. S. Olsson, J. Frellsen, W. Boomsma, K. V. Mardia and T. Hamelryck, Inference of Structure Ensembles of Flexible Biomolecules from Sparse, Averaged Data, *PLoS One*, 2013, **8**, e79439.
84. V. A. Voelz and G. Zhou, Bayesian inference of conformational state populations from computational models and sparse experimental observables, *J. Comput. Chem.*, 2014, **35**, 2215–2224.
85. P. Cossio and G. Hummer, Bayesian analysis of individual electron microscopy images: Towards structures of dynamic and heterogeneous biomolecular assemblies, *J. Struct. Biol.*, 2013, **184**, 427–437.
86. C. K. Fisher, A. Huang and C. M. Stultz, Modeling Intrinsically Disordered Proteins with Bayesian Statistics, *J. Am. Chem. Soc.*, 2010, **132**, 14919–14927.

87. S. Bottaro, T. Bengtsen and K. Lindorff-Larsen, Integrating Molecular Simulation and Experimental Data: A Bayesian/Maximum Entropy reweighting approach, *bioRxiv*, 2018, **457952**, DOI: 10.1101/457952.
88. L. G. Trabuco, E. Villa, K. Mitra, J. Frank and K. Schulten, Flexible Fitting of Atomic Structures into Electron Microscopy Maps Using Molecular Dynamics, *Structure*, 2008, **16**, 673–683.
89. K.-Y. Chan, L. G. Trabuco, E. Schreiner and K. Schulten, Cryo-Electron Microscopy Modeling by the Molecular Dynamics Flexible Fitting Method, *Biopolymers*, 2012, **97**, 678–686.
90. A. Singharoy, *et al.*, Molecular dynamics-based refinement and validation for sub-5 Å cryo-electron microscopy maps, *eLife*, 2016, **5**, e16105.
91. M. Igaev, C. Kutzner, L. V. Bock, A. C. Vaiana and H. Grubmüller, Automated cryo-EM structure refinement using correlation-driven molecular dynamics, *eLife*, 2019, **8**, e43542.
92. A. D. White, J. F. Dama and G. A. Voth, Designing Free Energy Surfaces That Match Experimental Data with Metadynamics, *J. Chem. Theory Comput.*, 2015, **11**, 2451–2460.
93. F. Marinelli and J. D. Faraldo-Gómez, Ensemble-Biased Metadynamics: A Molecular Simulation Method to Sample Experimental Distributions, *Biophys. J.*, 2015, **108**, 2779–2782.
94. J. Fennen, A. E. Torda and W. F. van Gunsteren, Structure refinement with molecular dynamics and a Boltzmann-weighted ensemble, *J. Biomol. NMR*, 1995, **6**, 163–170.
95. R. B. Best and M. Vendruscolo, Determination of Protein Structures Consistent with NMR Order Parameters, *J. Am. Chem. Soc.*, 2004, **126**, 8090–8091.
96. K. Lindorff-Larsen, R. B. Best, M. A. DePristo, C. M. Dobson and M. Vendruscolo, Simultaneous determination of protein structure and dynamics, *Nature*, 2005, **433**, 128.
97. K. S. Molnar, *et al.*, Cys-Scanning Disulfide Crosslinking and Bayesian Modeling Probe the Transmembrane Signaling Mechanism of the Histidine Kinase, PhoQ, *Structure*, 2014, **22**, 1239–1251.
98. T. O. Street, *et al.*, Elucidating the Mechanism of Substrate Recognition by the Bacterial Hsp90 Molecular Chaperone, *J. Mol. Biol.*, 2014, **426**, 2393–2404.
99. M. Bonomi, C. Camilloni, A. Cavalli and M. Vendruscolo, Metainference: A Bayesian inference method for heterogeneous systems, *Sci. Adv.*, 2016, **2**, e1501177.
100. R. Y.-R. Wang, *et al.*, Automated structure refinement of macromolecular assemblies from cryo-EM maps using Rosetta, *eLife*, 2016, **5**, e17219.
101. R. F. Alford, *et al.*, The Rosetta All-Atom Energy Function for Macromolecular Modeling and Design, *J. Chem. Theory Comput.*, 2017, **13**, 3031–3048.

102. S. Kirmizialtin, J. Loerke, E. Behrmann, C. M. T. Spahn and K. Y. Sanbonmatsu Chapter Sixteen – Using Molecular Simulation to Model High-Resolution Cryo-EM Reconstructions, in *Methods in Enzymology*, ed. S. A. Woodson and F. H. T. Allain, Academic Press, vol. 558, 2015, pp. 497–514.
103. S. Lindert and J. A. McCammon, Improved cryoEM-Guided Iterative Molecular Dynamics–Rosetta Protein Structure Refinement Protocol for High Precision Protein Structure Prediction, *J. Chem. Theory Comput.*, 2015, **11**, 1337–1346.
104. J. Ostmeyer, S. Chakrapani, A. C. Pan, E. Perozo and B. Roux, Recovery from slow inactivation in K+ channels is controlled by water molecules, *Nature*, 2013, **501**, 121–124.
105. N. R. Latorraca, *et al.*, Molecular mechanism of GPCR-mediated arrestin activation, *Nature*, 2018, **557**, 452.
106. R. O. Dror, *et al.*, Structural basis for modulation of a G-protein-coupled receptor by allosteric drugs, *Nature*, 2013, **503**, 295–299.
107. M. Landreh, *et al.*, Integrating mass spectrometry with MD simulations reveals the role of lipids in Na^+/H^+ antiporters, *Nat. Commun.*, 2017, **8**, 13993.
108. V. Leone, *et al.*, Interpretation of spectroscopic data using molecular simulations for the secondary active transporter BetP, *J. Gen. Physiol.*, 2019, **151**, 381–394.
109. S. A. Hollingsworth and R. O. Dror, Molecular Dynamics Simulation for All, *Neuron*, 2018, **99**, 1129–1143.

CHAPTER 8

Solid State Chemistry: Computational Chemical Analysis for Materials Science

ESTELINA LORA DA SILVA(iD 0000-0002-7093-3266),[a,b] SANDRA GALMARINI(iD 0000-0003-2183-3100),[c] LIONEL MAURIZI(iD 0000-0002-6346-7623),[d] MARIO JORGE CESAR DOS SANTOS(iD 0000-0002-3114-7473),[b] TAO YANG(iD 0000-0002-4053-124X),[e] DAVID J. COOKE(iD 0000-0001-5996-7900)[f] AND MARCO MOLINARI(iD 0000-0001-7144-6075)*[f]

[a] FIMUP, Departamento de Física e Astronomia da Faculdade de Ciências da Universidade do Porto, Rua do Campo Alegre, 687, 4169-007 Porto, Portugal; [b] Instituto de Diseño para la Fabricación y Producción Automatizada, and MALTA Consolider Team, Universitat Politècnica de València, Camí de Vera s/n, 46022 València, Spain; [c] Empa Materials Science and Technology, Dubendorf, ZH, Switzerland; [d] Laboratoire Interdisciplinaire Carnot de Bourgogne, UMR 6303 CNRS – Université Bourgogne Franche-Comté, BP 47870, F-21078 Dijon cedex, France; [e] College of New Materials and New Energies, Shenzhen Technology University, Shenzhen, 518118, China; [f] Department of Chemical Sciences, School of Applied Sciences, University of Huddersfield, Queensgate, Huddersfield HD1 3DH, United Kingdom
*Email: m.molinari@hud.ac.uk

Theoretical and Computational Chemistry Series No. 20
Computational Techniques for Analytical Chemistry and Bioanalysis
Edited by Philippe B. Wilson and Martin Grootveld

Published by the Royal Society of Chemistry, www.rsc.org

8.1 Introduction

The development of materials with physico-chemical properties for specific applications, especially for medical and biological purposes, has seen lightning growth in the last two decades. Among many promising functionalities, materials can be used to improve, cure or prevent human diseases or weaknesses. For instance, metallic materials provide functionalities in antimicrobial and therapeutic compounds such as coatings on medical devices or implants. Biocompatible materials can also repair or replace body parts such as bones and soft tissues. Whereas nanotechnology offers promising applications in diagnosis and therapeutics, the design and engineering of biocompatible materials is still in its infancy. An interplay between experimental and computational techniques is necessary to achieve a full understanding of biochemical analysis. Although this chapter aims to cover solid state chemistry within the biomedical remit, highlighting the advantages of computational analysis, its exploitation within the biological community is still quite limited. Therefore, we will cover those techniques that are currently at the forefront of materials science, but still have not seen their full application in biochemistry.

8.2 Computational Spectroscopy: Interaction Between Matter and Electromagnetic Radiation

This section will focus on the most relevant computational methodologies that can be applied to probe the spectroscopic properties of systems for medical and biological applications (biomaterials). Firstly, we will highlight relevant systems with exceptional properties for biomedical applications (Section 8.2.1), including zinc oxide and 2D graphene-like nanoparticles. Secondly, Section 8.2.2 will discuss the importance of computational methodologies to study the optical (linear and non-linear properties) and electronic properties of nanoparticles for biomedical applications. Discussion of the theoretical framework with relevant experimental spectroscopy data will be highlighted, alongside the importance of the dimensionality of the biomaterials. This is an important feature that affects the optical and electronic properties of such systems, and understanding such phenomena may enable the possibility of engineering the confined properties of interest for biomedical purposes. In Section 8.2.3, we will focus on inelastic scattering techniques, mainly describing lattice dynamics and Raman spectroscopy as efficient tools to probe the vibrational properties of biodegradable and biocompatible materials with piezoelectric effects, such as those found for bilayer graphene, which presents a plethora of applications, mainly for regenerative medicine. Finally, in Section 8.2.4, we will discuss the important resonance spectroscopy tools used to study defects and impurities in biomaterials. More explicitly the discussion will focus on muon spectroscopy, which is applied not only for direct biomedical usage, but also to mostly characterise the behaviour of hydrogen impurities in biomaterials, and positron lifetime spectroscopy, which together with positron emission tomography can be employed for medical diagnostics and serve as a non-destructive tool to probe the native defects and vacancies of biostructures.

8.2.1 Physical and Chemical Properties of Biomaterials

8.2.1.1 Bulk versus *Nanoparticle*

The importance of the properties that can be exploited for biomedical applications imposes the gathering of a complete characterisation of the nanodevices. This should be applied to both bulk and nanostructures, as their properties differ significantly from those of bulk materials (*e.g.* the confinement excitonic effects strongly affect the optical properties), by probing the corresponding structural, electronic, optical, magnetic and dynamical properties. This characterisation includes the study of the most common and abundant impurities and chemical defects found in these systems, for example hydrogen impurities and native defects (*i.e.* vacancies), as these alter the chemical properties of interest. An interesting example is observed for cerium oxide (ceria), which can exist in two forms in the bulk state: CeO_2 and Ce_2O_3, owing to the transformation of Ce^{4+} to Ce^{3+} being intermediated by oxygen vacancies. However, in the nanostructure, usually in the form of a nanoparticle (NP), a mixture of 3 + and 4 + states of Ce exist on the surface. When the diameter of the nanostructure decreases, the number of Ce^{3+} sites and O vacancies increases, enabling the nanoparticles (NPs) to absorb and release O owing to the redox-cycling between the two charged states. This property ensures that ceria NPs can be used in pharmacological and biological applications.[1] A more complete description, including the applications and computational techniques required to study these nanosystems will be detailed in Section 8.3.3.

8.2.1.2 Zinc Oxide Nanoparticles

Transition metal oxides are generally responsible for forming intermediate chemical bonding.[2] This variety of bonding configurations allows transition metal oxides to explore different crystal structures under specific temperature and pressure conditions; they may exhibit several phases including spin, charges and orbital states, which further enhances the diversity of applications.[3]

The plethora of applications can be evidenced with a simple example of nanoparticles of ZnO, which have been shown to exhibit excellent properties for orthopaedic and dental implants,[4] drug delivery, anti-cancer, anti-diabetic, antibacterial, antifungal and biological sensing applications.[5] Through electrohydrodynamic atomisation of ZnO NPs as a coating material, bacterial adhesion can be inhibited and osteoblast growth promoted.[4] ZnO NPs have also established their capability in biological imaging, owing to their high nonlinear optical coefficients; having been used for in-vitro imaging of living cells.[6] Moreover, as ZnO NPs possess large band-gaps (3.37 eV) with high exciton binding energies (60 meV), the promotion of a high catalytic efficiency is enabled, which is a desirable characteristic for biosensing.[5]

Development of novel nano-methodologies to promote alternative approaches for antimicrobial applications has led to the development of multifunctional agents with tailored physiochemical properties (shape, increased surface area and size, combined with their toxic nature towards prokaryotic organisms).[7] Alternatives to ZnO have been developed as dielectric films or laminar structures fabricated *via* deposition of an atomic layer of metal oxides, such as HfO_2, TiO_2, ZrO_2, Al_2O_3, and all have shown efficient antibacterial activity, comparable to the properties of ZnO.[3] The advantage of low temperature layer deposition (below 100 °C) has opened up the possibility of coating several types of perishable substrates, including soft tissue paper, cloth, and fabric for surgical masks and wound dressing, surgical instruments, door handles, as well as synthetic and organic materials for medicine, veterinary, health care and food industry supply.

8.2.1.3 Graphene-like Materials

Carbon nanotechnology is currently at the forefront of biomedical applications, although it is generally utilized in conjunction with existing biocompatible systems. Graphene-like systems (*i.e.* two or more layers of graphene) have proved efficient in terms of biomedical applications. Graphene and its derivatives have excellent mechanical and lubricant properties (decreasing friction),[8] enhanced capacity for antibacterial properties and biocompatibility.[9] Some applications include medical bone implants, mainly when coated onto ceramic biosurfaces, such as Ti surfaces embedded with SiC,[8] or hydroxyapatite (HaP).[10] HaP on its own is a biocompatible bioceramic used for hard tissue repair and regeneration, owing to its similarities to natural apatite. However, HaP has some drawbacks, which include fracture toughness, poor tensile strength and weak wear resistance. It is therefore important to employ reinforcement films such as graphene (or derivatives) that can be implemented in order to produce biocompatible composites with increased resistance and strength.[10] HaP also demonstrates a favourable affinity for bacterial adhesion[11] and therefore graphene-like coatings can provide the biosurface with antibacterial or bacteriostatic protection.[10]

Graphene has shown promising inexpensive, reliable, high-throughput sequencing with application in nanopore-based DNA sequencing.[12,13] A monoatomic thickness of 0.35 nm is similar to the DNA base spacing and graphene nanopores can be fabricated with a diameter of only 1.0 nm, which is about the size of a DNA sequence.

Further biological applications for graphene are found in neuroscience. Uncoated graphene can be used as a neuro-interface electrode without altering or damaging properties such as the signal strength or the formation of scar tissue. Graphene electrodes in the body are significantly more stable than electrodes of tungsten or silicon because of properties such as flexibility, biocompatibility and conductivity.[14]

When biosurfaces are coated with graphene, it is possible to generate a small electrical impulse through the piezoelectric effect, thus stimulating cellular and bone regeneration. In graphene bilayers, it is the existence of two stacking arrangements of individual layers that makes the generation of this electrical impulse possible.[15] As through the 'sliding' between these layers, transitioning from the AA-phase to the AB-phase (*e.g. via* the movement of a joint), electrical dipoles are generated. This effect can occur owing to tension and/or pressure applied between layers during the phase transition in which piezoelectric effects become dominant.[16] Furthermore, the application of strain between two graphene layers enables charge transfer between the layers,[15,17] enhancing the intensity of dipolar interactions between the graphene and the biosurface. The variety of distortions in graphene bilayers can also involve small angular distortions,[18] which generate systems known as twisted bilayer graphene. Similar to the sliding of graphene layers, the twisting enables the emergence of piezoelectric effects as the rotation of adjacent layers occurs in order to minimize the tension between the layers themselves.

The underlying idea of generating the piezoelectric effect by sliding or rotation of the graphene layers can also be applied when graphene is adsorbed on biosurfaces, for example graphene deposited on a Si and SiO_2 surface.[19] At the nanoscale, electric dipole moments (piezoelectric effect) are generated owing to the strain induced by the surface corrugations and the difference between the lattice topology of the two materials' surfaces.[19,20] Moreover, the addition of graphene to a host biomaterial has the potential to increase the conductivity if uniformly incorporated, even at a low value vol% owing to its high intrinsic conductivity. The unique properties of graphene can be transferred to the biosurface to enhance the physical and chemical properties of the host for biomechanical purposes.[21]

8.2.2 Absorption Spectroscopy: Opto-electronical Properties

Computer simulations of adsorption spectroscopy can provide insights into the mechanisms underlying the excitation processes at the atomistic level, and provide a fundamental understanding to direct the design and optimisation of materials for biomedical applications. This is generally done using *ab-initio* calculations within the framework of density functional theory (DFT).[3,22] This level of theory can be applied to both quantities related to the electronic ground-state (time-independent DFT) and to excited states (time-dependant DFT).

Computational spectroscopy allows for the prediction of accurate values of fundamental opto-electronical properties (gap, absorption spectra, excitons) and simulation of structural and dynamical properties of materials of interest for biomedical applications (nanostructured systems, large unit-cells, defects, doping, interfaces). For example, vibrational spectroscopy such as infrared (IR) absorption and Raman scattering has served as a fundamental tool for determining the weak interactions between guanine (DNA base adsorption) on ZnO model clusters (g-Zn_nO_n) and a comparison performed between DFT data. The adsorption of guanine with a ZnO cluster has been theoretically probed in terms of the geometry, binding energies, electronic and spectral properties, molecular orbitals (highest occupied molecular orbital (HOMO)-lowest unoccupied molecular orbital (LUMO)) charge distribution and occupation (*i.e.* Mulliken charges). Guanine favours physisorption with weak Zn-N bonds and tends to form on the active Zn^{2+} site.[23]

8.2.2.1 Linear and Non-linear Optical Properties: From Density Functional to Many-body Perturbation Methods

Based on first-principles calculations, a system of interacting electrons and nuclei in the presence of time-dependent external fields can be calculated by employing several different methods, depending on their accuracy. The most widely used are density based methods, such as the time-dependent extension of DFT to access neutral excitations,[24] Green's function based methods, and the many-body perturbation theory (GW[25] and BSE[26]).

Time-dependent (TD-)DFT allows the computation of excited state properties, as the respective Kohn–Sham scheme obeys a one-particle Schrödinger time-dependent equation. Within this approach, the absorption spectra are obtained through a linear change of the density produced by a small change in the external potential, which is the linear response theory. However, as with ground-state DFT, the time-dependent exchange-correlation (xc) and its first derivative, the exchange-correlation kernel (f_{xc}), encompass all non-trivial many-body effects, which have to be obtained by different approximations.[27] The most common approximation to the f_{xc} kernel is the adiabatic local-density approximation (ALDA), in which it is assumed that at time t, the kernel is equal to the ground-state local-density approximation (LDA) potential, obtained with the density $n(\mathbf{r}, t)$. ALDA fails to describe the absorption spectra of extended systems, especially for wide band-gap biomaterials, such as SiC, which possesses impressive properties for biosensing and glucose monitoring.[28] ALDA, does not satisfy the divergence of $1/\mathbf{k}^2$ for small wave-vectors, and instead converges to a constant value. Some successful kernels have currently been implemented (long-range corrected kernel, bootstrap kernel), which account for the correct divergence behaviour, however, these fail to capture excitonic effects.[28]

Many-body perturbation theory (MBPT) is a method based on the Green's function method, and provides a more accurate approach to obtaining excited state properties.[29] The one-particle Green's function, $G(\mathbf{r}, t; \mathbf{r}', t')$, can be thought of as being the probability amplitude for the propagation of an additional particle from ($\mathbf{r}'$, t') to ($\mathbf{r}$, t) in a many-body interacting system, thus taking into account an ensemble consisting of a particle interacting with surrounding opposite charges, quasiparticle (QP).[29] QP energies are calculated within the GW approximation, in which the expansion of the self-energy, Σ, in terms of G and the screened Coulomb interaction W, $\Sigma = GW$, is truncated after the first term. As the GW approximation considers the addition and/or removal of an electron (one-particle G) the one-particle excitation spectra can be computed. The experimental observable is related to the direct or inverse photoemission,[29] which provides information regarding the band-gap widths and characteristics of the valence and conduction band states.

Within MBPT, a method to compute the optical response is through the Bethe–Salpeter equation (BSE). As the BSE is a process involving two-particle excitations (two-particle G),[26,29] this method accounts for electron-hole interactions (excitonic effects), therefore prevailing over the non-interacting TD-DFT methodology. As optical absorption experiments create this electron-hole interaction (the exciton), a good agreement between the theory and experiment can be achieved by taking into account the exciton. This effect is important mainly when considering small-gap semiconductors and metals, as these materials screen excitons.[26] However, to obtain such theoretical accuracy, one has to consider that the two-particle nature of the BSE makes the calculations computationally expensive, as a four-point equation (owing to the propagation of two particles, starting and end points) has to be solved.[26] The GW method, combined with the solution of the BSE, is currently the most reliable approach for calculating optical response functions.

The optical properties of macroscopic (bulk) materials emerge from the induced polarisation when the material is subject to an external time-dependent electromagnetic field and macroscopic susceptibilities. By expanding the induced polarisation, the linear term becomes the macroscopic spontaneous polarisation and the non-linear effects come about from the second, third and higher order terms of the expansion.[30] For extended systems, the equations of motion and the coupling of the electrons with the external electric field are derived from the Berry-phase formulation of the dynamical polarisation.[31] It is then possible to calculate the second- and the third-harmonic generation of the systems under consideration. This formulation is developed upon the Green's function method. However, within TD-DFT, it is also possible to evaluate the second harmonic generation (SHG) in semiconductors which also takes into account the crystal local-field and excitonic effects.[32] However, the later method is limited to the treatment of the electron correlation of systems with weakly bound excitons.

It is also possible to express the optical properties of non-periodic systems, such as atoms, molecules, or clusters, within the dipole approximation (as the field-induced dipole polarisation[30]), in which the microscopic polarizability is defined by the linear term and the first and second hyperpolarizabilities are the non-linear contributions to the optical properties.[30,33] At the microscopic level the quadratic dependence on the field means that the magnitude of the quadratic hyperpolarizability tensor must be enhanced in order to produce improved SHG characteristics. This effect is produced when two photons create a single photon of twice the energy of the original two photons.[6] Materials displaying non-linear optical (NLO) properties are important not only for optoelectronics applications, but also for sensing and bioimaging.

8.2.2.2 Dependence of the Optical Properties on the Dimensionality of Biosystems

Time dependent density functional theory has been successfully applied to ZnO clusters to probe the respective optical properties (and validated by employing MBPT).[34] The absorption spectra shows an enhancement in the intensity with a blueshift of the spectral lines occurring with an increase in the cluster size. The observed optical gap has been demonstrated to be closely dependent on the size regime, geometrical shape, symmetry and dimensionality of the structures. From inspection of the molecular orbitals and eigenvalues, it was shown that the electronic transitions occur between the non-bonding O *p*-states and unoccupied Zn sp-hybrid orbitals.

ZnO NPs have also shown efficient NLO properties which enables efficient applications for *in vitro* imaging of living cells.[6] Moreover, it has been shown that ZnO NPs display a pronounced nonlinear doubling effect (SHG).[6]

Graphene provides similar functionality for biosensing applications and has been successfully exploited for protein binding sensor platforms.[35] In graphene, a further advantage is the tenability of the nonlinear third-harmonic generation (frequency tripling) using an electric gate voltage.[36] This technique has particular promise in biomedicine as this provides an alternative avenue for real-time imaging of histopathological quality; examples include the characterisation of tumour tissue *via* real-time optical biopsies.[37] A conventional technique that has been applied successfully to detect brain tumour tissue is stimulated Raman microscopy. However, such a technique is based on subtle differences in the vibrational spectra of the tumour tissue and healthy tissue, thus requiring extensive comparison of experimental spectra against libraries of reference spectra.[37]

8.2.3 Inelastic Scattering: Lattice Dynamics and Raman Spectroscopy

Carbon nanotechnology has shown great promise in biomedical applications. Over time, carbon nanotubes, fullerenes and graphene have all been shown to be promising nanozymes. Here, we show the computational application of Raman spectroscopy applied to graphene, which conceptually can be extended to any biochemical system that contains graphene. The application of Raman spectroscopy allows for the identification of monolayer or multi-layered graphenes in the Bernal (AB) configuration.[38] As graphene samples are produced in bulk both *via* exfoliation or epitaxial methods, with both techniques giving a variety of graphene rotational disorders,[38] each of these permutations will impact on the biochemical property of the systems. Hence, the characterization of such configurations becomes of paramount importance to control graphene behaviour in biomedical systems selectively.

Within the harmonic approximation, lattice dynamics enable the computation of the vibrational modes (phonons) at the zone-centre, which can be directly compared with experimentally derived Raman spectroscopy. Bilayer graphene consists of two coupled monolayers of carbon atoms, each with a honeycomb crystal structure. Although we have focused on biomedical applications, the very basis of this technique resides in the physics at the atomistic scale that describes the vibrational motion of the structure. In order to satisfy the translational and symmetrical properties of the Bravais lattice, the honeycomb graphene lattice can be seen as two sublattices, mathematically labelled as inequivalent A and B lattices, each containing two atoms. In the bilayer graphene there are two lattices (A and B), then there are four inequivalent carbon atoms that need a description within the unit cell, this leads to three acoustic (A) and nine ($3N-3$) optical (O) lattice vibrations or phonon branches.[15]

Comparison between the LDA calculated graphene phonon dispersions with the in-plane phonon dispersion of graphite obtained from inelastic X-ray scattering displays a reasonable agreement (Figure 8.1), although a small shift in the higher-frequency transverse optic (TO) and longitudinal optic (LO) modes is observed. LDA calculations often overestimate the energies of higher-frequency phonons, but despite this difference the characteristic features of the phonon dispersion are well reproduced. At low **q**-vectors, the in-plane transverse acoustic (TA) and longitudinal acoustic (LA) modes show linear dispersions. Although the doubly-degenerate LA mode has zero frequency at the Γ-point, the TA mode, also known as the shear-mode (Figure 8.2), has a non-zero frequency at the zone-centre ($\nu = 0.82$ THz).[39] The ZA mode is the flexural acoustic mode (Figure 8.2), which corresponds to out-of-plane, in-phase atomic displacements. In contrast to the TA and LA modes, the ZA branch shows a parabolic dispersion close to the Γ-point, indicating a low group velocity, a characteristic feature of layered structures.[39] The existence of a flexural mode is also a signature of 2D systems, and in particular it is a mode which is typically found in graphene-like systems. As the long-wave flexural mode has the lowest frequency, it is the easiest to excite.[40] The flexural mode is relevant for understanding of the intrinsic properties of graphene (*i.e.* electrical resistivity and thermal expansion coefficient). For example, for single layer graphene, the high thermal conductivity is a result of the vibrational morphology of the flexural mode.[40] When graphene-like systems are

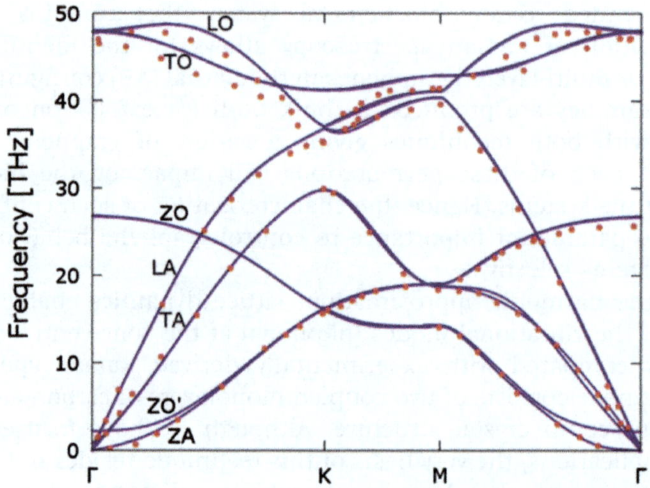

Figure 8.1 Phonon dispersion of AB-stacked bilayer graphene computed within the harmonic approximation (solid purple line). The unit cell contains four carbon atoms, leading to three acoustic (A) and nine optical (O) phonon branches. The calculated dispersions are compared to the in-plane phonon dispersion of graphite obtained from inelastic X-ray scattering (red dots). The phonon branches are marked with the labels assigned to the Γ-point phonons.

adsorbed on biosurfaces, this flexural mode tends to hybridize with the substrate, therefore making it possible to estimate the substrate induced changes of the thermal expansion coefficient and the temperature dependence of the electrical resistivity.[41]

At slightly higher frequencies, the out-of-plane ZO′ mode can be observed (Figure 8.2), which corresponds to interlayer motion along the *z*-axis (a layer-breathing mode). The other out-of-plane optical modes are characterised by the doubly degenerate ZO branch. At the Γ-point, the interlayer coupling causes the LO and TO modes to split into two doubly-degenerate branches, both of which correspond to the in-plane relative motion of atoms. With the exception of the ZA and ZO′ modes, all of the frequency branches have symmetry-imposed degeneracy at the zone-centre (Figure 8.1).

To describe the dispersion of the LO and the in-plane TO phonon branches near the Γ- and K-points correctly, it is important to consider the renormalization of the phonon energies, associated with a process in which a phonon can create an electron-hole pair.[38] These considerations are needed whether graphene is studied within materials science or in biochemistry. In order to understand the effects of the electron-phonon coupling one should resort to methodologies that go beyond the framework of the Born–Oppenheimer and density-functional approximations. One of the effects of the coupling interaction is the emergence of the Kohn anomaly,[42] causing the softening of certain Γ- and K-point phonons, leading to a discontinuity in its derivative. This effect can be caused by the strong inhomogeneities in

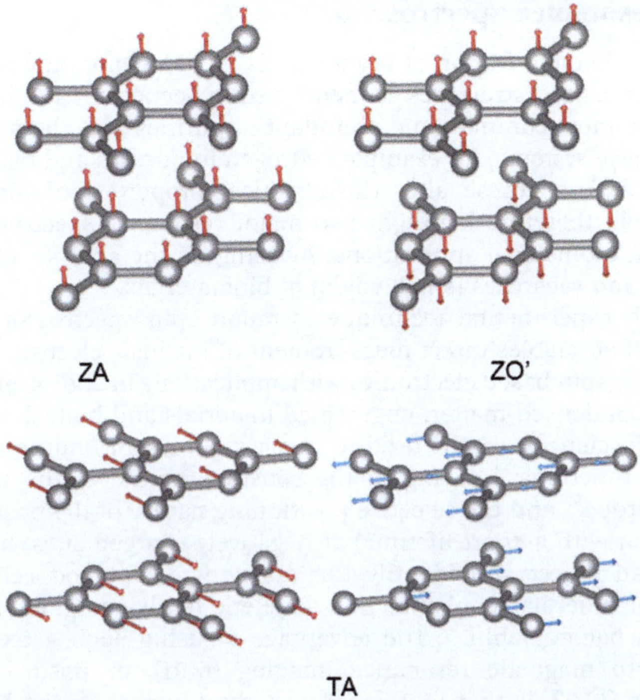

Figure 8.2 Eigenvectors corresponding to the vibrations in AB-stacked bilayer graphene. ZA (top left) and ZO′ (top right) correspond to the out-of-plane vibrations, and TA (bottom) denotes the degenerate in-plane transverse-acoustic modes.

the screening of lattice vibrations by the conduction electrons. It is shown that Kohn phonon vibrations cause the Fermi level oscillations and therefore a tuneable band gap opening.[43] For functionalised graphene (with group-IV elements and different functional groups, such as F, Cl, H_3C-, and H_2N-) the combination of an out-of-plane symmetry-breaking defect and a soft infrared-active phonon mode induces a large out-of-plane piezoelectric response into functionalized graphene.[44] It is therefore expected that the emergence of the Kohn anomaly on gated-graphene systems, will also affect the piezoelectric impulses, mentioned above, which are induced by the sliding and rotation of the adjacent graphene layers.

Regenerative medicine is currently a field of intensive interest, and thus the study of biodegradable and biocompatible materials with piezoelectric effects, such as graphene-like materials, is important. Not only can piezoelectric biomaterials be applied for bone and cartilage regeneration,[45] but also these effects have other biological applications, namely for diagnostic ultrasonography,[46] immunological biosensors[47] and pulmonary drug delivery.[48] In fact piezoelectric effects may occur naturally in tissue, playing a significant role in regulating and stimulating the continuous stress-induced modifications of the tissue (*i.e.* collagen).[49]

8.2.4 Resonance Spectroscopy

A complete characterisation of biomaterials includes, not only the study of bulk and nanosized structures as mentioned in Section 8.2.1.3, but also the study of the most common and abundant impurities and chemical defects found in these systems, for example hydrogen impurities and native defects (*i.e.* vacancies), as these alter the chemical properties of interest. The following will therefore highlight two major resonance spectroscopy techniques with biomedical applications, focusing on the analysis of hydrogen impurities and vacancies (small voids) of biomaterials.

One such experimental technique is muon spin spectroscopy (μSR), a technique that enables direct measurement of intrinsic electron spin relaxation rates in spin based electronics with applications including atomic-level studies in condensed matter, engineered materials and biotechnology.[50]

In the biosciences, μSR is used to image the human brain and to study new brain functions.[51] Owing to the sensitive nature of the muon as a magnetic probe[52] and the selective positioning nature of the positive muon (light proton with a short lifetime) at negatively charged sites, muon spectroscopy can detect and identify the magnetism of blood cells, namely haemoglobin met-haemoglobin, non-magnetic oxy-haemoglobin and magnetic deoxy-haemoglobin.[51] The advantage of using such a technique as compared to magnetic resonance imaging (MRI), or positron-emission tomography (PET), is that μSR can detect the function of the brain more efficiently (deeper probing, smaller volume, and short time-scale measurements) without the need to apply high magnetic fields. Moreover, muon spectroscopy can provide new aspects of the oxygenation reactions of haemoglobin as it can detect triplet excited states of oxy-hemoglobin.[51]

Implanted muons can sometimes bind to an electron to form a light isotope of hydrogen called muonium. By analysing the muonium behaviour inside a material one can learn about proton and hydrogen behaviour, which is important as the chemical and structural properties of the biomaterial will be affected.

Another technique of interest is positron annihilation lifetime spectroscopy, mostly employed in material science, which can provide information regarding atomic and molecular level free-volume and void sizes, molecular bonding, structures at depth-layers and phase transitions.[53] However, together with PET imaging, positron lifetime spectroscopy can serve as a multi-purpose detector and be applied for medical diagnostics, for example *in vivo* imaging of cell morphology,[54] allowing changes in the biomechanical parameters between normal and abnormal cells (*i.e.*, cancer cells) to be distinguished.[53]

Depending on the application of interest, and owing to the versatility of positron lifetime spectroscopy, this technique can serve in several fields with biotechnical applications (drug delivery formulations, drug encapsulation, and biocompatible use in medical products); macromolecular and membrane studies (macromolecular cavities, membranes, and conformational states affected by hydration, temperature and UV irradiation), and biological tissue research.[55]

8.2.4.1 *Muon Spin Rotation/Relaxation/Resonance Spectroscopy: Hydrogen Impurities*

Impurities and/or dopants may appear in materials from their initial growth conditions to the aging of fully developed devices,[56,57] affecting the device's properties and reliability. Among such impurities are hydrogen species that can be unintentionally incorporated during the growth environment.[56,58] Depending on the respective behaviour, hydrogen can be amphoteric, in which it counteracts the electric conductivity by passivating defective levels or it can enable a donor level close to the conduction band, thus inducing *n*-type conductivity.[56] Hydrogen may indeed influence the electrical, magnetic, optical and thermal properties of the host,[56] thus altering the respective properties, and hence the functionality. In graphene samples a variety of functional groups can be bound to the carbon network, however, the most common chemisorbed species is found to be the atomic hydrogen. Hydrogen impurities can generate localized spin states causing spin-half paramagnetism.[59,60]

Owing to the calculations being computationally simple, it is still ground-state electronic structure DFT calculations that are mostly applied to obtain several properties and extrapolate to provide insights into experimental spectroscopy data. Examples include the defect chemistry of solids to obtain defect formation energies and electronic configurations, which ultimately describe the electrical behaviour of the material. The position of defect levels within the band-gap can also be examined through the density of state calculations, which may provide insights if the defect level is located deep in the band-gap (amphoteric) or if it is a shallow defect and therefore resonant with the conduction states and positioned above the conduction band minimum.[58]

8.2.4.1.1 Suitability of Electronic Structure Methods Applied to Hydrogen Impurity Studies. First-principles studies of hydrogen in a wide range of oxides have been achieved[58] by employing mostly (semi-)local functionals within DFT. However, it is well known that these functionals do not calculate the band-gaps of bulk biomaterials accurately, and are also inadequate when considering strongly correlated systems, namely the d- and f-electron, because the incomplete cancellation of the Coulomb self-interaction favours delocalisation.[58] Non-local hybrid functionals, have received increased interest[61] and because of the mixing of a fraction of the generalized gradient approximation (GGA) exchange-energy with the exact Hartree–Fock exchange, they are able to determine the band-gap more accurately than the localized functionals. These functionals also have the advantage of being able to partially correct the self-interaction error, providing an improvement in the description of the electronic structure of f-electron systems. Another approach that can correct the severe self-interaction error of the (semi)-local functionals is to introduce a local Hubbard-U potential, characterized by the on-site Coulomb repulsion among the localised electrons. If the Hubbard correction is applied to spin-polarized GGA calculations, a good agreement for lattice parameters and the band gap can be achieved with a small value of U.[58]

GW is another alternative to DFT for calculating the electronic properties of biomaterials from first-principles, which corrects the systematic underestimation of the band-gap seen in DFT calculations. Most computationally affordable GW calculations employ the LDA eigenfunctions, as the starting point, to generate the self-energy (G_0W_0) from the LDA polarisation (or dielectric constant) one obtains the screened interaction, W_0.[26] When employing G_0W_0, it has been demonstrated that the fundamental band-gaps in sp^3 covalent materials show an improvement over LDA.[62] In spite of this, the one-shot GW band-gaps still underestimate band-gap, when compared to experiments, even for weakly correlated semiconductors.[63] This under estimation and the fact that the QP levels are closely related to the quality of the ground-state wave-functions (DFT exchange-correlation functional) makes one-shot GW methods an unsatisfactory alternative to standard methods.

Quasiparticle self-consistent GW (QSGW) applies a Hamiltonian that is found by optimisation and therefore calculations based on it predict reliable ground- and excited-states properties of a large number of weakly and moderately correlated materials.[63] Such calculations also result in reliable QP levels for a wide range of materials, not only in the description of the fundamental gaps in semiconductors, but also for the majority of the energy levels. In strongly correlated d- and f-electron systems, systematic errors larger than those observed in the experimental results are still predicted.[63] Despite the increased accuracy of QSGW, this method tends to slightly overestimate semiconductor band-gaps and underestimate dielectric constants.[62] The reason for this is because W does not include electron-hole correlation; the inclusion of the correlation energy would reduce the pair excitation energy in its intermediate states.[62,63]

Therefore, in order to obtain a more precise position of the defect levels inside the band-gap, care has to be taken to employ the correct functional or methodology and even the use of a hybrid functional, or (QS)GW will increase the band-gap width. However, the position of the defect levels will be more consistent and more reliable in these more advanced methods, enabling a better comparison with experimental data. These points are neatly illustrated by considering Y_2O_3,[64] which is a promising material for biological imaging applications.[65,66] Calculations have been performed with the hybrid non-local functional (HSE06). The results demonstrated the existence of metastable minimum-energy sites of which amphoteric behaviour was found for hydrogen after considering the lowest-energy structures for each charge state. For all neutral and negative configurations, localised defect levels were found inside the gap. The calculated acceptor transition level was observed to be near the midgap, and was consistent with the experimental data.[64] However, for the case of lanthanide oxides, relevant for applications such as magnetic resonance imaging, X-ray computed tomography and fluorescence imaging,[67] the *f*-states must also be considered. It therefore becomes difficult to perform many-body perturbation calculations, or even non-local calculations within DFT. Therefore, DFT + U can aid in localising the correlated f-states of the metal, and hence increase the band-gap width to slightly higher values when compared to the performance of (semi-)local functionals. The DFT + U approach is computationally as efficient as conventional DFT, and is well suited for defect studies, which require a large number of atoms, and has also proved reliable in recent studies of defects in f-electron solids. Examples of this approach can be found in studies of hydrogen impurities in Lu_2O_3 by employing the effective Hubbard-type potential to treat the on-site 4*f*-electron correlations.[58] By considering the lowest-energy configurations to obtain the charge-transitions, the results predict that hydrogen is an amphoteric impurity in Lu_2O_3, as observed experimentally.[58]

8.2.4.1.2 Experimental Probing of Hydrogen Impurities. Several experimental techniques exist that may extend the search for new information regarding the hydrogen impurities. These include, electron paramagnetic resonance (EPR), in which hydrogen is studied in the paramagnetic state or under non-equilibrium conditions (because interstitial hydrogen is not stable in the neutral charge state). Such a technique provides insights into redox biochemistry research, owing to its ability to distinguish and quantify the hydroxyl radicals and hydrogen atoms.[68] Nuclear magnetic resonance (NMR) is yet another method that can gather information about the bonding configurations of hydrogen and which is widely applied to biomedical and pharmaceutical research.

In spite of the advantages, the use of microscopic techniques, such as EPR or IR vibrational spectroscopy, show some drawbacks when gathering microscopic information about hydrogen configurations and electric defect levels, as these are limited to systems in which hydrogen is usually present in high concentrations.

8.2.4.1.3 Muon Spin Spectroscopy and Density-functional Theory Working Together. As they possess high mobilities and thus tend to pair with other defects, isolated hydrogen impurities are very difficult to study experimentally,[57,69] muon spin resonance can be used to study hydrogen in biomaterials,[70] providing very similar results to EPR. It is an experimental technique performed on muonium, which is an electron bound to a positive muon, being a pseudo-isotope of hydrogen with a similar chemical behaviour. From this spectroscopy technique it is possible to exploit the muon spin rotation, relaxation and resonance.[64] The short lifetime (2.2 μs) of muonium reflects short time-scale measurements under non-equilibrium conditions, thus favouring the observation of isolated defect centres, the centres responsible for the electrical activity.[64] Although the muon is only one ninth the mass of the proton, it is still possible to compare the respective properties to those of hydrogen in a wide range of materials. These properties include electronic structure, thermal stability and charge-state transitions,[64] and can often be compared with similar results from parallel *ab initio* computational studies.

Another advantage of muonium spectroscopy is that the technique can provide detailed information about the electronic structure of muonium/hydrogen through a hyperfine interaction (this refers to small shifts and splittings in the energy levels owing to the interaction between nuclei spins and electron density).[71] Such information is difficult to obtain for isolated hydrogen using any other technique. Calculation of the hyperfine tensors provides useful information for characterising the neutral impurity centres, and thus allows direct comparison with the experimentally obtained hyperfine constants measured using μSR.[58] It is important to be aware that when providing *ab-initio* results for comparison of the hyperfine tensors at the atomic sites, one needs to be aware of the exchange-correlation functional employed to compute the hyperfine tensor. It is well known that a pure Perdew–Burke–Ernzerhof (PBE) approach tends to delocalise the spin density of defect centres,[71] whereas hybrid functionals give more localised centres and generally provide results more comparable to experimental data. For the case of the studies performed on Lu_2O_3, discussed previously, calculation of the hyperfine constants for the neutral interstitial configurations show an isotropic hyperfine interaction with two distinct values at 926 and 1061 MHz for the Fermi-contact term which corresponds to two interstitial positions of hydrogen in the lattice. These high values are consistent with the muonium spectroscopy measurements, which also reveal a strongly isotropic hyperfine signature for the neutral muonium fraction with a slightly larger magnitude of 1130 MHz.[58]

From a theoretical point-of-view, through DFT calculations, the higher energy configurations (metastable states) that may exist for the different hydrogen charges, should be considered in order to compare with μSR. This will allow defining the charge transition levels for each structural configuration that hydrogen can adopt in a system. This procedure is more realistic when compared to non-equilibrium muonium measurements as, owing to the short life-time of muonium (~2.2 μs), it is not possible to obtain full equilibrium measurements. Instead higher energy metastable states can also be accessed together with their acceptor and donor levels individually.

The thermodynamics of a material depend on the formation energies of different types of defects and therefore the stable charge states and their transitions can be determined from this quantity.[72] For the case of the hydrogen impurity, the formation energies are evaluated to obtain the properties of interstitial hydrogen in the host material and this is defined as being the energy required to incorporate the impurity in three charge states (H^-, H^0, H^+) in the host lattice.[69,71]

Having determined an adequate number of minimum-energy hydrogen configurations it is also important to identify representative pathways for the migration of the impurity that connects the initial and final ionic configurations. This type of study can be carried out by employing the nudge-elastic band (NEB) method.[73] From this it is possible to compare the proton/muon sites and the activation energies between equivalent sites.[74]

Polarised muons have proved to be reliable and sensitive probes of local magnetic fields in matter and an important tool for the investigation of hydrogen reactions in low electron density materials (*i.e.* graphene). In defective graphene, muon spectroscopy aids in answering two debated questions: the possible onset of magnetism and the interaction with hydrogen.

It is possible to observe clear muon spin precession in graphene samples, contrary to the behaviour of graphite, and this has been demonstrated to originate from localised muon-hydrogen nuclear dipolar interactions, rather than from a hyperfine interaction with magnetic electrons, and reveals the formation of an extremely stable CHMu state (analogous to CH_2). These results rule out the formation of magnetism in chemically synthesized graphene samples. Moreover, in μSR spectroscopy electronic states of edges and defect sites have revealed radically different signals with respect to the standard case of graphite, or even of the pristine graphene plane.[60]

8.2.4.2 Positron Lifetime Spectroscopy: Vacancies

Vacancies are an example of one of the most common defects found in graphene and can be induced by proton irradiation.[60] When a C atom is removed from the lattice, breaking of the symmetry of the system induces a band-gap and the three σ-dangling sp^2 bonds introduce localised states in the mid-gap, which split owing to crystal field effects and a Jahn-Teller distortion.[75] Although two of the electrons bind to each other to form a pentagonal ring, the third dangling bond becomes a paramagnetic centre. One impurity may induce magnetisation in the graphene lattice owing to the formation of spin-polarised localised states. However, the presence of several vacancies may affect the magnetisation of the system owing to the correlation of the impurity position. Also, as the presence of only one vacancy allows symmetry breaking of the lattice (inducing a band-gap), two identical vacancies, located at the same sub-lattice, will increase the asymmetry, therefore further increasing the gap width. Hence, this breaking of the symmetry may define the magnetic properties of the lattice, allowing for the defects to react ferromagnetically or antiferromagnetically with each other, depending on the extension of the asymmetry. The Fermi level of graphene is sensitive to perturbation, and alterations may occur owing to the mismatch and subsequent strain at biosurfaces caused by impurities at the interface of graphene.[75]

It is therefore important to investigate the defects or chemical impurities at the interface of the multilayer or single layer graphene with biosurfaces. The choice of the substrate on which graphene is deposited plays an important role in the electronic, magnetic and chemical properties of the film.

It is relevant to employ a non-destructive probe, which can investigate graphene over layers on various substrates without extensive sample preparation or probe modification of its chemical or physical properties.

A well-developed experimental technique suited to probing point defects, mainly vacancy-type defects, is positron annihilation spectroscopy. By measuring positron lifetimes and momentum distributions during the annihilation process, it is possible to obtain information on the electronic structure in solids, mainly the open volumes and the chemical environments of the defects, such as vacancy-type defects.[59] The bound state of a positron and an electron is analogous to a hydrogen atom, in which the proton is replaced by the positron. The annihilation radiation contains information on the electron momentum distribution at the annihilation site owing to conservation of momentum during the annihilation process.

Recently, measurements using a low energy positron beam have been reported and applied for the study of multilayer graphene adsorbed on a polycrystalline Cu sample.[76] The efficient trapping of positrons on the surface state of graphene at kinetic energies as low as 2 eV has been used to obtain the respective Doppler broadened spectra. This is the first ever measurement of graphene thin-films (2–3 nm of thickness) adsorbed on a substrate with a depth resolved Doppler broadening spectroscopy at positron kinetic energies below 10 eV, to have been reported so far. These findings open up the possibility of employing positron beam spectroscopy to characterise defects and chemical structure in 2D biomaterials.

8.2.4.2.1 Two-component Density Functional Theory. From a computational point of view, it is possible to employ a two-component generalization of density functional theory (TCDFT), to probe the positron state and annihilation characteristics of a system. The positron-electron correlation-energy functional is usually calculated at the limit of the vanishing positron density. For a delocalised positron the respective density is small at every point of the lattice, not influencing the electronic structure, and hence the electronic structure of the perfect lattice is solved separately from the positron density. When the positron and electron densities are known, then the positron annihilation rate can be calculated.[72,77]

When a positron is localised at a lattice defect, one has to take into account the fact that the positron attracts electrons, and the average electron density increases near the defect. The short-range screening effects have hence to be taken into account by correlation functionals which depend on both the electron and the positron densities. Nevertheless, in most applications for positron states at defects, the full scheme is simplified by using the same procedure as that for delocalised positrons. This means that the zero-positron-density correlation energy and enhancement factors are used. The simplified scheme can be well justified by arguing that the positron, and the respective screening cloud, forms a neutral quasiparticle that enters the system without changing the average electron density. On the other hand, the two-component calculations also support the use of conventional schemes within the LDA or GGA approximations.[77]

In order to investigate two dimensional materials such as graphene, it is essentially important to trap the positrons on the surface state. If positrons are implanted at energies less than 10 eV then it is possible to obtain relevant information from the top atomic layers of the nanoscaled sample, as there is a greater probability of trapping the positrons into the surface state.[76]

8.3 Computational Techniques for Biocompatible Materials

Biological reactions are mainly induced by interactions between the surrounding environment and the surfaces of the biocompatible materials. Thus, the key parameter to focus on in order to reduce, or at least control, the influence of a material at the bio-interface is its surface. Biological reactions are not always well understood and this chapter will describe how computational analysis can provide excellent tools to explain the interactions between the surfaces of materials and biological media. We will explain the scientific questions and focus on the computational and analytical techniques that can measure these properties at the interface. In the first part, we will discuss which properties of the surface are crucial, before discussing the importance of the chemistry of the biocompatible materials and present the computational techniques that can be used to investigate these surface characteristics. The biological behavior of surfaces is controlled in a large part by interactions between a specific surface (specific properties are discussed in the first part of the chapter) and different (bio-)molecules present in the surrounding media. This will be the subject of the second part, in which techniques able to study and predict surface interactions with bio-molecules such as polymers or proteins will be presented. The last part will finally discuss the importance of size in the case of nanoparticles bio-interactions, which has a large influence on colloidal stability, the formation of the so-called corona of proteins or other molecules or on the formation of micelles.

8.3.1 Overview of Surface Properties

As most interactions between biocompatible materials and any biological system happen at the surface of the material, the surface properties are of paramount importance. From an experimental point of view, despite accurate control of synthesis processes, understanding the phenomena at the interface is often neglected leading to biased observations and conclusions. The types, number and arrangement of chemical groups at the surface will determine what species adsorb[78–82] and what surface reactions can take place.[83,84] In this section, we will therefore discuss how computational solid state techniques can help define those properties of the surfaces likely to be exploited in biomedicine. We will discuss the equilibrium properties of pure materials and how to study them using different computational techniques. Then, we will focus on the influence of defects, the environment and non-equilibrium structural features, such as growth features.

8.3.1.1 Amorphous Solids

Surfaces cost energy. The lowest energy (equilibrium) arrangement of the atoms is disturbed, leading to a higher energy of the atoms/molecules at the surface. The energy contribution of the surface is more important for small structures than for larger systems, owing to the different scaling of the surface (d^2) and the volume (d^3) with the typical dimension d. Therefore, the shape of small (typically submicrometric) structures are often dominated by the minimization of the surface energy. For amorphous solids, in which the surface energy is independent of the orientation of the surface, this typically means a spherical shape, optimizing the surface to volume ratio. For larger structures of amorphous solids, in which the contribution of the surface energy becomes minimal, the chemical groups present at the surface are those that will minimize the surface energy.

8.3.1.2 Ionic and Covalent Crystals

For the crystalline phases the energy of an interface depends on the orientation with respect to the crystalline lattice. The reason for this is that the orientation of the interface will determine the nature of the "bonds" present in the original crystal structure that will be broken in forming the surface. The dependence of the interfacial energy on the interface orientation with respect to the crystal lattice means that the shape, minimizing the total interfacial energy of a finite sized crystal, is no longer a sphere. Wulff developed a general method to predict the equilibrium morphology of a finite sized crystal based on the interfacial energy per unit area $\gamma(\phi,\theta)$ as a function of the orientation.[85] In fact, Wulff showed that, in order to minimize the overall interfacial energy, the distance of an interface from the center of mass should be proportional to $\gamma(\phi,\theta)$ (Figure 8.3). The Wulff construction demands knowledge of the interfacial energy with respect to the interface orientation. However, $\gamma(\phi,\theta)$ is not easily accessible with experimental methods.

In the case of ionic crystals, the low energy interfaces are usually parallel to a crystal plane $(h\ k\ l)$, in which h, k and l are the Miller indices of the plane indicating the orientation of the plane with respect to the crystal lattice (Figure 8.3b). For a crystalline plane, h, k and l are inversely proportional to the intercept of the plane with the vectors of the crystal lattice, and for most lattices they are equivalent to the indices of the surface normal. The continuous function $\gamma(\phi,\theta)$ can thus be reduced to a certain number of surface energies of different crystallographic planes $(h\ k\ l)$. Atomistic simulation techniques can be used to determine the enthalpic contribution to the surface energy of different atomistic planes, taking into account the relaxation of the

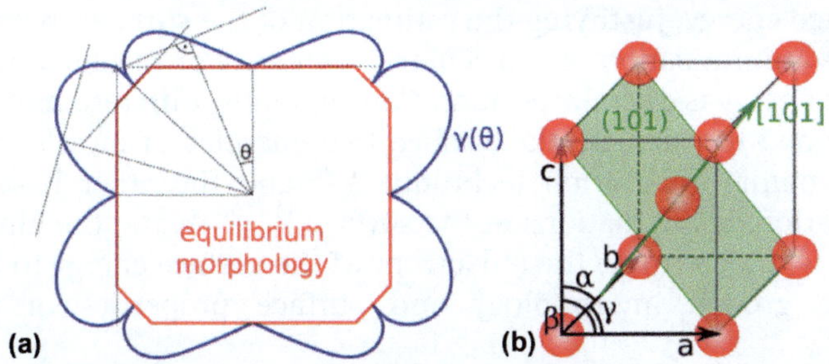

Figure 8.3 (a) Schematic illustration of the Wulff construction in 2D, the interfacial energy curve is shown in blue, the perpendicular lines to the θ-direction at the position of $\gamma(\theta)$ used for the Wulff construction in black and the resulting equilibrium morphology corresponding to the inner envelope to all perpendicular lines in red. (b) Schematic illustration of the cell parameters $\vec{a}$, $\vec{b}$, $\vec{c}$ and α, β, γ. Also shown is the (101) surface with the surface normal (101), in which the numbers in parenthesis are the Miller indices indicating the orientation of the plane.

position of the atoms at the surface. This can be done either using DFT[86–88] or classical atomistic force fields (potentials).[89,90] The energy of a surface is calculated by comparing the energy of the relaxed surface to that of the bulk crystal structure. Usually crystals are represented by periodic boundary conditions, or in other words, by a small part of the crystal that is thought to repeat in all three dimensions to build up a much larger bulk structure. However, in a direction normal to the surface, the periodicity of the crystal is disrupted and periodic boundary conditions can no longer be applied in this direction. There are two main methods to deal with this 1D lack of periodicity: (i) if an energy minimization or structural relaxation technique is used to find the lowest enthalpy structure, the bulk structure is often represented by a region with pre-relaxed, fixed atomic positions adjacent to the surface layer which is allowed to relax;[89] (ii) if molecular dynamic calculations are performed instead, the surface is often represented by a slab of material, terminated on both sides by a surface.[91] While the surface energy $\gamma(\phi,\theta)$ depends on the surface directions, there might be different surface terminations to take into account, resulting from the different nature of the broken bonds through the crystal at different depths, which share the same surface normal. To get the correct surface energy, the termination with the lowest energy has to be considered.[92] Additionally, some surfaces might present a large surface reconstruction (local rearrangement of the atoms close to the surface), something that has to be considered explicitly.[93]

8.3.1.3 Metals

So far, we have only discussed the enthalpic contribution to the surface energy. For ionic crystals this is generally by far the dominant contribution to the surface energy, justifying the estimation of the surface energy directly from energy minimization or molecular dynamics. For metals, however, the surface free energy is often important. Consequently, different methods have been developed to calculate the surface free energies of metals, one being a thermodynamic integration technique[94–96] and the other based on the characterization of fluctuations at the surface.[97–100] Again, the aim of these techniques is to determine the anisotropy of the surface energy to be able to predict the growth, morphology and surface properties of the final material.[101]

8.3.1.4 *Influence of the Environment*

The environment (melt, solution, gas phase) around the solid, both during the formation/synthesis and in the end application, also has a significant influence on the surface properties. Although the solid is growing, the interactions with the species from a melt or a solution for example, can change the relative surface energies and consequently the equilibrium shape.[79,91] The same is true if the material is surrounded by an aqueous environment. For calcium hydroxide, the stronger interaction of water with the higher energy surfaces compared to the low energy basal plane is likely to be the reason why the particles precipitated from pure, supersaturated $Ca(OH)_2$ solutions do not have the shape of hexagonal platelets observed for naturally occurring minerals.[91] As these interactions are generally transient and changing constantly, as well as the fact that liquids in contact with surfaces generally show density variations perpendicular to the surface in the length scale of 0.5–1 nm,[81,102] only simplified systems with the adsorption of a few molecules are generally studied with DFT or *ab-initio* methods,[83,87,103,104] while for the study of the larger scale systems, the preferred method is classical atomistic simulations.[91,102] If the surface is reacting with species from the environment on the other hand, then DFT is more frequently used in these studies, as chemical reactions are more easily described using DFT. For oxides for example, DFT in combination with thermodynamic calculations allow the prediction of the conditions, under which surfaces are likely to be hydrogenated or hydroxylated,[83,86,87] or it can be used to estimate the acidity of surface groups.[105] Especially for metals, surface reactions, such as oxidation, can lead to the formation of surface layers, which again can be studied using DFT methods.[106]

Defects and species in solution can have a higher or lower affinity to the interface region than to the bulk (Figure 8.4a), leading to a concentration gradient within the material normal to the surface. The energetics of this segregation of defects or adsorption of surrounding species can be estimated using atomistic simulations.[79,82,107] The easiest approach is to calculate the adsorption or segregation enthalpy, by comparing the formation energy of the defect in the bulk and at the interface. However, care has to be taken to consider all relevant positions at the interface. To estimate the free energy

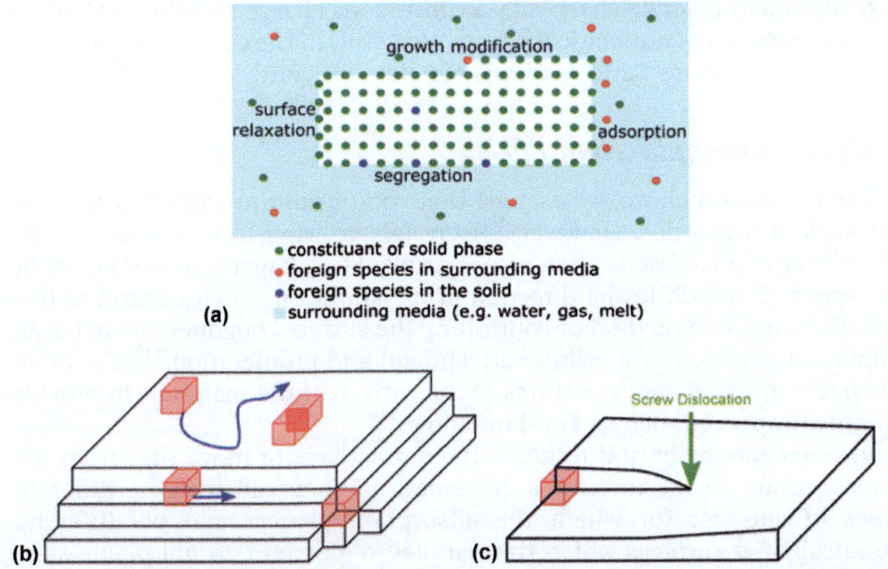

Figure 8.4 Schematic illustration of possible influences on the morphology of solids (a), different types of surface sites and surface diffusion (b) and defect growth mechanisms (c).

difference more advanced methods, such as NEB, metadynamics or other free energy methods may be needed.[81,82] An additional complexity is the distribution and equilibrium concentration of defects. For the estimation of this, additional methods are needed. In the case of weak surface adsorption from solution, direct molecular dynamics simulation of the system over sufficiently long times allows estimation of the distribution of the species in solution.[108] However, owing to the limited possible size and length scale of the simulations, this approach is limited to systems with a weak adsorption and fast dynamics, as well as to large concentrations ($\gtrsim$1 mol L^{-1}). This is of particular interest for biological interactions as biological fluids interacting with materials are usually highly concentrated. For other systems, either additional thermodynamic calculations[109] or Grand Canonical (or Metropolis) Monte Carlo calculations[110] are needed. Within this context single surfaces are also often extended to porous media. A special case arises when the species considered are charged, which is the norm in biological media. In this case the adsorption/segregation will lead to a spatial separation of charges normal to the interface. In the case of solids, one generally speaks of a space charge region, in the case of adsorption from the solution or an electrical double layer. For the latter a charge separation can also be caused by surface reactions such as acid/base reactions of surface groups. For both segregation[111,112] and adsorption[113,114] the distribution of the charges is often calculated using continuum methods (that is approximating the distribution of discrete charges by a continuous charge distribution), often using the results of atomistic methods However, in the case of the electrical double layer, Monte Carlo methods are also frequently used.[105,115,116]

8.3.1.5 Non-equilibrium Effects

In the discussion above, we assume that equilibrium properties determine the surface properties. However, for crystals growing from a solution, the prevalence of a surface is often also controlled to a larger or lesser extent by the speed of growth in the direction of the surface normal. Control of this growth is also one method of controlling the surface roughness which is an important parameter for cellular attachment and proliferation.[117] The main method is to grow nanostructures at the surface of the materials to provide specific properties such as anti-biofouling.[118]

In order for the crystal to grow, the atoms have to move first from the bulk solution to the interfaces. Schematically one can imagine different types of surfaces for which the adsorption process can be different: atomically flat surfaces which run parallel to a certain crystal plane, stepped surfaces with only one direction in common with a crystal plane and finally kinked surfaces that have no direction in common with any crystalline planes (Figure 8.4b). An atom that adsorbs at a kink site will immediately become part of the crystal and indistinguishable from other surface sites. An atom that adsorbs at a step site still retains a certain mobility in one direction, owing to the reduced number of bonds formed with the surface compared with a kink site. In order to definitely become part of the crystal it would have to associate with one or more atoms, forming a one dimensional nucleus, terminated on both ends by a kink site that can then grow easily. Alternatively, an atom adsorbed at a step site can move along the step until it finds a kink site where it can be integrated into the crystal. An atom adsorbed at a flat surface is even less strongly bound to the surface and even more mobile. It can undergo 2D surface diffusion until it either encounters a step or a kink site or other surface atoms to form a 2D layer nucleus. Generally, the relative surface energies also give an indication of the relative growth rates as they are governed by similar properties, a kinked surface will clearly grow faster than a stepped surface and much faster than an atomically flat surface and at the same time can be expected to have a much higher surface energy. Similarly, solvent or solute species adsorbing at flat, stepped or kinked surfaces will preferentially slow down growth in that direction, owing to the competition of the adsorbed species with the growth species, and will reduce the relative surface energy. However, there are a few exceptions in which kinetic effects should be considered.

First of all, for a growth species to adsorb, it will have to move through the electrical double layer and through the structured water layers[102,119] and it will have to lose its hydration shell. These processes are often associated with an energetic barrier which can vary from surface to surface. Such effects can be studied using the same free energy methods that can be used to study the adsorption of other species from solution.[81] Additionally, the changing diffusion coefficient of species in the structured water layers close to the surface,[102] as well as the surface diffusion coefficient,[120] can be important for growth. Diffusion coefficients can either be estimated from activation energies[121] or from direct molecular dynamics calculations, however, especially for the latter great care has to be taken to take into account the correct time scale that allows for large scale diffusion and to not be limited to fast, local movements.[122,123]

Growth in a certain direction can be accelerated by the presence of defects at the surface, namely screw dislocations. A screw dislocation is a line defect at which, at a certain line in the crystal, the atomic planes are partially shifted by an interatomic distance parallel to the defect line. If the screw dislocation reaches a surface, step sites will be created over part of the crystal (Figure 8.2c). This step can now grow in a spiral fashion without ever reaching the edge of the surface and being annihilated. This means that owing to the presence of screw dislocations, a flat surface oriented parallel to an atomic plane can grow continuously without needing to periodically go through a layer nucleation step. This process is well known to influence growth[124] and to be of great importance for the growth of nanoscale structures such as nanorods or nanotubes.[125,126] The energetics and the atomic structure, and even the kinetics (using coarse grained kinetic Monte Carlo) of the growth of these dislocations can be studied using atomistic simulations.[127,128]

Finally, growth can also happen by ordered attachment or aggregation of preformed nanocrystalline building blocks, especially if mediated by organic molecules,[129] a process that can also lead to the formation of screw dislocations that will promote further growth[130] and that has already been studied using atomistic simulation.[131]

8.3.1.6 Example of Application

After having discussed in detail different ways in which computational solid state methods can and have been used to understand the properties of surfaces relevant for biocompatible systems, let us look at a specific application and how the different techniques were applied for that system. A good example of the application of the techniques described above is the study of TiO_2 for photocatalytic water treatment. TiO_2 is one of the most promising candidates for advanced oxidation processes (AOPs), which are based on the generation of highly reactive oxygen species (*i.e.* H_2O_2, OH, O_2^-, O_3) at the surface of particles suspended in the water *via* photocatalysis. These reactive species then lead to the mineralization of refractory organic compounds, water pathogens and disinfection by-products, without creating any secondary pollution. However, photocatalytic processes will improve the production of reactive oxygen species (ROS) leading to potential environmental toxicity.[132] It is crucial to have a good understanding of the surface properties to understand, predict and improve the efficiency of the surface reactions in order to decrease their potential undesired side effects on the living environment. Thus, computational methods, in combination with experimental techniques, have been used extensively to show which surfaces are low-energy and thus present on the particles,[133,134] how these surfaces are modified by reconstruction,[135] surface reactions[87] or defects[136] and finally how the photocatalytic synthesis of ROS occurs at the different surfaces.[83,84]

8.3.2 Surface Adsorption and Surface Functionalization

As discussed above, the properties of surfaces are of paramount importance for biological systems and are often tailored to selectively control the properties of materials. Within this context surface properties are modified by the steered adsorption or grafting of organic molecules providing additional surface functionalities. Typical examples are implants, stents or antibacterial surfaces, which are often specifically functionalized to tailor their biological interactions. For example, surface functionalization can prevent, on different engineered surfaces (silica-, titanium- or polymeric-based for example), the proliferation of micro-organisms such as bacteria or fungi.[137]

The functionalization can also change the charges at the surface. The surface functionalization is mainly done by adsorption of small chemical groups providing new charges or new surface groups without completely masking the surface properties of the original material. For example, small molecules containing amino ($-NH_2$) and/or carboxylic ($-COOH$) groups often have a positive and negative surface charge, enabling further functional molecules to be grafted.[138,139] Another approach consists of adsorbing large molecules such as polymers onto the material surfaces to change their biological interactions or protect the integrity of the bulk material. Polyethyleneglycol (PEG) or polyvinylalcohol (PVA) are used to protect the surfaces of materials from biofouling to colloidally stabilize particles. In the special case of particles being injected or added to biological systems, either *in vitro* or *in vivo*, the surface properties will determine, amongst other things, the cellular uptake and their biodistribution.[140,141] In addition to this steered or voluntary adsorption, in biological media the interaction between the surface and the media generally leads to the formation of a layer of adsorbed bio-molecules. Recently, it has become clear that, to understand the final behavior of bio-interfaces, the adsorption of proteins onto the surface plays a crucial role, something that is still being studied intensively.[142–145]

Thus, whether steered or unsteered, the adsorption of organic molecules is a very important process for biocompatible materials. In this chapter, we will discuss different phenomena that are important for the interaction between surfaces and organic molecules and we will discuss different computational methods that can be used to study them.

8.3.2.1 Influence of the Functional Groups of Adsorbed Molecules

The organic molecules in question possess different functional groups, the molecules (polymers or proteins) can be very large and they are often charged. These three aspects are often studied with different techniques. The interaction of different functional groups with the surface are generally studied for small molecules, peptides or oligomers with either classical atomistic methods[146–148] or DFT[149,150] (Figure 8.3a), although the increasing system sizes required means the use of DFT rapidly becomes prohibitive in terms of computer time.[151] For surface functionalization with small molecules, such simulations, taking into account the charges, can be sufficient. The limitation of the size of the studied molecules does not only come from the system size, but also from the time scale of the movement and the change of conformation of larger molecules, which is much larger than the approximately 100 ns accessible with full atomistic molecular dynamics.[152] Even using advanced techniques, such as metadynamics, full atomistic simulations of the adsorption of realistically sized larger molecules on surfaces, such as many proteins and polymers, are currently not accessible. Especially if the system of interest is not only one molecule adsorbed at the surface, but a more or less compact layer of adsorbed proteins/polymers, in which entanglement and confinement (Figure 8.5b and c) phenomena become important.

8.3.2.2 Influence of the Conformation of Adsorbed Molecules

To study the conformation of polymers or large proteins adsorbed or grafted to a surface, different techniques are used. Closely related to the atomistic methods discussed above are coarse grained methods in which instead of taking into account each atom, only the positions of and the interactions between groups of atoms are described.[153,154] Additionally, the solvent is not always considered explicitly, but can be approximated by adapting the interactions between and the dynamics of the coarse grained entities. Approaches to adapt the dynamics of a system taking into account an implicit solvent are Brownian dynamics[155,156] and Langevin dynamics.[157–159] The advantage of this approach is not only the reduced number of entities within the system, but also the fact that, owing to the larger mass and slower dynamics of the coarse grained entities, a larger timestep can be used and thus larger timescales become accessible. Such coarse grained simulations have, for instance, been used to study end-grafted polymers under shear flow to understand their tribological properties.[155,158] In addition to molecular dynamics, Monte Carlo approaches are also used frequently to study the conformation of adsorbed polymers.[160,161] Finally, owing to the complexity of polymeric systems, a number of thermodynamics and mean field theories have been developed to describe the adsorption of polymers on surfaces and, at least for linear polymers[156,163] and some well-defined more complex

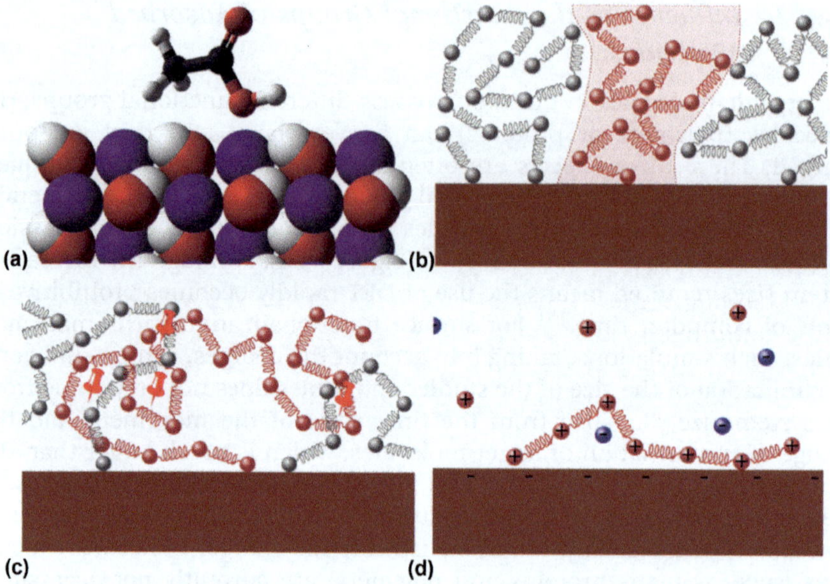

Figure 8.5 Different aspects of surface adsorption of organic molecules: (a) atomistic interactions between the surface groups and functional groups of the organic molecule; (b) confinement effects owing to the restricted space available within an adsorbed layer of molecules; (c) entanglement effects partially pinning the different molecules and restricting its free movement; and (d) interdependence of the adsorption of charged molecules and the ionic distribution within an electrolyte.

systems,[164,165] give a valuable first insight into the phenomenon governing their conformation and behavior. As an example, these theories have been used to explain the mechanisms of thermoresponsive cell culture substrates.[166,167]

8.3.2.3 Influence of the Charges of Adsorbed Molecules

In theory, both the full atomistic and coarse grained methods mentioned above take into account charges on the molecules. However, a further complexity arises when looking at charged molecules adsorbing onto a charged surface within an electrolyte solution. In this case, the conformation and adsorption of the organic molecules is influenced by the distribution of the charges in the electrolyte and the resulting electrical potential, which in turn is influenced by the presence of the molecules. This phenomenon is most commonly studied using coarse grained Metropolis or Grand Canonical Monte Carlo methods, in which the organic molecules are treated as polyelectrolytes[168,169] or using mean field theory.

An example of this field of study is the development of anti-fouling surfaces. The anti-fouling surfaces are specifically designed to avoid unwanted cellular adhesion and proliferation, as well as the unspecific adsorption of biomolecules. A common strategy to render a surface non-biofouling is the formation of a hydrophilic or zwitterionic polymer-layers either through adsorption, coating, chemical grafting or the formation of a self-assembled monolayer.[170] The first theoretical studies into the effectiveness of these layers used mean field theories to study the effect of surface density and polymer length.[171,172] Further atomistic[173,174] and DFT calculations[175] showed the importance of the hydration layer on the polymer surface to understand the non-fouling properties, something that is consistent with experimental observations[176,177] and significantly increased the theoretical understanding of these systems. Subsequent studies looked at the influence of the conformation and arrangement of the molecules in the non-fouling layer.[178,179]

8.3.3 Special Case of Nanoscale Materials

Nanotechnology is a research area which has received growing interest for the last twenty years. The possibility of using the singular properties of nanomaterials to solve long standing problems is becoming more and more promising. NPs have become the most important components in nanotechnology. Their size (usually below 100 nm in at least one direction) with specific physico-chemical characteristics, opens the way for the development of novel solutions for catalysis, the food industry, medicine and many other applications.[180–182] The small size of the particles has different effects compared to those of the bulk solid, as already mentioned above. The physical solid state properties of a nanoscale structure can be different from that of larger structures owing to nano-confinement, additionally the surface properties can be different owing to the influence of the higher concentration of edges and kinks. Also, the specific geometry of the surface (Figure 8.4) changes both the mechanisms of adsorption of large molecules and, in the case of charged particles in the presence of electrolytes, the charge distribution close to the surface. Finally, for any application of nanoparticles dispersed in an aqueous environment, such as most biological media, and thus for biomedical applications in particular, the small size and change in adsorption behaviors influence the stability and colloidal interactions of the nanoparticles. Thus, we will discuss the special case of nanoparticles and other nanoscale materials, as well as their interactions, and will present computational methods that can be used to study them, as well as their current limitations.

8.3.3.1 Nanoparticles

Many nanostructures exhibit unusual electrical, optical or magnetic properties which are directly linked to their size. These properties are either the effect of the high surface to volume ratio and the ensuing high influence of the surface (*e.g.* lower melting temperature, different phase diagram *etc.*) or due to the intermediate nature between the molecule and continuous

3D solid (*e.g.* band structure, optical properties *etc.*). For the latter, quantum mechanics methods (*e.g.* DFT) are of most interest; however, the systems cannot be treated using the methods used for bulk solid-state materials such as periodic boundaries. On the other hand, many of the nanostructures used are too large to be studied using DFT, thus DFT studies are generally restricted to small nanoparticles or clusters[183,184] or to nanoparticle model structures.[185] Using classical atomistic methods, larger structures can be studied to gain insight into changes in the relative stability of different phases or other thermodynamic properties for example.[186,187] Another field of study in which classical atomistic methods are frequently used is the study of the edge and kink-site dominated growth and equilibrium morphologies of nanoparticles which can deviate significantly from that of the large particles discussed above.[186,188]

The small volume of nanoparticles compared to macro particles of nanoparticles has a significant effect on the volume available to the adsorbed species and thus notably influences the adsorption of macromolecules (Figure 8.6). To some extent mean field theory can be used to describe the adsorption of linear molecules,[162] however, for atomistic methods, the adsorption of larger molecules on flat surfaces is already a challenge, as discussed in the previous section. For nanoparticles, now the systems to be studied are even larger, as periodic boundary conditions cannot be applied in the same way as for flat surfaces. Consequently, in the vast majority of studies on polymer/protein adsorption on nanoparticles, coarse grained methods are applied.[157,189–192] In addition, most of the time oligomers, rather than larger molecules, have been studied,[189,193,194] or the nanoparticle has been simplified.[190,191,195] Other coarse grained studies specifically deal with the adsorption of a single charged polyelectrolyte.[196,197]

8.3.3.2 Other Nanoscale Materials

There are two related fields that consider similar phenomena as the adsorption of large organic molecules on nanoparticle surfaces. One is the study of polymer nanocomposites,[198–201] looking at the changes in the conformation of polymers in a polymer melt around a nanoparticle.

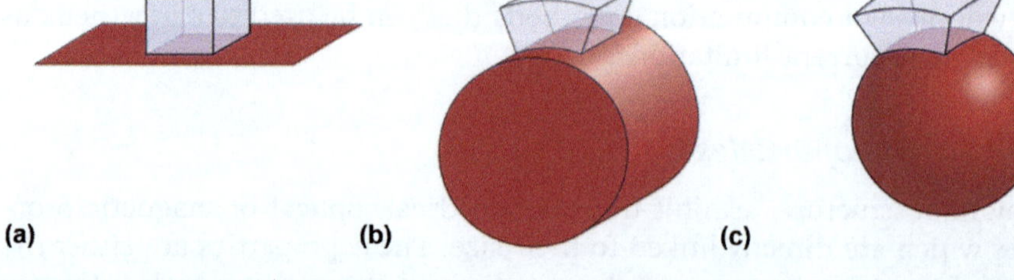

Figure 8.6 Schematic comparison between the volume available to macromolecules adsorbed on (a) a planar surface (large particle), (b) on a nanorod and (c) on a nanoparticle.

Although this phenomenon is also dependent on the interactions of polymers with nanoparticles, the difference between these systems and surface adsorption is that the polymers are confined in the polymer melt, as well as close to the nanoparticle surface. For the study of polymer nanocomposites, generally coarse grained methods similar to those mentioned above are used. A second field of study, which has many similarities to surface adsorption, is the formation of micelles, which, depending on their size, can also be considered as nanostructures, however, instead of being confined by a surface, the systems are confined by hydrophilic/hydrophobic forces acting on different parts of the molecules. As there are only the organic molecules to simulate, and not both the organic molecules and a nanoparticle, many all-atom (or united atom, treating all but the hydrogen atoms explicitly) simulations of micelles can be found,[202,203] although coarse grained simulations on larger micelle structures exist as well.[204,205]

8.3.3.3 Interactions of Nanoscale Structures

In most applications, nanostructures and particles have a tendency to interact, agglomerate and even coalesce or sinter, thus changing their properties and behavior. This process has been studied by several authors for the case of gaseous phase synthesis in which the influence of the surrounding phase on the interaction between particles is negligible and the interaction between small nanoparticles can be simulated in a vacuum.[131,206] Other systems, such as nanoparticles in suspension, are more complicated. Typical biological media for instance are complex electrolytes containing a high concentration of organic molecules. Thus, the interaction between nanoparticles are not only the result of van der Waals and, depending on the material, magnetic forces, mitigated by the surrounding media, but also of electrostatic forces including the electrical double layer and the contribution of any organic molecules adsorbed at the surface of the nanoparticle. Owing to lack of alternatives, the electrical double layer and the electrostatic interaction of nanoparticles in an electrolyte are often described by analytical theories developed for large particles in which the surface can be considered to be flat locally.[113,114,207,208] However, both mean field calculations[209,210] and coarse grained Monte Carlo simulations[211,212] have shown that the ionic distribution around a nanoparticle can differ significantly from that of a flat surface. Nevertheless, only two of the studies mentioned above also look at how the changes in the ionic distribution around the nanoparticle effect the interparticle forces[209] and there are still no readily available models that would be generally applicable to describe the electrostatic forces between nanoparticles. The situation is similar for the effect of adsorbed organic molecules on the interparticle forces. There are some models available,[113] however, they have been developed for larger particles and it is unclear how this is applicable to nanoparticles. The coarse grained studies that exist in the literature concerning the interaction of polymer decorated nanoparticles,[157,213] whilst giving insight into the particular systems considered, do not permit a general view of the contribution of adsorbed organic molecules on the interaction of nanoparticles.

8.4 Summary and Conclusions

Although the field of solid-state computational chemistry is quite established, its application to nanoscale biocompatible materials is still a comparatively young field, but it has high potential and is rapidly growing. The need to control the behavior of materials for biomedical applications drives the study of both bulk and surface structures and properties.

In this chapter, we have shown that computational methodologies can be used as powerful tools to probe the spectroscopic properties of materials applied to medicine. Computational spectroscopy allows for the prediction of opto-electronical properties of the materials and can complement experimental characterization. This gives structural, electronic, optical, magnetic, mechanical and dynamical features at the atomistic level, providing a fundamental understanding to aid the optimization of the design and engineering of biocompatible materials, which include dimensionality of the system and structural defects and impurities. Although adsorption computational spectroscopy can provide information on the mechanisms underlying the excitation processes at the atomistic level, inelastic scattering spectroscopy can probe the vibrational properties of the materials and allow for the analysis of dynamical stability at different conditions of temperature, pressure, and electric fields, and resonance spectroscopy can provide a full picture of chemical impurities and/or native defects found in the biocompatible materials.

The interactions between materials and biological systems are mainly controlled by surface properties. The morphology, surface composition and adsorption events are therefore crucial considerations to take into account when designing the biocompatible material. Typically, biological media are strong electrolytes with large organic molecules, such as proteins, in suspension. The adsorption of proteins and other biomolecules at the surface will largely determine the biological interactions. The adsorption of large molecules is determined by the interactions between the surface and different functional groups of the molecules, confinement and entanglement effects of the large molecules at the surface, as well as by electrostatic forces. The computational study of the first two aspects is well established, using for example atomistic methods for the first, and coarse grained for the second. However, the interdependence between the electrical double layer and the surface adsorption of large biological molecules is often neglected. To take this into account, there are two main methods: (i) coarse grained Metropolis or Grand Canonical Monte Carlo methods, in which the organic molecules are treated as polyelectrolytes; and (ii) mean field theory.

It is clear that for computational techniques to gain a full picture of the behavior of biocompatible materials, a fully predictive multiscale approach is needed. Perhaps a more systematic approach will be needed, such as the one implemented in the Materials' Project,[214] which aims to accelerate the discovery of new technological materials through advanced scientific computing. However, for biocompatible materials, this should not be limited to bulk properties, but be extended to surface properties.

References

1. A. Dhall and W. Self, Cerium Oxide Nanoparticles: A Brief Review of Their Synthesis Methods and Biomedical Applications, *Antioxidants*, 2018, 7(8), 97.
2. G. M. Rignanese, Dielectric Properties of Crystalline and Amorphous Transition Metal Oxides and Silicates as Potential High-K Candidates: The Contribution of Density-Functional Theory, *J. Phys.: Condens. Matter*, 2005, **17**(7), R357.
3. M. Godlewski, S. Gierałtowska, Ł. Wachnicki, R. Pietuszka, B. S. Witkowski, A. Słońska, Z. Gajewski and M. M. Godlewski, High-K Oxides by Atomic Layer Deposition—Applications in Biology and Medicine, *J. Vac. Sci. Technol., A*, 2017, **35**(2), 021508.
4. K. Memarzadeh, A. S. Sharili, J. Huang, S. C. F. Rawlinson and R. P. Allaker, Nanoparticulate Zinc Oxide as a Coating Material for Orthopedic and Dental Implants, *J. Biomed. Mater. Res., Part A*, 2015, **103**(3), 981–989.
5. Y. Zhang, T. R. Nayak, H. Hong and W. Cai, Biomedical Applications of Zinc Oxide Nanomaterials, *Curr. Mol. Med.*, 2013, **13**(10), 1633–1645.
6. B. E. Urban, P. Neogi, K. Senthilkumar, S. K. Rajpurohit, P. Jagadeeshwaran, S. Kim, Y. Fujita and A. Neogi, Bioimaging Using the Optimized Nonlinear Optical Properties of ZnO Nanoparticles, *IEEE J. Sel. Top. Quantum Electron.*, 2012, **18**(4), 1451–1456.
7. M. Rai, A. Yadav and A. Gade, Silver Nanoparticles as a New Generation of Antimicrobials, *Biotechnol. Adv.*, 2009, **27**(1), 76–83.
8. H. S. Dong and S. J. Qi, Realising the Potential of Graphene-Based Materials for Biosurfaces – a Future Perspective, *Biosurf. Biotribol.*, 2015, **1**(4), 229–248.
9. S. Kumar and K. Chatterjee, Comprehensive Review on the Use of Graphene-Based Substrates for Regenerative Medicine and Biomedical Devices, *ACS Appl. Mater. Interfaces*, 2016, **8**(40), 26431–26457.
10. M. Li, P. Xiong, F. Yan, S. Li, C. Ren, Z. Yin, A. Li, H. Li, X. Ji, Y. Zheng and Y. Cheng, An Overview of Graphene-Based Hydroxyapatite Composites for Orthopedic Applications, *Bioact. Mater.*, 2018, **3**(1), 1–18.
11. Y. Y. Shi, M. Li, Q. Liu, Z. J. Jia, X. C. Xu, Y. Cheng and Y. F. Zheng, Electrophoretic Deposition of Graphene Oxide Reinforced Chitosan–Hydroxyapatite Nanocomposite Coatings on Ti Substrate, *J. Mater. Sci.: Mater. Med.*, 2016, **27**(3), 48.
12. M. Xu, D. Fujita and N. Hanagata, Perspectives and Challenges of Emerging Single-Molecule DNA Sequencing Technologies, *Small*, 2009, **5**(23), 2638–2649.

13. N. Yang and X. Jiang, Nanocarbons for DNA Sequencing: A Review, *Carbon*, 2017, **115**, 293–311.
14. A. Fabbro, D. Scaini, V. León, E. Vázquez, G. Cellot, G. Privitera, L. Lombardi, F. Torrisi, F. Tomarchio, F. Bonaccorso, S. Bosi, A. C. Ferrari, L. Ballerini and M. Prato, Graphene-Based Interfaces Do Not Alter Target Nerve Cells, *ACS Nano*, 2016, **10**(1), 615–623.
15. C. Tayran, S. Aydin, M. Çakmak and Ş. Ellialtıoğlu, Structural and Electronic Properties of AB and AA-Stacking Bilayer-Graphene Intercalated by Li, Na, Ca, B, Al, Si, Ge, Ag, and Au Atoms, *Solid State Commun.*, 2016, **231–232**, 57–63.
16. P. V. Santos, T. Schumann Jr., M. H. Oliveira, J. M. J. Lopes and H. Riechert, Acousto-Electric Transport in Epitaxial Monolayer Graphene on SiC, *Appl. Phys. Lett.*, 2013, **102**(22), 221907.
17. M. V. D. Donck, C. D. Beule, B. Partoens, F. M. Peeters and B. V. Duppen, Piezoelectricity in Asymmetrically Strained Bilayer Graphene, *2D Mater.*, 2016, **3**(3), 035015.
18. S.-Y. Li, K.-Q. Liu, L.-J. Yin, W.-X. Wang, W. Yan, X.-Q. Yang, J.-K. Yang, H. Liu, H. Jiang and L. He, Splitting of Van Hove Singularities in Slightly Twisted Bilayer Graphene, *Phys. Rev. B*, 2017, **96**(15), 155416.
19. G. da Cunha Rodrigues, P. Zelenovskiy, K. Romanyuk, S. Luchkin, Y. Kopelevich and A. Kholkin, Strong Piezoelectricity in Single-Layer Graphene Deposited on SiO_2 Grating Substrates, *Nat. Commun.*, 2015, **6**, 7572.
20. H. Yan, Z.-D. Chu, W. Yan, M. Liu, L. Meng, M. Yang, Y. Fan, J. Wang, R.-F. Dou, Y. Zhang, Z. Liu, J.-C. Nie and L. He, Superlattice Dirac Points and Space-Dependent Fermi Velocity in a Corrugated Graphene Monolayer, *Phys. Rev. B: Condens. Matter Mater. Phys.*, 2013, **87**(7), 075405.
21. K. Ahmad, C. Wan, M. A. Al-Eshaikh and A. N. Kadachi, Enhanced Thermoelectric Performance of Bi_2Te_3 Based Graphene Nanocomposites, *Appl. Surf. Sci.*, 2019, **474**, 2–8.
22. P. Hohenberg and W. Kohn, Inhomogeneous Electron Gas, *Phys. Rev.*, 1964, **136**(3B), B864–B871.
23. V. L. Chandraboss, B. Karthikeyan and S. Senthilvelan, Experimental and First-Principles Study of Guanine Adsorption on ZnO Clusters, *Phys. Chem. Chem. Phys.*, 2014, **16**(42), 23461–23475.
24. M. A. L. Marques and E. K. U. Gross, Time-Dependent Density Functional Theory, *Annu. Rev. Phys. Chem.*, 2004, **55**(1), 427–455.
25. L. Hedin, New Method for Calculating the One-Particle Green's Function with Application to the Electron-Gas Problem, *Phys. Rev.*, 1965, **139**(3A), A796–A823.
26. E. E. Salpeter and H. A. Bethe, A Relativistic Equation for Bound-State Problems, *Phys. Rev.*, 1951, **84**(6), 1232–1242.
27. S. E. Saddow, Silicon Carbide Materials for Biomedical Applications, in *Silicon Carbide Biotechnology* ed. S. E. Saddow, Elsevier, 2nd edn, ch. 1, 2016, pp. 1–25.

28. S. Rigamonti, S. Botti, V. Veniard, C. Draxl, L. Reining and F. Sottile, Estimating Excitonic Effects in the Absorption Spectra of Solids: Problems and Insight from a Guided Iteration Scheme, *Phys. Rev. Lett.*, 2015, **114**(14), 146402.
29. A. Marini, C. Hogan, M. Grüning and D. Varsano, Yambo: An Ab Initio Tool for Excited State Calculations, *Comput. Phys. Commun.*, 2009, **180**(8), 1392–1403.
30. B. F. Milne, Enhancement of Nonlinear Optical Properties in Late Group 15 Tetrasubstituted Cubanes, *Dalton Trans.*, 2014, **43**(17), 6333–6338.
31. C. Attaccalite and M. Grüning, Nonlinear Optics from an Ab Initio Approach by Means of the Dynamical Berry Phase: Application to Second- and Third-Harmonic Generation in Semiconductors, *Phys. Rev. B: Condens. Matter Mater. Phys.*, 2013, **88**(23), 235113.
32. E. Luppi, H. Hübener and V. Véniard, Ab Initio Second-Order Nonlinear Optics in Solids: Second-Harmonic Generation Spectroscopy from Time-Dependent Density-Functional Theory, *Phys. Rev. B: Condens. Matter Mater. Phys.*, 2010, **82**(23), 235201.
33. H. A. Kurtz, J. J. P. Stewart and K. M. Dieter, Calculation of the Nonlinear Optical Properties of Molecules, *J. Comput. Chem.*, 1990, **11**(1), 82–87.
34. J. N. Balasaheb, C. Sunil and B. Vaishali, Study of Electronic and Optical Properties of ZnO Clusters Using TD-DFT Method, *Mater. Res. Express*, 2017, **4**(10), 106304.
35. J. Peña-Bahamonde, H. N. Nguyen, S. K. Fanourakis and D. F. Rodrigues, Recent Advances in Graphene-Based Biosensor Technology with Applications in Life Sciences, *J. Nanobiotechnol.*, 2018, **16**(1), 75.
36. T. Jiang, D. Huang, J. Cheng, X. Fan, Z. Zhang, Y. Shan, Y. Yi, Y. Dai, L. Shi, K. Liu, C. Zeng, J. Zi, J. E. Sipe, Y.-R. Shen, W.-T. Liu and S. Wu, Gate-Tunable Third-Order Nonlinear Optical Response of Massless Dirac Fermions in Graphene, *Nat. Photonics*, 2018, **12**(7), 430–436.
37. N. V. Kuzmin, P. Wesseling, P. C. d. W. Hamer, D. P. Noske, G. D. Galgano, H. D. Mansvelder, J. C. Baayen and M. L. Groot, Third Harmonic Generation Imaging for Fast, Label-Free Pathology of Human Brain Tumors, *Biomed. Opt. Express*, 2016, 7(5), 1889–1904.
38. L. M. Malard, M. A. Pimenta, G. Dresselhaus and M. S. Dresselhaus, Raman Spectroscopy in Graphene, *Phys. Rep.*, 2009, **473**(5), 51–87.
39. V. N. Popov, Two-Phonon Raman Bands of Bilayer Graphene: Revisited, *Carbon*, 2015, **91**, 436–444.
40. J. Jin-Wu, W. Bing-Shen, W. Jian-Sheng and S. P. Harold, A Review on the Flexural Mode of Graphene: Lattice Dynamics, Thermal Conduction, Thermal Expansion, Elasticity and Nanomechanical Resonance, *J. Phys.: Condens. Matter*, 2015, **27**(8), 083001.
41. B. Amorim and F. Guinea, Flexural Mode of Graphene on a Substrate, *Phys. Rev. B: Condens. Matter Mater. Phys.*, 2013, **88**(11), 115418.

42. W. Kohn, Image of the Fermi Surface in the Vibration Spectrum of a Metal, *Phys. Rev. Lett.*, 1959, **2**(9), 393–394.
43. I. Milošević, N. Kepčija, E. Dobardžić, M. Damnjanović, M. Mohr, J. Maultzsch and C. Thomsen, Kohn Anomaly in Graphene, *Mater. Sci. Eng., B*, 2011, **176**(6), 510–511.
44. K. E. El-Kelany, P. Carbonnière, A. Erba, J.-M. Sotiropoulos and M. Rérat, Piezoelectricity of Functionalized Graphene: A Quantum-Mechanical Rationalization, *J. Phys. Chem. C*, 2016, **120**(14), 7795–7803.
45. N. More and G. Kapusetti, Piezoelectric Material – a Promising Approach for Bone and Cartilage Regeneration, *Med. Hypotheses*, 2017, **108**, 10–16.
46. J.-Y. Chapelon, D. Cathignol, C. Cain, E. Ebbini, J.-U. Kluiwstra, O. A. Sapozhnikov, G. Fleury, R. Berriet, L. Chupin and J.-L. Guey, New Piezoelectric Transducers for Therapeutic Ultrasound, *Ultrasound Med. Biol.*, 2000, **26**(1), 153–159.
47. G. N. M. Ferreira, A.-C. da-Silva and B. Tomé, Acoustic Wave Biosensors: Physical Models and Biological Applications of Quartz Crystal Microbalance, *Trends Biotechnol.*, 2009, **27**(12), 689–697.
48. A. Rodzinski, R. Guduru, P. Liang, A. Hadjikhani, T. Stewart, E. Stimphil, C. Runowicz, R. Cote, N. Altman, R. Datar and S. Khizroev, Targeted and Controlled Anticancer Drug Delivery and Release with Magnetoelectric Nanoparticles, *Sci. Rep.*, 2016, **6**, 20867.
49. C. A. L. Bassett and R. O. Becker, Generation of Electric Potentials by Bone in Response to Mechanical Stress, *Science*, 1962, **137**(3535), 1063.
50. L. Nuccio, L. Schulz and A. J. Drew, Muon Spin Spectroscopy: Magnetism, Soft Matter and the Bridge between the Two, *J. Phys. D: Appl. Phys.*, 2014, **47**(47), 473001.
51. K. Nagamine, Muon Application to Advanced Bio- and Nano-Sciences, *AIP Conf. Proc.*, 2008, **981**(1), 375–377.
52. M. H. Dehn, D. J. Arseneau, P. Böni, M. D. Bridges, T. Buck, D. L. Cortie, D. G. Fleming, J. A. Kelly, W. A. MacFarlane, M. J. MacLachlan, R. M. L. McFadden, G. D. Morris, P.-X. Wang, J. Xiao, V. M. Zamarion and R. F. Kiefl, Communication: Chemisorption of Muonium on Gold Nanoparticles: A Sensitive New Probe of Surface Magnetism and Reactivity, *J. Chem. Phys.*, 2016, **145**(18), 181102.
53. L. V. Elnikova, On the Application of Positron Annihilation Spectroscopy for Studying the Photodestruction of Biological Membranes, *J. Surf. Invest.: X-ray, Synchrotron Neutron Tech.*, 2015, **9**(2), 257–262.
54. P. Moskal, D. Kisielewska, C. Curceanu, E. Czerwiński, K. Dulski, A. Gajos, M. Gorgol, B. Hiesmayr, B. Jasińska, K. Kacprzak, Ł. Kapłon, G. Korcyl, P. Kowalski, W. Krzemień, T. Kozik, E. Kubicz, M. Mohammed, Sz. Niedźwiecki, M. Pałka, M. Pawlik-Niedźwiecka, L. Raczyński, J. Raj, S. Sharma, Shivani, R. Y. Shopa, M. Silarski, M. Skurzok, E. Stępień, W. Wiślicki and B. Zgardzińska, Feasibility Study of the Positronium Imaging with the J-PET Tomograph, *Phys. Med. Biol.*, 2019, **64**(5), 055017.
55. H. M. Chen, J. D. van Horn and Y. C. Jean, Applications of Positron Annihilation Spectroscopy to Life Science, *Defect Diffus. Forum*, 2012, **331**, 275–293.

56. C. G. V. d. Walle and J. Neugebauer, Hydrogen in Semiconductors, *Annu. Rev. Mater. Res.*, 2006, **36**(1), 179–198.
57. E. L. Silva, A. G. Marinopoulos, R. C. Vilão, R. B. L. Vieira, H. V. Alberto, J. Piroto Duarte and J. M. Gil, Hydrogen Impurity in Yttria: Ab Initio and μSr Perspectives, *Phys. Rev. B: Condens. Matter Mater. Phys.*, 2012, **85**(16), 165211.
58. E. L. da Silva, A. G. Marinopoulos, R. B. L. Vieira, R. C. Vilão, H. V. Alberto, J. M. Gil, R. L. Lichti, P. W. Mengyan and B. B. Baker, Electronic Structure of Interstitial Hydrogen in Lutetium Oxide from DFT + U Calculations and Comparison Study with μSr Spectroscopy, *Phys. Rev. B*, 2016, **94**(1), 014104.
59. I. Makkonen, M. Hakala and M. J. Puska, First-Principles Calculation of Positron States and Annihilation at Defects in Semiconductors, *Phys. B*, 2006, **376–377**, 971–974.
60. M. Riccò, D. Pontiroli, M. Mazzani, M. Choucair, J. A. Stride and O. V. Yazyev, Muons Probe Strong Hydrogen Interactions with Defective Graphene, *Nano Lett.*, 2011, **11**(11), 4919–4922.
61. S. Park, B. Lee, S. H. Jeon and S. Han, Hybrid Functional Study on Structural and Electronic Properties of Oxides, *Curr. Appl. Phys.*, 2011, **11**(3, Supplement), S337–S340.
62. M. S. Hybertsen and S. G. Louie, Electron Correlation in Semiconductors and Insulators: Band Gaps and Quasiparticle Energies, *Phys. Rev. B: Condens. Matter Mater. Phys.*, 1986, **34**(8), 5390–5413.
63. M. van Schilfgaarde, T. Kotani and S. Faleev, Quasiparticle Self-Consistent GW Theory, *Phys. Rev. Lett.*, 2006, **96**(22), 226402.
64. S. F. J. Cox, J. L. Gavartin, J. S. Lord, S. P. Cottrell, J. M. Gil, H. V. Alberto, J. P. Duarte, R. C. Vilão, N. A. d. Campos, D. J. Keeble, E. A. Davis, M. Charlton and D. P. v. d. Werf, Oxide Muonics: II. Modelling the Electrical Activity of Hydrogen in Wide-Gap and High-Permittivity Dielectrics, *J. Phys.: Condens. Matter*, 2006, **18**(3), 1079.
65. T. Andelman, S. Gordonov, G. Busto, P. V. Moghe and R. E. Riman, Synthesis and Cytotoxicity of Y_2O_3 Nanoparticles of Various Morphologies, *Nanoscale Res. Lett.*, 2009, **5**(2), 263.
66. D. den Engelsen, G. R. Fern, T. G. Ireland, P. G. Harris, P. R. Hobson, A. Lipman, R. Dhillon, P. J. Marsh and J. Silver, Ultraviolet and Blue Cathodoluminescence from Cubic Y_2O_3 and Y_2O_3:Eu^{3+} Generated in a Transmission Electron Microscope, *J. Mater. Chem. C*, 2016, **4**(29), 7026–7034.
67. W. Xu, Y. Chang and G. H. Lee, Biomedical Applications of Lanthanide Oxide Nanoparticles, *J. Biomater. Tissue Eng.*, 2017, **7**(9), 757–769.
68. I. Spasojević, Electron Paramagnetic Resonance - a Powerful Tool of Medical Biochemistry in Discovering Mechanisms of Disease and Treatment Prospects, *J. Med. Biochem.*, 2010, **29**(3), 175–188.
69. C. G. Van de Walle, Hydrogen as a Cause of Doping in Zinc Oxide, *Phys. Rev. Lett.*, 2000, **85**(5), 1012–1015.

70. H. Li and J. Robertson, Behaviour of Hydrogen in Wide Band Gap Oxides, *J. Appl. Phys.*, 2014, **115**(20), 203708.
71. K. Szász, T. Hornos, M. Marsman and A. Gali, Hyperfine Coupling of Point Defects in Semiconductors by Hybrid Density Functional Calculations: The Role of Core Spin Polarization, *Phys. Rev. B: Condens. Matter Mater. Phys.*, 2013, **88**(7), 075202.
72. M. N. Risto, Issues in First-Principles Calculations for Defects in Semiconductors and Oxides, *Modell. Simul. Mater. Sci. Eng.*, 2009, **17**(8), 084001.
73. H. Jónsson, G. Mills and K. W. Jacobsen, Nudged Elastic Band Method for Finding Minimum Energy Paths of Transitions, in *Classical and Quantum Dynamics in Condensed Phase Simulations*, 1998, pp. 385–404.
74. A. G. Marinopoulos, Protons in Cubic Yttria-Stabilized Zirconia: Binding Sites and Migration Pathways, *Solid State Ionics*, 2018, **315**, 116–125.
75. B. R. K. Nanda, M. Sherafati, Z. S. Popović and S. Satpathy, Electronic Structure of the Substitutional Vacancy in Graphene: Density-Functional and Green's Function Studies, *New J. Phys.*, 2012, **14**(8), 083004.
76. V. A. Chirayath, M. D. Chrysler, A. D. McDonald, R. W. Gladen, A. J. Fairchild, A. R. Koymen and A. H. Weiss, Investigation of Graphene Using Low Energy Positron Annihilation Induced Doppler Broadening Spectroscopy, *J. Phys.: Conf. Ser.*, 2017, **791**(1), 012032.
77. M. J. Puska and R. M. Nieminen, Theory of Positrons in Solids and on Solid Surfaces, *Rev. Mod. Phys.*, 1994, **66**(3), 841–897.
78. P. Billemont, B. Coasne and G. De Weireld, Adsorption of Carbon Dioxide, Methane, and Their Mixtures in Porous Carbons: Effect of Surface Chemistry, Water Content, and Pore Disorder, *Langmuir*, 2013, **29**(10), 3328–3338.
79. J. P. Allen, W. Gren, M. Molinari, C. Arrouvel, F. Maglia and S. C. Parker, Atomistic Modelling of Adsorption and Segregation at Inorganic Solid Interfaces, *Mol. Simul.*, 2009, **35**(7), 584–608.
80. U. Aschauer, F. Jones, W. R. Richmond, P. Bowen, A. L. Rohl, G. M. Parkinson and H. Hofmann, Growth Modification of Hematite by Phosphonate Additives, *J. Cryst. Growth*, 2008, **310**(3), 688–698.
81. S. Galmarini and P. Bowen, Atomistic Simulation of the Adsorption of Calcium and Hydroxyl Ions onto Portlandite Surfaces - Towards Crystal Growth Mechanisms, *Cem. Concr. Res.*, 2016, **81**, 16–23.
82. S. Galmarini, A. Kunhi Mohamed and P. Bowen, Atomistic Simulations of Silicate Species Interaction with Portlandite Surfaces, *J. Phys. Chem. C*, 2016, **120**(39), 22407–22413.
83. A. Tilocca and A. Selloni, Reaction Pathway and Free Energy Barrier for Defect-Induced Water Dissociation on the (101) Surface of TiO_2-Anatase, *J. Chem. Phys.*, 2003, **119**(14), 7445–7450.
84. U. Aschauer, Y. He, H. Cheng, S.-C. Li, U. Diebold and A. Selloni, Influence of Subsurface Defects on the Surface Reactivity of TiO_2: Water on Anatase (101), *J. Phys. Chem. C*, 2010, **114**(2), 1278–1284.

85. G. Wulff, Zur Frage Der Geschwindigkeit Des Wachstums Und Der Auflösung Der Krystallflächen, *Z. Krystallogr. Mineral.*, 1901, **34**, 449–530.
86. A. Marmier and S. C. Parker, Ab Initio Morphology and Surface Thermodynamics of α–Al_2O_3, *Phys. Rev. B: Condens. Matter Mater. Phys.*, 2004, **69**(11), 115409.
87. C. Arrouvel, M. Digne, M. Breysse, H. Toulhoat and P. Raybaud, Effects of Morphology on Surface Hydroxyl Concentration: A DFT Comparison of Anatase–TiO_2 and Γ-Alumina Catalytic Supports, *J. Catal.*, 2004, **222**(1), 152–166.
88. L. Vitos, A. V. Ruban, H. L. Skriver and J. Kollár, The Surface Energy of Metals, *Surf. Sci.*, 1998, **411**(1), 186–202.
89. P. W. Tasker, Surface Energies, Surface Tensions and Surface-Structure of the Alkali-Halide Crystals, *Philos. Mag. A*, 1979, **39**(2), 119–136.
90. U. Aschauer, P. Bowen and S. C. Parker, Atomistic Modeling Study of Surface Segregation in Nd:YAG, *J. Am. Ceram. Soc.*, 2006, **89**(12), 3812–3816.
91. S. Galmarini, A. Aimable, N. Ruffray and P. Bowen, Changes in Portlandite Morphology with Solvent Composition: Atomistic Simulations and Experiment, *Cem. Concr. Res.*, 2011, **41**(12), 1330–1338.
92. N. H. de Leeuw and S. C. Parker, Surface Structure and Morphology of Calcium Carbonate Polymorphs Calcite, Aragonite, and Vaterite: An Atomistic Approach, *J. Phys. Chem. B*, 1998, **102**(16), 2914–2922.
93. J. Schardt, J. Bernhardt, U. Starke and K. Heinz, Crystallography of the (3×3) Surface Reconstruction of 3*C* – SiC (111), 4*H* – SiC (0001), and 6*H* – SiC (0001) Surfaces Retrieved by Low-Energy Electron Diffraction, *Phys. Rev. B: Condens. Matter Mater. Phys.*, 2000, **62**(15), 10335–10344.
94. J. Q. Broughton and G. H. Gilmer, Molecular Dynamics Investigation of the Crystal–Fluid Interface. VI. Excess Surface Free Energies of Crystal–Liquid Systems, *J. Chem. Phys.*, 1986, **84**(10), 5759–5768.
95. R. L. Davidchack and B. B. Laird, Crystal Structure and Interaction Dependence of the Crystal-Melt Interfacial Free Energy, *Phys. Rev. Lett.*, 2005, **94**(8), 086102.
96. G. Grochola, S. P. Russo, I. Yarovsky and I. K. Snook, "Exact" Surface Free Energies of Iron Surfaces Using a Modified Embedded Atom Method Potential and Λ Integration, *J. Chem. Phys.*, 2004, **120**(7), 3425–3430.
97. M. Asta, J. J. Hoyt and A. Karma, Calculation of Alloy Solid-Liquid Interfacial Free Energies from Atomic-Scale Simulations, *Phys. Rev. B: Condens. Matter Mater. Phys.*, 2002, **66**(10), 100101.
98. C. A. Becker, D. Olmsted, M. Asta, J. J. Hoyt and S. M. Foiles, Atomistic Underpinnings for Orientation Selection in Alloy Dendritic Growth, *Phys. Rev. Lett.*, 2007, **98**(12), 125701.
99. R. L. Davidchack, J. R. Morris and B. B. Laird, The Anisotropic Hard-Sphere Crystal-Melt Interfacial Free Energy from Fluctuations, *J. Chem. Phys.*, 2006, **125**(9), 094710.

100. J. R. Morris, M. I. Mendelev and D. J. Srolovitz, A Comparison of Crystal–Melt Interfacial Free Energies Using Different Al Potentials, *J. Non-Cryst. Solids*, 2007, **353**(32), 3565–3569.
101. M. Asta, C. Beckermann, A. Karma, W. Kurz, R. Napolitano, M. Plapp, G. Purdy, M. Rappaz and R. Trivedi, Solidification Microstructures and Solid-State Parallels: Recent Developments, Future Directions, *Acta Mater.*, 2009, **57**(4), 941–971.
102. S. Kerisit, D. J. Cooke, D. Spagnoli and S. C. Parker, Molecular Dynamics Simulations of the Interactions between Water and Inorganic Solids, *J. Mater. Chem.*, 2005, **15**(14), 1454–1462.
103. M. Claus, H. Jürgen, K. Walter and B. Kurt, Water Adsorption on Amorphous Silica Surfaces: A Car–Parrinello Simulation Study, *J. Phys.: Condens. Matter*, 2005, **17**(26), 4005.
104. Y. Foucaud, M. Badawi, L. O. Filippov, I. V. Filippova and S. Lebègue, Surface Properties of Fluorite in Presence of Water: An Atomistic Investigation, *J. Phys. Chem. B*, 2018, **122**(26), 6829–6836.
105. S. V. Churakov, C. Labbez, L. Pegado and M. Sulpizi, Intrinsic Acidity of Surface Sites in Calcium Silicate Hydrates and Its Implication to Their Electrokinetic Properties, *J. Phys. Chem. C*, 2014, **118**(22), 11752–11762.
106. M. Aykol and K. A. Persson, Oxidation Protection with Amorphous Surface Oxides: Thermodynamic Insights from Ab Initio Simulations on Aluminum, *ACS Appl. Mater. Interfaces*, 2018, **10**(3), 3039–3045.
107. S. Galmarini, U. Aschauer, P. Bowen and S. C. Parker, Atomistic Simulation of Y-Doped Alpha-Alumina Interfaces, *J. Am. Ceram. Soc.*, 2008, **91**(11), 3643–3651.
108. S. Kerisit, D. J. Cooke, A. Marmier and S. C. Parker, Atomistic Simulation of Charged Iron Oxyhydroxide Surfaces in Contact with Aqueous Solution, *Chem. Commun.*, 2005, **24**, 3027–3029.
109. W. C. Mackrodt and P. W. Tasker, Segregation Isotherms at the Surfaces of Oxides, *J. Am. Ceram. Soc.*, 1989, **72**(9), 1576–1583.
110. J. A. Purton, J. C. Crabtree and S. C. Parker, Dl_Monte: A General Purpose Program for Parallel Monte Carlo Simulation, *Mol. Simul.*, 2013, **39**(14–15), 1240–1252.
111. S. B. Desu and D. A. Payne, Interfacial Segregation in Perovskites: I, Theory, *J. Am. Ceram. Soc.*, 1990, **73**(11), 3391–3397.
112. M. F. Yan, R. M. Cannon and H. K. Bowen, Space Charge, Elastic Field, and Dipole Contributions to Equilibrium Solute Segregation at Interfaces, *J. Appl. Phys.*, 1983, **54**(2), 764–778.
113. O. Burgos-Montes, R. Moreno and P. Bowen, Hamaker 2: A Toolkit for the Calculation of Particle Interactions and Suspension Stability and Its Application to Mullite Synthesis by Colloidal Methods Au - Aschauer, Ulrich, *J. Dispersion Sci. Technol.*, 2011, **32**(4), 470–479.
114. S. H. Behrens and M. Borkovec, Electrostatic Interaction of Colloidal Surfaces with Variable Charge, *J. Phys. Chem. B*, 1999, **103**(15), 2918–2928.

115. J. Forsman, B. Jönsson, C. E. Woodward and H. Wennerström, Attractive Surface Forces Due to Liquid Density Depression, *J. Phys. Chem. B*, 1997, **101**(21), 4253–4259.
116. C. Labbez, I. Pochard, B. Jönsson and A. Nonat, C-S-H/Solution Interface: Experimental and Monte Carlo Studies, *Cem. Concr. Res.*, 2011, **41**(2), 161–168.
117. L. R. Freschauf, J. McLane, H. Sharma and M. Khine, Shrink-Induced Superhydrophobic and Antibacterial Surfaces in Consumer Plastics, *PLoS One*, 2012, **7**(8), e40987.
118. H. Wang, L. Wang, P. Zhang, L. Yuan, Q. Yu and H. Chen, High Antibacterial Efficiency of Pdmaema Modified Silicon Nanowire Arrays, *Colloids Surf., B*, 2011, **83**(2), 355–359.
119. P. Zarzycki, S. Kerisit and K. M. Rosso, Molecular Dynamics Study of the Electrical Double Layer at Silver Chloride – Electrolyte Interfaces, *J. Phys. Chem. C*, 2010, **114**(19), 8905–8916.
120. H. J. Fan, M. Knez, R. Scholz, D. Hesse, K. Nielsch, M. Zacharias and U. Gösele, Influence of Surface Diffusion on the Formation of Hollow Nanostructures Induced by the Kirkendall Effect: The Basic Concept, *Nano Lett.*, 2007, **7**(4), 993–997.
121. K. S. Smirnov and D. Bougeard, A Molecular Dynamics Study of Structure and Short-Time Dynamics of Water in Kaolinite, *J. Phys. Chem. B*, 1999, **103**(25), 5266–5273.
122. U. Aschauer, P. Bowen and S. C. Parker, Oxygen Vacancy Diffusion in Alumina: New Atomistic Simulation Methods Applied to an Old Problem, *Acta Mater.*, 2009, **57**(16), 4765–4772.
123. A. Tewari, U. Aschauer and P. Bowen, Atomistic Modeling of Effect of Mg on Oxygen Vacancy Diffusion in A-Alumina, *J. Am. Ceram. Soc.*, 2014, **97**(8), 2596–2601.
124. F. C. Frank, The Influence of Dislocations on Crystal Growth, *Discuss. Faraday Soc.*, 1949, **5**(0), 48–54.
125. L. Chen, B. Liu, A. N. Abbas, Y. Ma, X. Fang, Y. Liu and C. Zhou, Screw-Dislocation-Driven Growth of Two-Dimensional Few-Layer and Pyramid-Like WSe_2 by Sulfur-Assisted Chemical Vapor Deposition, *ACS Nano*, 2014, **8**(11), 11543–11551.
126. Y. Chu, P. Chen, J. Tang and P. Rao, Engineer in Situ Growth of A-Al_2O_3 Whiskers by Axial Screw Dislocations, *Cryst. Growth Des.*, 2017, **17**(4), 1999–2005.
127. F. Ding, A. R. Harutyunyan and B. I. Yakobson, Dislocation Theory of Chirality-Controlled Nanotube Growth, *Proc. Natl. Acad. Sci. U. S. A.*, 2009, **106**(8), 2506.
128. S. Piana, M. Reyhani and J. D. Gale, Simulating Micrometre-Scale Crystal Growth from Solution, *Nature*, 2005, **438**, 70.
129. H. Cölfen and S. Mann, Higher-Order Organization by Mesoscale Self-Assembly and Transformation of Hybrid Nanostructures, *Angew. Chem., Int. Ed.*, 2003, **42**(21), 2350–2365.

130. R. L. Penn and J. F. Banfield, Imperfect Oriented Attachment: Dislocation Generation in Defect-Free Nanocrystals, *Science*, 1998, **281**(5379), 969.
131. D. Spagnoli, J. F. Banfield and S. C. Parker, Free Energy Change of Aggregation of Nanoparticles, *J. Phys. Chem. C*, 2008, **112**(38), 14731–14736.
132. N. von Moos, V. B. Koman, C. Santschi, O. J. F. Martin, L. Maurizi, A. Jayaprakash, P. Bowen and V. I. Slaveykova, Pro-Oxidant Effects of Nano-TiO_2 on Chlamydomonas Reinhardtii During Short-Term Exposure, *RSC Adv.*, 2016, **6**(116), 115271–115283.
133. M. Ramamoorthy, D. Vanderbilt and R. D. King-Smith, First-Principles Calculations of the Energetics of Stoichiometric TiO_2 Surfaces, *Phys. Rev. B: Condens. Matter Mater. Phys.*, 1994, **49**(23), 16721–16727.
134. H. Perron, C. Domain, J. Roques, R. Drot, E. Simoni and H. Catalette, Optimisation of Accurate Rutile TiO_2 (110), (100), (101) and (001) Surface Models from Periodic DFT Calculations, *Theor. Chem. Acc.*, 2007, **117**(4), 565–574.
135. S. D. Elliott and S. P. Bates, Assignment of the (1×2) Surface of Rutile TiO_2(110) from First Principles, *Phys. Rev. B: Condens. Matter Mater. Phys.*, 2003, **67**(3), 035421.
136. H. Cheng and A. Selloni, Surface and Subsurface Oxygen Vacancies in Anatase TiO_2 and Differences with Rutile, *Phys. Rev. B: Condens. Matter Mater. Phys.*, 2009, **79**(9), 092101.
137. B. R. Coad, S. E. Kidd, D. H. Ellis and H. J. Griesser, Biomaterials Surfaces Capable of Resisting Fungal Attachment and Biofilm Formation, *Biotechnol. Adv.*, 2014, **32**(2), 296–307.
138. M. Yamaura, R. L. Camilo, L. C. Sampaio, M. A. Macêdo, M. Nakamura and H. E. Toma, Preparation and Characterization of (3-Aminopropyl)Triethoxysilane-Coated Magnetite Nanoparticles, *J. Magn. Magn. Mater.*, 2004, **279**(2), 210–217.
139. M. A. Dobrovolskaia, A. K. Patri, J. Zheng, J. D. Clogston, N. Ayub, P. Aggarwal, B. W. Neun, J. B. Hall and S. E. McNeil, Interaction of Colloidal Gold Nanoparticles with Human Blood: Effects on Particle Size and Analysis of Plasma Protein Binding Profiles, *Nanomedicine: NBM*, 2009, **5**(2), 106–117.
140. F. Alexis, E. Pridgen, L. K. Molnar and O. C. Farokhzad, Factors Affecting the Clearance and Biodistribution of Polymeric Nanoparticles, *Mol. Pharmaceutics*, 2008, **5**(4), 505–515.
141. L. Maurizi, A.-L. Papa, L. Dumont, F. Bouyer, P. Walker, D. Vandroux and N. Millot, Influence of Surface Charge and Polymer Coating on Internalization and Biodistribution of Polyethylene Glycol-Modified Iron Oxide Nanoparticles, *J. Biomed. Nanotechnol.*, 2015, **11**(1), 126–136.
142. L. Vroman, Effect of Adsorbed Proteins on the Wettability of Hydrophilic and Hydrophobic Solids, *Nature*, 1962, **196**, 476.
143. N. Bertrand, P. Grenier, M. Mahmoudi, E. M. Lima, E. A. Appel, F. Dormont, J.-M. Lim, R. Karnik, R. Langer and O. C. Farokhzad,

Mechanistic Understanding of in Vivo Protein Corona Formation on Polymeric Nanoparticles and Impact on Pharmacokinetics, *Nat. Commun.*, 2017, **8**(1), 777.

144. S. Galmarini, U. Hanusch, M. Giraud, N. Cayla, D. Chiappe, N. von Moos, H. Hofmann and L. Maurizi, Beyond Unpredictability: The Importance of Reproducibility in Understanding the Protein Corona of Nanoparticles, *Bioconjugate Chem.*, 2018, **29**(10), 3385–3393.
145. U. Sakulkhu, L. Maurizi, M. Mahmoudi, M. Motazacker, M. Vries, A. Gramoun, M.-G. Ollivier Beuzelin, J.-P. Vallée, F. Rezaee and H. Hofmann, Ex Situ Evaluation of the Composition of Protein Corona of Intravenously Injected Superparamagnetic Nanoparticles in Rats, *Nanoscale*, 2014, **6**(19), 11439–11450.
146. U. Aschauer, J. Ebert, A. Aimable and P. Bowen, Growth Modification of Seeded Calcite by Carboxylic Acid Oligomers and Polymers: Toward an Understanding of Complex Growth Mechanisms, *Cryst. Growth Des.*, 2010, **10**(9), 3956–3963.
147. M. J. Penna, M. Mijajlovic and M. J. Biggs, Molecular-Level Understanding of Protein Adsorption at the Interface between Water and a Strongly Interacting Uncharged Solid Surface, *J. Am. Chem. Soc.*, 2014, **136**(14), 5323–5331.
148. T. Utesch, G. Daminelli and M. A. Mroginski, Molecular Dynamics Simulations of the Adsorption of Bone Morphogenetic Protein-2 on Surfaces with Medical Relevance, *Langmuir*, 2011, **27**(21), 13144–13153.
149. D. Bonvin, U. J. Aschauer, J. A. M. Bastiaansen, M. Stuber, H. Hofmann and M. Mionić Ebersold, Versatility of Pyridoxal Phosphate as a Coating of Iron Oxide Nanoparticles, *Nanomaterials*, 2017, **7**(8), 202.
150. D. Bonvin, U. Aschauer, D. T. L. Alexander, D. Chiappe, M. Moniatte, H. Hofmann and M. Mionić Ebersold, Protein Corona: Impact of Lymph Versus Blood in a Complex in Vitro Environment, *Small*, 2017, **13**(29), 1700409.
151. O. Bertran, B. Zhang, A. D. Schlüter, M. Kröger and C. Alemán, Modeling Nanosized Single Molecule Objects: Dendronized Polymers Adsorbed onto Mica, *J. Phys. Chem. C*, 2015, **119**(7), 3746–3753.
152. T. Wei, M. A. Carignano and I. Szleifer, Lysozyme Adsorption on Polyethylene Surfaces: Why Are Long Simulations Needed?, *Langmuir*, 2011, **27**(19), 12074–12081.
153. C. M. Hassan, P. Trakampan and N. A. Peppas, Water Solubility Characteristics of Poly(Vinyl Alcohol) and Gels Prepared by Freezing/Thawing Processes, in *Water Soluble Polymers: Solutions Properties and Applications*, ed. Z. Amjad, Springer US, Boston, MA, 2002, pp. 31–40.
154. S. Wei, L. S. Ahlstrom and C. L. Brooks III, Exploring Protein–Nanoparticle Interactions with Coarse-Grained Protein Folding Models, *Small*, 2017, **13**(18), 1603748.
155. P. S. Doyle, E. S. G. Shaqfeh and A. P. Gast, Rheology of Polymer Brushes: A Brownian Dynamics Study, *Macromolecules*, 1998, **31**(16), 5474–5486.

156. S. Qi and F. Schmid, Dynamic Density Functional Theories for Inhomogeneous Polymer Systems Compared to Brownian Dynamics Simulations, *Macromolecules*, 2017, **50**(24), 9831–9845.
157. K. J. Modica, T. B. Martin and A. Jayaraman, Effect of Polymer Architecture on the Structure and Interactions of Polymer Grafted Particles: Theory and Simulations, *Macromolecules*, 2017, **50**(12), 4854–4866.
158. M. K. Singh, P. Ilg, R. M. Espinosa-Marzal, M. Kröger and N. D. Spencer, Effect of Crosslinking on the Microtribological Behavior of Model Polymer Brushes, *Tribol. Lett.*, 2016, **63**(2), 17.
159. J. Sheng and K. Luo, Conformation and Adsorption Transition on an Attractive Surface of a Ring Polymer in Solution, *RSC Adv.*, 2015, **5**(3), 2056–2061.
160. W. Janke and W. Paul, Thermodynamics and Structure of Macromolecules from Flat-Histogram Monte Carlo Simulations, *Soft Matter*, 2016, **12**(3), 642–657.
161. S. Förster, E. Kohl, M. Ivanov, J. Gross, W. Widdra and W. Janke, Polymer Adsorption on Reconstructed Au(001): A Statistical Description of P3HT by Scanning Tunneling Microscopy and Coarse-Grained Monte Carlo Simulations, *J. Chem. Phys.*, 2014, **141**(16), 164701.
162. A. Halperin, M. Tirrell and T. P. Lodge, Tethered Chains in Polymer Microstructures, in *Macromolecules: Synthesis, Order and Advanced Properties*, Springer Berlin Heidelberg, Berlin, Heidelberg, 1992, pp. 31–71.
163. S. T. Milner, Polymer Brushes, *Science*, 1991, **251**(4996), 905.
164. M. Daoud and J. P. Cotton, Star Shaped Polymers: A Model for the Conformation and Its Concentration Dependence, *J. Phys. France*, 1982, **43**(3), 531–538.
165. M. Kröger, O. Peleg and A. Halperin, From Dendrimers to Dendronized Polymers and Forests: Scaling Theory and Its Limitations, *Macromolecules*, 2010, **43**(14), 6213–6224.
166. A. Halperin and M. Kröger, Theoretical Considerations on Mechanisms of Harvesting Cells Cultured on Thermoresponsive Polymer Brushes, *Biomaterials*, 2012, **33**(20), 4975–4987.
167. A. Halperin and M. Kröger, Thermoresponsive Cell Culture Substrates Based on PANIPAM Brushes Functionalized with Adhesion Peptides: Theoretical Considerations of Mechanism and Design, *Langmuir*, 2012, **28**(48), 16623–16637.
168. J. McNamara, C. Y. Kong and M. Muthukumar, Monte Carlo Studies of Adsorption of a Sequenced Polyelectrolyte to Patterned Surfaces, *J. Chem. Phys.*, 2002, **117**(11), 5354–5360.
169. G. Luque-Caballero, A. Martín-Molina and M. Quesada-Pérez, Polyelectrolyte Adsorption onto Like-Charged Surfaces Mediated by Trivalent Counterions: A Monte Carlo Simulation Study, *J. Chem. Phys.*, 2014, **140**(17), 174701.
170. S. Chen, L. Li, C. Zhao and J. Zheng, Surface Hydration: Principles and Applications toward Low-Fouling/Nonfouling Biomaterials, *Polymer*, 2010, **51**(23), 5283–5293.

171. I. Szleifer, Protein Adsorption on Surfaces with Grafted Polymers: A Theoretical Approach, *Biophys. J.*, 1997, **72**(2, Part 1), 595–612.
172. F. Fang and I. Szleifer, Controlled Release of Proteins from Polymer-Modified Surfaces, *Proc. Natl. Acad. Sci. U. S. A.*, 2006, **103**(15), 5769.
173. R. Nagumo, K. Akamatsu, R. Miura, A. Suzuki, H. Tsuboi, N. Hatakeyama, H. Takaba and A. Miyamoto, Assessment of the Antifouling Properties of Polyzwitterions from Free Energy Calculations by Molecular Dynamics Simulations, *Ind. Eng. Chem. Res.*, 2012, **51**(11), 4458–4462.
174. M. Zolk, F. Eisert, J. Pipper, S. Herrwerth, W. Eck, M. Buck and M. Grunze, Solvation of Oligo(Ethylene Glycol)-Terminated Self-Assembled Monolayers Studied by Vibrational Sum Frequency Spectroscopy, *Langmuir*, 2000, **16**(14), 5849–5852.
175. R. L. C. Wang, H. J. Kreuzer and M. Grunze, Molecular Conformation and Solvation of Oligo(Ethylene Glycol)-Terminated Self-Assembled Monolayers and Their Resistance to Protein Adsorption, *J. Phys. Chem. B*, 1997, **101**(47), 9767–9773.
176. S. Herrwerth, W. Eck, S. Reinhardt and M. Grunze, Factors That Determine the Protein Resistance of Oligoether Self-Assembled Monolayers – Internal Hydrophilicity, Terminal Hydrophilicity, and Lateral Packing Density, *J. Am. Chem. Soc.*, 2003, **125**(31), 9359–9366.
177. D. Nagasawa, T. Azuma, H. Noguchi, K. Uosaki and M. Takai, Role of Interfacial Water in Protein Adsorption onto Polymer Brushes as Studied by SFG Spectroscopy and QCM, *J. Phys. Chem. C*, 2015, **119**(30), 17193–17201.
178. A. J. Pertsin and M. Grunze, Computer Simulation of Water near the Surface of Oligo(Ethylene Glycol)-Terminated Alkanethiol Self-Assembled Monolayers, *Langmuir*, 2000, **16**(23), 8829–8841.
179. J. Zheng, Y. He, S. Chen, L. Li, M. T. Bernards and S. Jiang, Molecular Simulation Studies of the Structure of Phosphorylcholine Self-Assembled Monolayers, *J. Chem. Phys.*, 2006, **125**(17), 174714.
180. A. K. Gupta and M. Gupta, Synthesis and Surface Engineering of Iron Oxide Nanoparticles for Biomedical Applications, *Biomaterials*, 2005, **26**(18), 3995–4021.
181. W. J. Stark, P. R. Stoessel, W. Wohlleben and A. Hafner, Industrial Applications of Nanoparticles, *Chem. Soc. Rev.*, 2015, **44**(16), 5793–5805.
182. M. Scotter, J. Blackburn, B. Ross, A. Boxall, L. Castle, R. Aitken and R. Watkins, Applications and Implications of Nanotechnologies for the Food Sector Au - Chaudhry, Qasim, *Food Addit. Contam., Part A*, 2008, **25**(3), 241–258.
183. D.-H. Lim and J. Wilcox, DFT-Based Study on Oxygen Adsorption on Defective Graphene-Supported Pt Nanoparticles, *J. Phys. Chem. C*, 2011, **115**(46), 22742–22747.
184. R. Lazzari, J. Goniakowski, G. Cabailh, R. Cavallotti, N. Trcera, P. Lagarde and J. Jupille, Surface and Epitaxial Stresses on Supported Metal Clusters, *Nano Lett.*, 2016, **16**(4), 2574–2579.

185. F. De Angelis, A. Tilocca and A. Selloni, Time-Dependent Dft Study of $[Fe(CN)_6]^{4-}$ Sensitization of TiO_2 Nanoparticles, *J. Am. Chem. Soc.*, 2004, **126**(46), 15024–15025.
186. P. K. Naicker, P. T. Cummings, H. Zhang and J. F. Banfield, Characterization of Titanium Dioxide Nanoparticles Using Molecular Dynamics Simulations, *J. Phys. Chem. B*, 2005, **109**(32), 15243–15249.
187. A. G. Bembel, On the Size Dependences of the Metallic Nanoparticle Evaporation and Sublimation Heats: Thermodynamics and Atomistic Modeling, *Russ. Phys. J.*, 2017, **59**(10), 1567–1574.
188. G. Grochola, S. P. Russo and I. K. Snook, On Morphologies of Gold Nanoparticles Grown from Molecular Dynamics Simulation, *J. Chem. Phys.*, 2007, **126**(16), 164707.
189. Y. Li, M. Kröger and W. K. Liu, Endocytosis of Pegylated Nanoparticles Accompanied by Structural and Free Energy Changes of the Grafted Polyethylene Glycol, *Biomaterials*, 2014, **35**(30), 8467–8478.
190. T.-Y. Wei, S.-Y. Lu and Y.-C. Chang, Transparent, Hydrophobic Composite Aerogels with High Mechanical Strength and Low High-Temperature Thermal Conductivities, *J. Phys. Chem. B*, 2008, **112**(38), 11881–11886.
191. A. J. Trazkovich, M. F. Wendt and L. M. Hall, Effect of Copolymer Sequence on Structure and Relaxation Times near a Nanoparticle Surface, *Soft Matter*, 2018, **14**(28), 5913–5921.
192. N. Kashyap and B. Rai, Molecular Mechanism of Transdermal Co-Delivery of Interferon-Alpha Protein with Gold Nanoparticle – a Molecular Dynamics Study Au - Gupta, Rakesh, *Mol. Simul.*, 2018, **44**(4), 274–284.
193. J. Lin, H. Zhang, Z. Chen and Y. Zheng, Penetration of Lipid Membranes by Gold Nanoparticles: Insights into Cellular Uptake, Cytotoxicity, and Their Relationship, *ACS Nano*, 2010, **4**(9), 5421–5429.
194. C. Rossner, Q. Tang, M. Müller and G. Kothleitner, Phase Separation in Mixed Polymer Brushes on Nanoparticle Surfaces Enables the Generation of Anisotropic Nanoarchitectures, *Soft Matter*, 2018, **14**(22), 4551–4557.
195. T. Mandal, N. V. Konduru, A. Ramazani, R. M. Molina and R. G. Larson, Effect of Surface Charge and Hydrophobicity on Phospholipid-Nanoparticle Corona Formation: A Molecular Dynamics Simulation Study, *Colloid Interface Sci. Commun.*, 2018, **25**, 7–11.
196. P. Chodanowski and S. Stoll, Polyelectrolyte Adsorption on Charged Particles: Ionic Concentration and Particle Size Effects—a Monte Carlo Approach, *J. Chem. Phys.*, 2001, **115**(10), 4951–4960.
197. P. Chodanowski and S. Stoll, Polyelectrolyte Adsorption on Charged Particles in the Debye – Hückel Approximation. A Monte Carlo Approach, *Macromolecules*, 2001, **34**(7), 2320–2328.
198. S. K. Kumar, V. Ganesan and R. A. Riggleman, Perspective: Outstanding Theoretical Questions in Polymer-Nanoparticle Hybrids, *J. Chem. Phys.*, 2017, **147**(2), 020901.

199. P. J. Dionne, C. R. Picu and R. Ozisik, Adsorption and Desorption Dynamics of Linear Polymer Chains to Spherical Nanoparticles: A Monte Carlo Investigation, *Macromolecules*, 2006, **39**(8), 3089–3092.
200. J. S. Smith, D. Bedrov and G. D. Smith, A Molecular Dynamics Simulation Study of Nanoparticle Interactions in a Model Polymer-Nanoparticle Composite, *Compos. Sci. Technol.*, 2003, **63**(11), 1599–1605.
201. J. Sampath and L. M. Hall, Influence of a Nanoparticle on the Structure and Dynamics of Model Ionomer Melts, *Soft Matter*, 2018, **14**(22), 4621–4632.
202. S. J. Marrink, D. P. Tieleman and A. E. Mark, Molecular Dynamics Simulation of the Kinetics of Spontaneous Micelle Formation, *J. Phys. Chem. B*, 2000, **104**(51), 12165–12173.
203. D. Yordanova, E. Ritter, T. Gerlach, J. H. Jensen, I. Smirnova and S. Jakobtorweihen, Solute Partitioning in Micelles: Combining Molecular Dynamics Simulations, Cosmomic, and Experiments, *J. Phys. Chem. B*, 2017, **121**(23), 5794–5809.
204. S. Bogusz, R. M. Venable and R. W. Pastor, Molecular Dynamics Simulations of Octyl Glucoside Micelles: Structural Properties, *J. Phys. Chem. B*, 2000, **104**(23), 5462–5470.
205. A. K. Soper and K. J. Edler, Coarse-Grained Empirical Potential Structure Refinement: Application to a Reverse Aqueous Micelle, *Biochim. Biophys. Acta, Gen. Subj.*, 2017, **1861**(6), 1652–1660.
206. M. R. Zachariah and M. J. Carrier, Molecular Dynamics Computation of Gas-Phase Nanoparticle Sintering: A Comparison with Phenomenological Models, *J. Aerosol Sci.*, 1999, **30**(9), 1139–1151.
207. B. V. Derjaguin, A Theory of the Heterocoagulation, Interaction and Adhesion of Dissimilar Particles in Solutions of Electrolytes, *Discuss. Faraday Soc.*, 1954, **18**, 85–98.
208. J. Forsman, A Simple Correlation-Corrected Poisson – Boltzmann Theory, *J. Phys. Chem. B*, 2004, **108**(26), 9236–9245.
209. H. Ohshima, T. W. Healy and L. R. White, Accurate Analytic Expressions for the Surface Charge Density/Surface Potential Relationship and Double-Layer Potential Distribution for a Spherical Colloidal Particle, *J. Colloid Interface Sci.*, 1982, **90**(1), 17–26.
210. C. Pfeiffer, C. Rehbock, D. Hühn and C. Carrillo-Carrion, D. Jimenez de Aberasturi, V. Merk, S. Barcikowski and W. J. Parak, Interaction of Colloidal Nanoparticles with Their Local Environment: The (Ionic) Nanoenvironment around Nanoparticles Is Different from Bulk and Determines the Physico-Chemical Properties of the Nanoparticles, *J. R. Soc., Interface*, 2014, **11**(96), 20130931.
211. J. J. López-García, J. Horno and C. Grosse, Ion Size Effects on the Dielectric and Electrokinetic Properties in Aqueous Colloidal Suspensions, *Curr. Opin. Colloid Interface Sci.*, 2016, **24**, 23–31.
212. B. Svensson and B. Jönsson, The Interaction between Charged Aggregates in Electrolyte Solution. A Monte Carlo Simulation Study, *Chem. Phys. Lett.*, 1984, **108**(6), 580–584.

213. V. Pryamitsyn and V. Ganesan, Pair Interactions in Polyelectrolyte-Nanoparticle Systems: Influence of Dielectric Inhomogeneities and the Partial Dissociation of Polymers and Nanoparticles, *J. Chem. Phys.*, 2015, **143**(16), 164904.
214. A. Jain, S. P. Ong, G. Hautier, W. Chen, W. D. Richards, S. Dacek, S. Cholia, D. Gunter, D. Skinner, G. Ceder and K. A. Persson, Commentary: The Materials Project: A Materials Genome Approach to Accelerating Materials Innovation, *APL Mater.*, 2013, **1**(1), 011002.

CHAPTER 9

Electron Spin Resonance for the Detection of Paramagnetic Species: From Fundamentals to Computational Methods for Simulation and Interpretation

INOCENCIO MARTÍN,[a] LEO MARTIN,[a] ANWESHA DAS,[b] MARTIN GROOTVELD,[b] VALENTIN RADU,[c] MELISSA L. MATHER[c] AND PHILIPPE B. WILSON*[d]

[a] Universita de la Laguna, Calle Padre Herrera, s/n, 38200 San Cristóbal de La Laguna, Santa Cruz de Tenerife, Spain; [b] Faculty of Health and Life Sciences, De Montfort University, The Gateway, Leicester LE1 9BH, UK; [c] Optics and Photonics Research Group, Faculty of Engineering, University Park, University of Nottingham, Nottingham NG7 2RD, UK; [d] School of Animal, Rural and Environmental Sciences, Nottingham Trent University, Brackenhurst Campus, Brackenhurst Lane, Southwell, Nottinghamshire NG25 0QF, UK
*Email: philippe.wilson@ntu.ac.uk

9.1 Introduction

Techniques such as electron spin/paramagnetic resonance (ESR/EPR) and optically-detected magnetic resonance (ODMR) spectroscopy employ microwave radiation and a static magnetic field in order to investigate analytes with

Theoretical and Computational Chemistry Series No. 20
Computational Techniques for Analytical Chemistry and Bioanalysis
Edited by Philippe B. Wilson and Martin Grootveld

Published by the Royal Society of Chemistry, www.rsc.org

one or more unpaired electrons.[1] This allows us to investigate systems including transition metal ion complexes and free radicals, the latter including those usually found in biosystems such as superoxide anions and hydroxyl radicals ($^{\bullet}$OH). Whilst providing detailed information on such systems, ESR is arguably a more complex technique in terms of data extraction compared to some alternatives considered earlier on in this volume.[2] Computational techniques are applied in order to extract information from such systems, which also allow molecular 'fingerprinting' to take place.[3]

Unpaired electrons are found in the valence shell, such as the d orbitals in transition metals, or the highest occupied molecular orbitals (HOMOs) of other paramagnetic species, including free radicals.[4] As a form of absorption spectroscopy, ESR describes a series of excited electronic states over substantial ranges in frequency (from approximately 1 to 170 GHz). Although the electronic transitions between these states are probed using UV/Visible radiation, if a magnetic field is applied, the ground electronic state is split into finely distributed energy levels, with the ΔE, or the spacing between energy levels, being determined by the strength of the applied magnetic field; therefore, small field strengths lead to transitions in the order of 1–5 GHz. As ESR as a technique probes these transitions, their subtle nature leads to a molecularly-probing method with a substantial level of analytical sensitivity.

NMR and ESR spectroscopy share a number of common features in both their theory and application. Both techniques are dependent on the magnetic moment, μ, which is associated with a particle possessing an angular momentum, such as an electron or nucleus.[5] A moiety with a spin $= 1/2$ exhibits two states when in the presence of a static magnetic field, B, with the energy difference between these states, ΔE, corresponding to eqn (9.1).

$$\Delta E = g\mu B \tag{9.1}$$

Within eqn (9.1), g corresponds to the so-called g-value of a moiety or particle. For instance, the g-value of an electron is 2.002319, whilst that of a proton is 5.5856. Owing to the difference in mass between electrons and nuclei, their magnetic moments differ substantially, according to eqn (9.2).

$$\mu_e = g_e\beta_e S = -\frac{g_e e\hbar S}{2m_e} = \gamma_e S\hbar \tag{9.2}$$

In which γ is the gyromagnetic ratio, and acts as a proportionality constant for the intrinsic magnetic moment of the electron and intrinsic angular momentum, S.[4]

Therefore, for the electron, β_e, also known as the Bohr magneton is equal to 9.274×10^{-24} J T^{-1}, and the nuclear magneton μ_N corresponding to the magnetic moment of the proton, is of the order of 5.051×10^{-27} J T^{-1}.[6] As these g-values and magnetic moments differ between the electron and proton, they constitute the difference in characteristic frequencies, and differences in sensitivity between the nuclear magnetic and electron spin resonance techniques, according to the well-known relationship shown in eqn (9.3).

$$\Delta E = h\nu \tag{9.3}$$

The magnitude of the *g*-value of an electron will differ based on its environment. These values will deviate from the Bohr magneton for unpaired electrons, and are sensitive to the environmental effects in paramagnetic systems.

Within this chapter, we will cover the basic theory of ESR spectrometry, and recent applications of computational techniques to the interpretation and simulation of spectra acquired on paramagnetic systems.

9.2 Theory

Although ESR and NMR share a number of common features, and are grounded in similar theoretical constructs, the main difference is the nature of the system being probed. Indeed, ESR experiments directly probe the spin properties of the electron, whilst NMR spectroscopic strategies investigate the nuclear spin. ESR experiments are based on identifying the magnetic field at which unpaired electrons come into resonance with microwave radiation of a given frequency. Spectrometers are available, operating at a range of microwave bands; however, X-band (approximately 9 GHz) spectrometers are now common. Higher frequency instruments have provided an increase in resolution and diminution of second-order effects. Notably, with relaxation taking place with $t_{1/2}$ values in the region of approximately 10^{-9} s, line sharpening may be observed.[4] The sweep range of microwave frequencies is generally selected to cover the band of interest, whilst a standard with a known *g*-value is commonly used as a reference. Diphenylpicrylhydrazyl (DPPH) is often used as such a reference, with a *g*-value of 2.0036 and a pitch of 2.0028. The *g*-value is recorded on the basis of eqn (9.4):

$$g_{\text{sample}} = \frac{(g.B)_{\text{standard}}}{B_{\text{sample}}} \tag{9.4}$$

Spectral noise can be reduced by employing a modulation strategy, that is, cyclically varying the magnetic field. The signal is then detected by inspecting the phase of the wave, with spectra displayed as their first derivative forms, or d*I*/d*B* instead of explicit absorption spectra.[1] The maximum absorption can therefore be determined as the point at which the first-derivative trace crosses the baseline, with the signal width being the distance between the minima and maxima of adjacent signals (Figure 9.1). In second-derivative spectra, minima correspond to the maxima visible in the zero-order spectra.

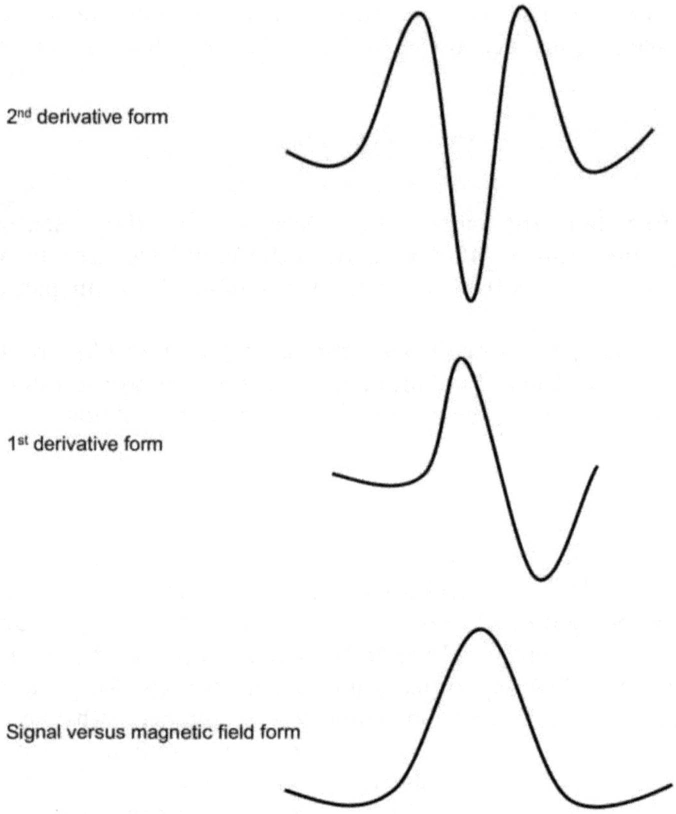

Figure 9.1 Example of EPR spectral lineshapes for idealised absorbance, first and second derivative forms.

Continuous wave spectra are acquired with a small voltage in the resonator; there is a small net magnetisation in the X-band, and B_1 perturbs the magnetisation by only small amounts. There will therefore be a very small signal which will need to be distinguished from spectral noise, leading to the use of modulation of B_0 with phase-sensitive detection, which greatly enhances the signal to noise (S : N).[3] With B_0 applied to an area within the ESR signal, the field leads to the signal oscillating at the same frequency; therefore, the first derivative of the absorption is proportional to the amplitude of the oscillating detected signal.

As described above, a free electron in a vacuum has a *g*-value of 2.002319. In chemical systems, the *g*-value reflects the environment in which the electron is located, specifically an orbital which itself can be more or less delocalised across the molecule. In free radical species, *g* is close to 2.0023, although in more complex systems, *g* will vary depending on the effects of the orbital motion to the angular momentum of the electron, in which the *g*-value can increase to values of greater than 10.0. Effects owing to orbital interactions, leading to such changes in the *g*-value, tend to be observed in systems with transition metal ions.[7] Orbitals possess angular momentum, which can be described by the magnetic quantum number, m_l. For d-orbitals, m_l values for electrons in the d_{xy} and $d_{x^2-y^2}$ orbital systems are ±2. The d_{xz} and d_{yz} orbitals have a single m_l value of ±1, whilst the electrons in the d_{z^2} orbital have an angular momentum of zero about the *z*-axis. The orbital angular momentum quantum number, *l*, is equal to the sum of the individual m_l values for electrons within the system.[8] Assuming that electrons will preferentially populate the orbitals with the highest m_l values leads to an expression for *L* in a d^4 system:

$$L = (+2) + (+1) + (0) + (-1) = 2$$

The spin of the system is given by $S = n/2$, in this case *n* corresponds to the number of unpaired electrons, and it is this, along with the angular momentum, which determines the total angular momentum, *J*. The total angular momentum comprises $|l \pm S|$, with the *g*-value determined using eqn (9.5), also known as the modified Landé formula.[9]

$$g = 1 + [J(J+1) + (S(S+1) - (L(L+1)]\frac{1}{[2J(J+1)]} \tag{9.5}$$

Let us now consider the electronic spin for a single particle; the electron possesses an intrinsic angular momentum or spin of 1/2, as described above. As a vector quantity, its magnitude is determined from eqn (9.6).

$$|S| = [S(S+1)\hbar]^{-1/2} \tag{9.6}$$

Therein, two eigenstates can be projected on the quantisation axis (*z*), that is $m_s = +1/2$ or $-1/2$. According to eqn (9.2), γ_e is negative, indicating that μ_e is antiparallel to spin.

The magnetic moments of the electron spins in a magnetic field B_0 differ according to eqn (9.1), in which B_0 is applied in the *z* axis. The splitting of the energy levels of these spins in a magnetic field is known as the Zeeman effect, which breaks the degeneracy of the spin states. In the applied magnetic field, the spin-*down* state will have a lower energy than the spin-up state. Indeed, according to eqn (9.7) and substituting into eqn (9.2), we obtain eqn (9.8):

$$E = \mu\beta_0 \tag{9.7}$$

$$\begin{aligned} E_\uparrow &= +\frac{g_e\beta_e B_0}{2} \\ E_\downarrow &= -\frac{g_e\beta_e B_0}{2} \end{aligned} \tag{9.8}$$

The electronic spin has no restrictions on orientation, the evolution of a spin state with time can first be described according to eqn (9.9), in terms of its state; the phase of this state is described by eqn (9.10).[3]

$$\left|\Psi > = \cos\left(\frac{\theta}{2}\right)\right|0 > + \sin\left(\frac{\theta}{2}\right)\exp(i\phi)|1 > \tag{9.9}$$

$$\phi(t) = \gamma B_t \tag{9.10}$$

Eqn (9.9) is also known as the Bloch vector, and describes the spin orientations within a spherical representation, in which θ represents the *z*-axis components, and ϕ in the *xy* plane. The spin therefore precesses around the applied magnetic field, B_0 (Figure 9.2) according to eqn (9.11).[10]

$$\omega_L = \gamma B_0 \tag{9.11}$$

In eqn (9.11), ω_L is the Larmor frequency, which is also commonly encountered in NMR spectroscopy.

The application of an oscillating microwave field, which is of significantly smaller magnitude than B_0 will lead the spin to precess around this additional field, B_1 with its frequency described as ω_1. The microwave field precesses orthogonally to the Larmor motion of the electron, leading to a complex overall motion with the spin precessing around B_0 and B_1 (Figure 9.3).[11]

We may consider that if the Larmor and microwave frequencies are equivalent, the fields B_0 and B_1 act in phase, leading to small changes in the microwave field engendering large variations in the magnetic moment. This equality of ω_L and ω_1 is described as the resonance condition, in which the magnetic moment of a spin originally precessing about the *z*-axis now *nutates* about $\pm z$. The physical result of this is that the applied microwave field is

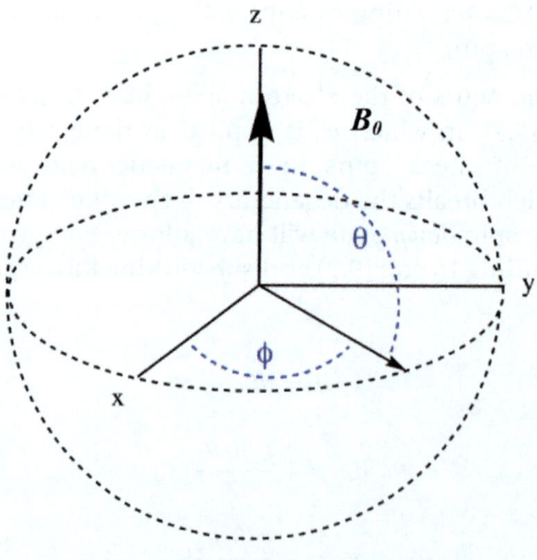

Figure 9.2 Bloch sphere illustrating the spin precessing in a magnetic field, B_0.

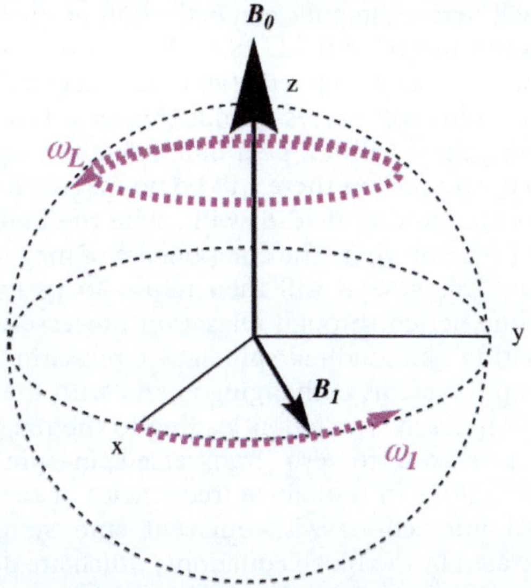

Figure 9.3 Bloch sphere representation of orthogonal fields on electronic spin properties.

driving the transitions we observe in ESR. Indeed, from eqn (9.2) and (9.11), we determine the nutation frequency (ω_n) as:

$$\omega_n = \frac{g\beta_e B_1}{\hbar} \tag{9.12}$$

Considering that the microwave field, B_1 scales in magnitude to the root of the applied microwave power, there is also a linear dependency of the nutation frequency on the applied microwave field.[12]

Those *au fait* with NMR spectroscopy will be familiar with the notion of the rotating frame; conceptually simplifying the motions of a complex vector such as that described above. We can consider the rotating frame using the Bloch sphere represented in Figure 9.3. Instead of the cartesian axes being located orthogonally to each other, these can be considered as the axes of the rotating frame, in which the entire coordinate system will rotate at frequency ω_n about z. Therefore, upon satisfying the resonance condition, the magnetic moment is static in only the applied magnetic field B_0.[13] This effectively eliminates the effects of B_0 and B_1, and thereby becomes static in the x axis of the rotating frame, with the magnetic moment of the electron precessing at a frequency of ω_n about B_1. Conversely, when the resonance condition is satisfied, and in the absence of the applied microwave field B_1, then the magnetic moment is also static, whilst if an inequality arises between the Larmor and microwave frequencies, the magnetic moment will precess about B_0 within the rotating frame at a frequency-corrected value:

$$\Omega = \omega_L - \omega_1 \tag{9.13}$$

Much as NMR will detect the bulk magnetisation of nuclear spins in the presence of a powerful magnetic field, ESR follows a similar theory.[14]

In the absence of an applied magnetic field, bulk magnetisation = 0; once this is applied in z, spins will precess about this axis. However, under the influence of B_0 alone, there will not be a bulk effect to magnetisation, and therefore at thermal equilibrium there will be no magnetisation in the x or y axes.[7] Our applied microwave field B_1 will rotate the bulk magnetisation out of equilibrium (in z), in which the components of magnetisation in x, y, and z are non-zero. This system will then return to its equilibrium state within a defined time period through relaxation processes, such as the return to z magnetisation (longitudinal/spin–lattice relaxation, T_1). This takes place through the spin systems exchanging energy with the 'spin bath' and the environment. Conversely, relaxation leading to the magnetisation components in x and y relaxing to zero (transverse/spin–spin relaxation) originates from the variations in resonance frequencies of each spin, and will take place through interactions with adjacent spin systems. Relaxation processes are illustrated by the Bloch equations, which are developed first in the absence of the microwave field, that is prior to its application:

$$\frac{\mathrm{d}M_x}{\mathrm{d}t} = -\Omega M_y - \frac{M_x}{T_2} \tag{9.14}$$

$$\frac{\mathrm{d}M_z}{\mathrm{d}t} = -\frac{M_z - M_0}{T_1} \tag{9.15}$$

Eqn (9.14) translates to the y component relaxation by substitution of x with y and *vice-versa*. M refers to the bulk magnetisation, and T_1 and T_2 to the spin–lattice and spin–spin relaxation processes respectively.[15]

However, to what extent do these processes influence the spectral data obtained in ESR experiments? If relaxation is efficient, then T_1 will be appropriately small; should this be of a more significant magnitude, it may be necessary for the power of the microwaves to be attenuated in the experiment in order to avoid saturation.

9.3 Fine Structure in ESR Spectroscopy

Fine structure in ESR spectra can be defined as the appearance of multiple signals; species with spin S have a total of $2S+1$ energy states described by the quantum number M_s. For M_s non-zero states in the absence of B_0, these will be double degenerate, the signals separated by the electric fields produced by neighbouring atoms through spin–orbit coupling, which is governed by the structure of the species investigated. Applying B_0 removes redundant degeneracy, and we can therefore observe transitions between states according to the rule $\Delta M_s = \pm 1$ (Figure 9.4).[16] Indeed, $2S$ transitions may take place, and in cases in which $S > 1/2$, the extent to which these transitions can be observed, and their g-values, are dependent on the electronic environment, zero-field splitting and B range of the instrument.

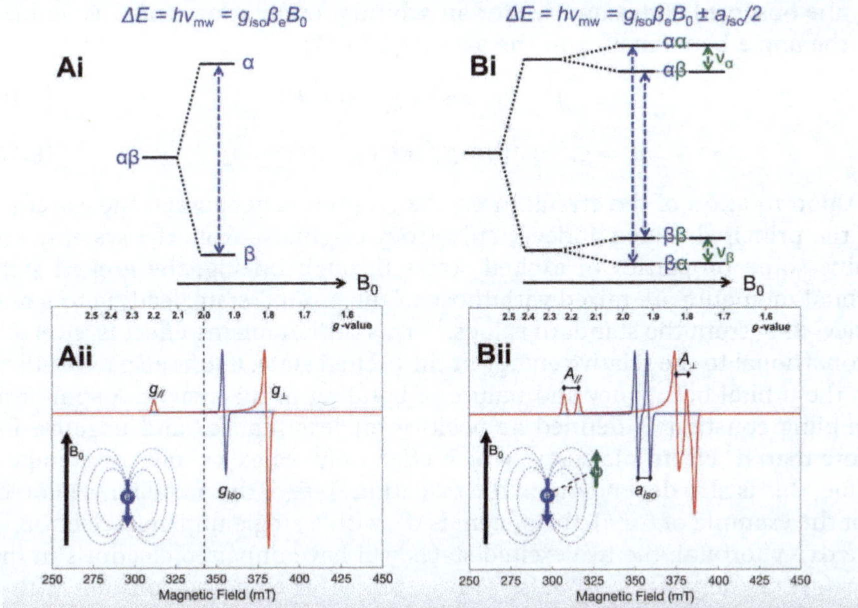

Figure 9.4 B_0 is scanned across the leading axis giving rise to the energies of the two spin states of an unpaired electron breaking degeneracy; at the value of B_0, at which $\Delta E = h\nu$, the spins absorb energy, which we define as resonance. *A* represents the Zeeman term for a $S = 1/2$ system and one unpaired electron, although *B* illustrates the hyperfine interaction with a ^{1}H nucleus. The ESR transitions are denoted with blue dotted lines in parts (i) for isotropic systems, and correspond to the blue lines in parts (ii). The position corresponding to g_e in *Aii* is shown by the dotted line. Reproduced from ref. 17, with permission from the Royal Society of Chemistry.

Some systems will be highly symmetric and therefore their properties are independent of the direction of an applied magnetic field. However, the majority of systems are classified as *anisotropic*, or have properties dependent on the orientation of B_0. This links the direction of the applied magnetic field to variations in the separation of energy levels, and the magnitude of linked properties. In anisotropic systems, *g* will depend on the orientation of B_0 *cf.* the molecular symmetry axis. The common illustration of this is in a number of Cu(II) systems, this example is from a previously published study.[4] If the *x* and *y* axes are equivalent in terms of magnetic field directionality, but differ from the *z* component, then the system behaves as one with C_3 symmetry; two *g* values dominate, that of the field parallel ($g_\parallel$) and that perpendicular ($g_\perp$) to the axis of the spin angular momentum. For crystalline samples with all molecules with the same orientation at θ to B_0, the expression for *g* is shown in eqn (9.16). Conversely, if the symmetry is lower than this, such as in many systems, the *g* values must be defined based on the observed cartesian axes for an arbitrary orientation, and θ is defined as the angle between B_0 and the axes (eqn (9.17)).

$$g^2 = g_{||}{}^2\cos^2 + g_{\perp}{}^2\sin^2\theta \tag{9.16}$$

$$g^2 = g_x{}^2\cos^2\theta_x + g_y{}^2\cos^2\theta_y + g_z{}^2\cos^2\theta_z \tag{9.17}$$

Upon rotation of the crystal in B_0, the g values vary between the extremes of the principal values. Indeed, anisotropy originates from the system, and gains some properties of excited states through mixing; the excited state orbital momenta are mixed with those of the ground state, leading to μ and g deviating from the standard values.[17] This orbital mixing effect is inversely proportional to the relative energy of the excited state, and is also dependant on the orbital occupancy and nature of bonding in the system. A spin-orbit coupling constant is defined as positive for less than d^5 and negative for more than d^5 electronic states, which effectively scales g about the standard value; this is also dependant on the oxidation state of the metal ion explored. For the example of Cu(II), the system is d^9, with a single unpaired electron in the dx^2–y^2 orbital; the two excited states will have unpaired electrons in the dxy ($g_{\|}$) or dxz/yz ($g_{\perp}$) orbitals; with a spin–orbit coupling constant with a negative magnitude for d^9, this leads to $g_{\|} > g_{\perp}$.[4]

Conversely, in a solid sample matrix, in solution the molecules are in a constant state of motion at a rate substantially larger than the spectrometer frequency, and these anisotropic motions are averaged during the time taken to acquire the spectrum. This will therefore lead to the appearance of isotropicity and a uniform value of g. The effect of solidifying a solution will not change the likelihood of all possible orientations being present; however, these then become fixed in space leading to each molecule possessing a characteristic g-value and producing a summative spectrum over all orientations.[16] Indeed, it is statistically more likely that a greater number of symmetry axes will be aligned perpendicular rather than parallel to B_0, which when considered relative to the angular dependency of the intensity, leads to a derivative spectrum such as that shown in Figure 9.5.

Additional information on the structures of paramagnetic species can be obtained from interactions between electrons and nuclei therein. Indeed, atomic nuclei possess magnetic moments which themselves induce local magnetic fields in the electrons, linking into the concept of electronic shielding in NMR. Interactions between the electrons and nuclei are defined as the *hyperfine interaction*. The magnetic moment of the nucleus induces a magnetic field on the electron B_e, which is either orthogonal or parallel to the applied magnetic field, with parallel interactions leading to an enhancement of it, decreasing the magnitude of the applied field to establish resonance by B_e.[18] Let us consider the interaction of an unpaired electron spin with an $S = 1/2$ nucleus such as ^1H. The ESR signal is split into two, each being shifted by B_e from the signal, with a hyperfine splitting constant, a_n, in which n refers to the nucleus interacting with the unpaired electron (in this case, H). The hyperfine splitting constant is equivalent to the

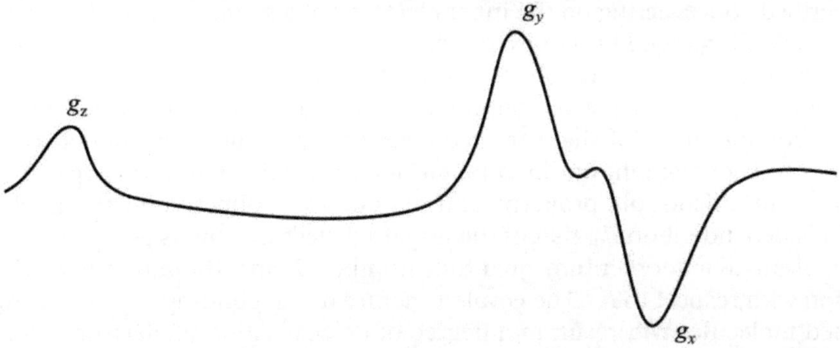

Figure 9.5 The first derivative form of an ESR spectrum for a non-cubic system (lower symmetry) such as rhombic. It can be difficult to identify the difference between g_x and g_y for such systems; however, if these three major values of g are substantially different, each will lead to a well-defined signal in the spectrum.

separation of the two resonance lines, or $2 \times B_e$. The presence of further $S = 1/2$ nuclei will lead to a further splitting in the signal. A general rule for such nuclei is to expect 2^n EPR signals for n nuclei of $S = 1/2$, similar to NMR with the $2nI + 1$ rule for the fine structure of coupled spectral signals.[6]

The intensity of the ESR signal allows us to measure species concentrations, and follows a similar integral-based methodology to NMR spectroscopy, being proportional to the concentration of unpaired electrons in the sample under normal conditions. These conditions also include a dependence on microwave power; should this be particularly low, signal intensity will increase with the root of power, as described above, whilst the inverse is true. We describe this phenomenon as *saturation*, with an increased broadening and decreased intensity observed at high power levels. Saturation is therefore detrimental, and prevents the acquisition of accurate linewidths, shapes, intensities and intricate hyperfine splittings. Should microwave power be decreased on a spectrometer, then signal intensity proportionately decreases *via* a square root relationship, and only then can saturation be deemed to be absent.[19]

Let us consider the hyperfine *coupling* constant, A, in some detail. This quantity is dependent on the interactions between the nucleus and the unpaired electron, with a number of principal components governing its magnitude, and facilitates derivation of the hyperfine splitting constant, a. The orbital term is dependent on the angular momentum of the electron; an electron with $l = 0$ has a finite probability of being located close to the nucleus, and this interaction is referred to as the Fermi contact term, which has a positive contribution to A, and suggests a low system symmetry. Electron spin exchange effects refer to the polarising influence of unpaired electron spins in the valence orbitals on the electronic density in the inner orbitals, leading to a non-uniform distribution of spin in the core orbitals. These effects influence A through the interaction of the orbital spin systems described above, acting on the inner electrons of a system and leading to the spins coupling together. For example, the electron in a 2s orbital interacting with a 3d unpaired electron will have an antiparallel spin to the d orbital system leading to negative coupling.[10] The A term is also affected by the magnetic moments of the unpaired electrons and nuclei, or dipolar coupling, and is dependent on the orientation of the spins and their separation. This is an anisotropic property, and although the direction of the nuclear spin is dependent on B_0, that of the unpaired electron spin is governed by its orbital angular momentum quantum number, l, and the molecular orientation with respect to B_0. The covalent nature of the bonding within organic-based molecules will result in a degree of delocalisation of electrons, being inversely proportional to A.[17] The combined A value is a summation of these differing contributing terms. A behaves similarly to the g-value for crystals as outlined above, whilst for solutions the anisotropic effects become averaged in view of the random motion of the system, and therefore the expression for the isotropic A value becomes:

$$A_{\text{iso}} = \frac{(A_{\parallel} + 2A_{\perp})}{3} \tag{9.18}$$

This relationship can be further split in order to expand the Fermi-contact term and the spin densities of the associated nuclei (eqn (9.19)).

$$A_{\text{iso}}(N) = \frac{4\pi}{3}\beta_e\beta_N \cdot g_e g_N \cdot \sqrt{\langle S_Z \rangle \cdot \rho_N^{\alpha-\beta}} \tag{9.19}$$

The first b terms represent the product of the Bohr and nuclear magneton, whilst the g-values are those for the free electron and nuclear variants, respectively. The third or root term refers to the expectation value of the electronic spin z component, whilst the fourth term describes the spin density in the nucleus. These approximations are complicated in the anisotropic case, and readers are directed to other sources for a full mathematical derivation thereof.

9.4 Quantum Mechanical Consideration of ESR Theory

It is also possible to consider spin-orbit coupling quantum mechanically, and these interactions can be modelled as such. We have thus far mainly considered the basis of ESR spectroscopy classically. However, in order to obtain a robust understanding of the theory and applications (including computational simulations thereof), it is necessary to attain some grasp of the quantum mechanical background of the technique.

Eqn (9.7) expressed the energy of μ in B_0 as a scalar product. However, we may also consider B as magnetic induction in Tesla, in which H, the operator representing the sum of energies of the system, is represented in eqn (9.20), whilst including the proportionality between the electronic angular momentum and the magnetic moment expressed in eqn (9.2):

$$\hat{H} = -\hat{\mu}B = \gamma_e B\hat{l} = \gamma_e B_0 \hat{l}_z = g_e \gamma_e B_0 \hat{s}_z \tag{9.20}$$

In this equation, in which l is the orbital angular momentum, and s the spin angular momentum operator, we note that the Hamiltonian for the spin system in a magnetic field of magnitude B_0 is aligned along the z-axis of the system. We now develop this approach to consider the variation of the g value for the ground state of systems with zero angular momentum. The following section is adapted from the excellent work of Weil and Bolton,[20] and serves to demonstrate a derivation of the g-value and its deviation from β_e.

As we have briefly alluded to above, the total angular momentum operator for the electron is the vector sum of contributions from both the spin angular and orbital angular momenta. The Zeeman–Hamiltonian operator can be obtained, and demonstrates the coupling of both spin and orbital angular momentum arising from the spin–orbit effect, in which L is the ground state electronic orbital angular momentum operator. This, once added to the electronic terms, yields an expression for the spin–orbit interaction term in view of the spin μ and B couplings arising from the orbital angular momentum (eqn (9.21)).

$$\hat{H}(r) = \hat{H}_z + \hat{H}_{so} = \beta_e B^T(\hat{L} + g_e\hat{S}) + \lambda\hat{L}^T\hat{S} \tag{9.21}$$

In this Equation, subscripts Z and SO denote the electronic Zeeman and spin–orbit energies respectively, the Hamiltonian describing the energy from the above couplings. This can be applied to consider a system, such as a ground state with non-degenerate orbitals. The ground state, $|G, M_s\rangle$, is represented by the wavefunction G, and the spin state M_s. For a spin in the ground state, the first order energy can be described as:

$$E_G = \langle G, M_s|g_e\beta_e B_z\hat{S}_z|G, M_s\rangle + \langle M_s|\beta_e B_z + \lambda\hat{S}_z|M_s\rangle \tag{9.22}$$

Eqn (9.22) takes a number of equivalencies into account: the first term describes the so-called *spin-only* electron Zeeman energy, whilst the second has been somewhat simplified from eqn (9.21) by considering that the total electronic orbital angular momentum is equal to zero for a non-degenerate orbital state in which there is no action of spin–orbit coupling. Each element in the Hamiltonian can be corrected to the second order according to eqn (9.23), in which the sum includes all orbital states. The elements from the spin-only electron Zeeman energy term in eqn (9.22) can be removed from the second-order correction as $|G/n\rangle = 0$,

$$H_{M_s,M_s'} = \sum_{n \neq G} \frac{[\langle M_s|\beta_e B + \lambda\hat{S}|M_s'\rangle\langle G|\hat{L}|n\rangle\langle n|\hat{L}|G\rangle\langle M_s'|\beta_e B + \lambda\hat{S}|M_s\rangle]}{E_n^0 - E_G^0} \tag{9.23}$$

and with the corresponding factors grouped, yields the matrix expressed in eqn (9.24), with the ijth elements of it determined by the final equality, in which the orbital angular momentum operators, L, correspond here to directions on the three cartesian axes.

$$-\sum_{n\neq G}\frac{\langle G|\hat{L}|n\rangle\langle n|\hat{L}|G\rangle}{U_n^0-U_G^0}=\begin{bmatrix}\Lambda_{xx} & \Lambda_{xy} & \Lambda_{xz}\\ \Lambda_{xy} & \Lambda_{yy} & \Lambda_{yz}\\ \Lambda_{xz} & \Lambda_{yz} & \Lambda_{zz}\end{bmatrix}=\Lambda \therefore \Lambda_{ij}=-\sum_{n\neq G}\frac{\langle G|\hat{L}_i|n\rangle\langle n|\hat{L}_j|G\rangle}{U_n^0-U_G^0}. \quad (9.24)$$

Substitution of eqn (9.24) into eqn (9.23), which is then combined with the spin operator

$$g_e\beta_e B^T \cdot \hat{S}$$

yields the spin Hamiltonian, which can be expressed as eqn (9.25):

$$\hat{H}=\beta_e B^T \cdot g \cdot \hat{S}+\hat{S}^T \cdot \lambda^2\Lambda \cdot \hat{S} \quad (9.25)$$

In eqn (9.25), the term describing the product of the matrix term from eqn (9.24) and the eigenvalue of the state are often combined as described using the quantity, D. Should the angular momentum of a system only be dependent on the spin angular momentum, then g will be isotropic and equivalent to g_e. Anisotropy, which is observed in many real systems, comes out of the matrix described in eqn (9.24), in view of the contributions of the orbital angular momenta from excited states.[17] This fundamental description of the quantum mechanical nature of ESR spectroscopy will be built upon in the discussion of density functional theory (DFT) and its applications to the simulation of ESR spectra further on in this chapter. We have now surveyed the theory of ESR spectroscopy in some detail; this theory is also somewhat transferrable to optically-detected magnetic resonance, in which the signal is read-out optically, but excitations occur in an equivalent manner according to the methodology described herein. It is therefore appropriate to offer a limited description of spectral interpretation prior to describing the excellent recent work in the field of computational simulation of spectra, and an analysis thereof.

9.5 Interpretation

As described earlier in this chapter, ESR data are conventionally observed as first-derivative spectra of the absorption intensity against B_0. We can derive the g-values from comparisons of the field at the signal position with that of a standard according to eqn (9.4).

As noted above, there are numerous factors which affect the linewidth, shape, and position of signals in ESR spectra. Although the first-order effects, such as the principal A and g-values can be obtained with relative ease, those corresponding to hyperfine interactions can be trickier to derive and interpret. For accurate evaluations of complex experimental spectra, it is often necessary to use one or more of the multitude of simulation software modules available in order to compare a computed spectrum against that obtained experimentally.[21] This is often carried out much as in *ab initio* quantum chemical calculations, which use a trial value of the wavefunction to gradually improve the approximation until a reasonable solution for the system energy is obtained. In ESR spectral simulations, trial values can be employed and modified until correspondence is achieved with experimental data. Comparisons of signal maxima-to-minima in the first-derivative spectra (*i.e.* maximum minus minimum dI/dν values, in which I and ν represent the intensity and frequency parameters), can help to compute relative intensities for simple analytes. However, as shown in Figure 9.1, a second-derivative spectrum can improve determinations of relative intensities further.

Systems with a single unpaired electron, such as radicals, often yield well-defined signals with *g*-values close to that of g_e. Although ESR is commonly employed in research, it receives far less attention in undergraduate courses in view of the perceived complexity of experimental set-ups and interpretation. In order to demonstrate the basic interpretative principles, we survey recent pedagogical literature available on this topic. Students have, for example, been tasked with a series of experiments, firstly involving spectrometer calibrations followed by the measurement of a *g*-value for a standard, leading to the estimation of the *g*-value for more complex analytes.[22] The organic radical DPPH noted above can be utilised for the study of a basic spectrum. The resonance frequency of DPPH will be mediated by the magnitude of B_0 leading to the measurement of *g*, which gives a quantitative estimation of the magnetic moment of the system. Following this, ESR spectra of $Cu(acac)_2$ and $VO(acac)_2$ were obtained. DPPH has a single unpaired electron, delocalised across the system; from Eqn (9.3) and (9.8), we can determine the *g*-value for the single electron spin from the resonance frequency and B_0. As a calibration, students are tasked to measure the DPPH signal at up to ten field strengths; the transition frequency against B_0 allows for the calculation of the *g*-value of DPPH (**1**). This is carried out in an analogous manner through the absorption measurements of a scanned *B*, the first derivative of the spectrum determining *g*. Once this is determined for DPPH, students analyse a d^9 complex ($Cu(acac)_2$) (**2**) and a d^1 complex ($VO(acac)_2$) (**3**), which both have one unpaired electron spin, and spectra showing hyperfine interactions. The spin behaviours of the Cu(II) and oxo-V(IV) complexes differ; whilst the V(IV) system has an unpaired electron, the Cu(II) complex shows the behaviour of an unpaired positively-charged 'hole' (Scheme 9.1). The samples were prepared in a 40% : 60% $CHCl_3$: toluene solvent mixture, and DPPH was included as in internal reference within a capillary tube within the ESR sample tube. Once the spectrometer cavity is tuned, and the signal optimised, derivate spectra were obtained.

Spectra were obtained for this experiment on a Varian E4 spectrometer, and students were requested to identify the number of transitions and assign the hyperfine structure, followed by determining the *g*-value relative to that of the DPPH internal standard. The recorded spectra from this exercise by Butera and Waldeck are shown in Figure 9.6.

Eqn (9.26) can be employed to estimate the g-values of the complexes, which are inherently different to g_e, with one of the complexes having $g > g_e$, and the other $g < g_e$, which are determined by variations in spin-orbit couplings, as described further above.

Indeed, Basu has also communicated an experiment involving the interpretation of ESR spectra obtained from paramagnetic transition metal complexes.[23] Therein, the electronic structure of transition metal ion-tetraphenylporphyrin complexes is deduced, again including one d^9 (Cu(II)) and one d^1 (Mo(V)) complex, both with $S = 1/2$. Spectra were recorded on a Bruker ESP 300 E X-band instrument.

The work by Basu is detailed in its description of the experimental procedure, and importantly delivers the complex interpretation necessary for the spectra observed. We can observe a principal central signal for the Mo(V) complex with six satellites, arising from the natural abundance of stable Mo isotopes, 75% of the abundance being for $S = 0$ (large central line), the remainder being $S = 5/2$; therefore, $2nS + 1 = 6$, corresponding to the six satellite lines. Similarly, Figure 9.7c displays the room temperature spectrum of Cu(TPP) with four principal signals: ^{63}Cu and ^{65}Cu, both with $S = 3/2$, yield four resonances. A was estimated for Mo and Cu from these spectra, being 48 and 95 G, respectively; and g values of 1.971 and 2.117 respectively. These deviate from g_e by 0.1147 for Cu(II) and −0.0313 for Mo(V). We can relate the superhyperfine interactions seen in a frozen solution of TPP to eqn (9.18) above; indeed, the complex spectral artefacts show anisotropic g-values parallel and perpendicular to the field ($g_{\parallel}$ and $g_{\perp}$, equivalent to 2.195 and 2.078 respectively). We can explain the substantial superhyperfine structure from the interaction of the unpaired electron with the N atoms in the TPP chelator; the N isotope of highest abundance is ^{14}N at 99.63% and $S = 1$. With $m = 0.4038$ for ^{14}N, this is less than those of the Cu(II) or Mo(V) ions, leading to a lesser interaction *cf* to these transition metal ions. Four N atoms

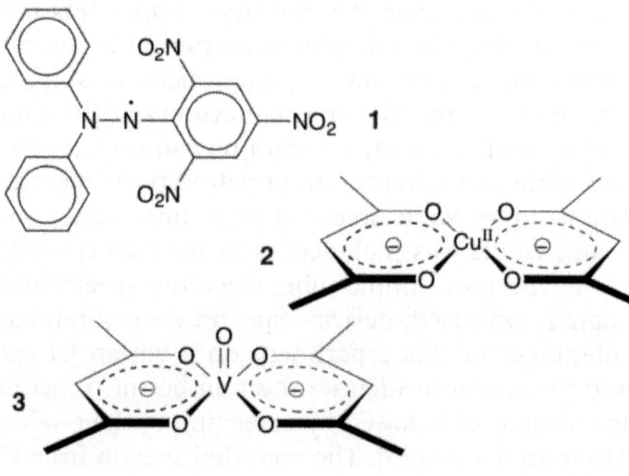

Scheme 9.1

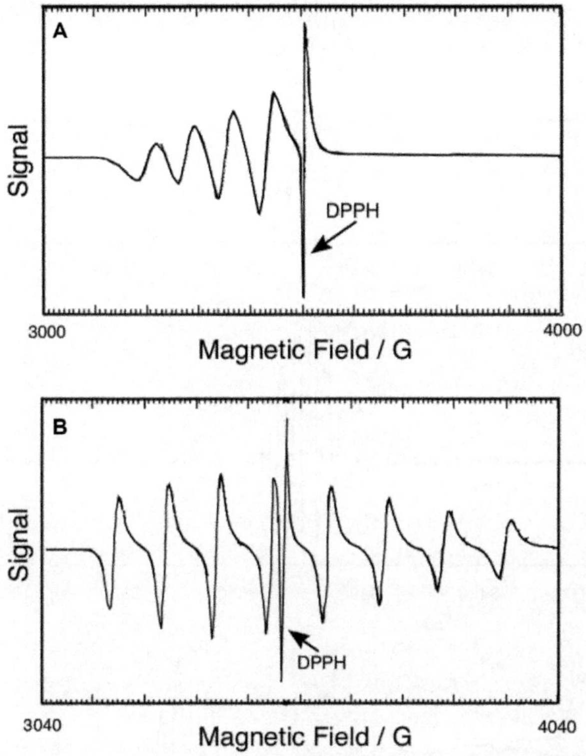

Figure 9.6 ESR spectra recorded using a Varian E4 spectrometer for (A) $Cu^{II}(acac)_2$ and (B) $V^{IV}O(acac)_2$, with DPPH as an internal reference standard. Reproduced from ref. 22 with permission from American Chemical Society, Copyright 2000.

in the porphyrin rings could lead to nine signals, which are observed in the room temperature spectrum on Cu(TPP); however, this is not clear from the first-derivative Mo(v) complex spectrum, improved somewhat in Figure 9.7b. It is possible to deduce the mean *a* coupling with N, a_N as 14 G for the Cu(II) and 2.7 G for the Mo(v) complex; these are attributable to the more diffuse nature of the Cu(II) 4d orbital *cf.* the 3d orbital of Mo(v).

Basu also describes links between the molecular structure and super-hyperfine constant; we assume that the unpaired electron in the Cu(II) complex should be located in the $3d(x^2-y^2)$ orbital planar to the N atoms on the *x* and *y* axes, whilst the $4d_{xy}$ orbital of the Mo(v) complex is orthogonal to the latter. Hence, the interactions we observe between the unpaired electron in the Cu(II) and Mo(v) complexes differ, an effect which is ascribable to the degree of interaction with the N atoms in the TPP macrocycle.[23]

In this current section, we have surveyed a limited number of pedagogical studies in order to briefly provide an overview of spectral interpretation. This, combined with the initial methodological and theoretical portions of

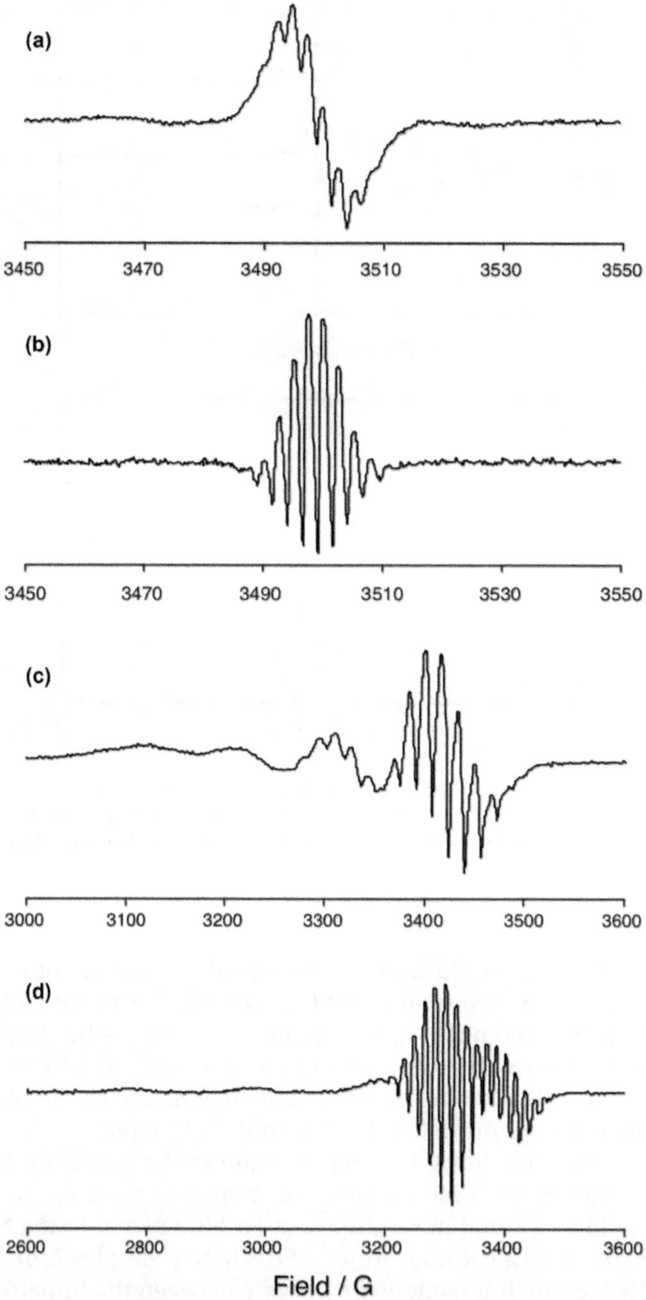

Figure 9.7 Continuous wave ESR spectra of (a) MoOCl(TPP) in the first-derivative, and (b) in the second-derivative form. (c) Room temperature spectrum of Cu(TPP), and (d) 77 K spectrum of Cu(TPP).
Reproduced from ref. 23 with permission from American Chemical Society, Copyright 2001.

this chapter, now allows us to progress to simulating ESR spectra and computational tools as aids to interpretation.

9.6 Early Work in the Simulation of ESR Spectral Parameters

The simulation of ESR parameters was originally developed in tandem with experimental work in view of the realisation that computed spectra could assist elucidation and derive the complexity of the experimental data obtained.[24] The main terms necessary for the simulation of ESR data were described in 1960 by Griffith,[25] and an excellent summary is provided by Neese and Munzarová.[26] Firstly, the *g*-matrix describes the Zeeman splittings arising from the total electronic μ with B_0; the hyperfine tensor incorporates the interactions between the electronic and nuclear spins; the zero-field splitting describes the interactions which break the degeneracy of $2S+1$ signals in a spin multiplet. Additionally, the interactions of the nuclear quadrupole moment and nuclear Zeeman interactions arising from nuclear magnetic dipoles with B are necessary for sufficiently parameterised models to simulate spin systems.[26] This is all based on the concept of the spin Hamiltonian described earlier in the theoretical sections of this chapter, and further terms are added in order to account for the electronic exchange and dipolar interactions[27] observed for groups of chemical systems, such as transition metal ion complexes and organic radicals.

Early computing of spin populations employed the configuration interaction method based on simple molecular orbital frameworks, whilst McLachlan modified the Hückel MO theory to include spin polarisation terms, generating a fundamental and widely-applied model.[28] The work of John Pople, who is known for his Gaussian package,[29] involved implantation of the above theory with semi-empirical methods. Indeed, with early computing resources it was impractical to implement Hartree–Fock (HF) theory level calculations for π-radicals, for example, and therefore a Hartree–Fock-intermediate neglect of differential overlap (HF-INDO) implementation was successfully demonstrated to describe the isotropic hyperfine coupling in numerous radical systems.[30] This difficulty in *ab initio* simulations employing the methodology at the time meant that HF led to ESR simulations to be somewhat overlooked in the beginning. Original iterations of the HF theory were far more suited to calculate parameters and optimised structures for small organic molecules rather than the more electronically-complex transition metal ions, in which, for example, solution-state spectra of organic radicals were commonly recorded, any *g*-matrix anisotropy or HFC were found to average out as described earlier, leading to anisotropic hyperfine coupling or *g*-values to be measured only rarely.[26] The difficulty in the calculation of the *g*-tensor was related to the very small changes from g_e in systems such as organic radicals, in which precise measurement of this deviation is highly complex; Neese points out that *g* is also a second-order property, which will not behave simply in terms of its expectation value across the ground state, and an unrestricted Hartree–Fock wavefunction.[31] As these shifts of *g* are relatively small *cf.*

to the magnitude, these arise from the similarly small electron ground state angular momentum described above, a consequence of spin–orbit coupling constants of most light atoms. From the 1970s onwards, it became clear that quantum chemical techniques employing terms for electronic correlation, vibronic and environmental coupling yielded the most quantitatively-proximate simulated data for ESR parameters.[32] Once DFT became more widely implemented[33,34] in the late 20th century, the approximations implemented therein led to estimations of hyperfine couplings of systems with unpaired electrons which were comparable in accuracy to those obtained from variations of configuration interaction or coupled cluster theory; evidently they were functional-dependent.[35]

9.7 Progress in Density Functional Theory Approaches

Although early calculations employed HF theory for the estimation of hyperfine couplings, it became clear that consideration of parameters such as electronic correlation was necessary in order to obtain quantitatively-comparable data from simulations. Unfortunately, at the time in which HF theory was the mainstay of quantum chemical methods, this was impractical for the present computational resource; development of configurational interaction and correlated techniques improved the scope, although they were more computationally-expensive. The development of DFT methods opened up the simulation of larger, more electronically-complex systems for the calculation of ESR parameters. Within DFT, the total energy is minimised by applying the concept of the many-particle ground state as a function of the electron density. This became practical with the implementation of work by Kohn and Sham[36] in defining a reference to which the 'real' system can be compared. The electronic correlation term remains in the DFT approximation, and it is this which includes higher level terms. The hyperfine coupling tensor is therefore calculated based on this theory, implementing eqn (9.20)–(9.25) in the context of the above approximation.

Density functional theory methods will vary in applicability for the calculation of ESR parameters based on the electronic correlation potential, and the terms employed. Indeed, isotropic coupling constants are substantially sensitive to the choice of the functional and basis set when compared to anisotropic constants, this being further confirmed in the case of systems with unpaired electrons in *s* orbitals. This is complicated in higher orbitals in which the unpaired spin transfer to the nucleus is undertaken by spin polarisation, the mechanism of which is difficult to reproduce in calculations of hyperfine coupling constants.[37] For paramagnetic metal ions, it is recommended that large basis sets are employed, particularly in order to account for π interactions. The B3LYP functional combined with the

EPR-III,[38] IGLO-III or EXT basis sets show good agreement with model experiments.[37] Hyperfine coupling constants are more challenging to estimate for unsaturated sigma radicals at the HF level, whereas DFT appears to show good accuracy, with the PBE0 performing well, and BP86 and B3LYP computed isotropic couplings being within 10% of the experimental values.[37]

The importance of relativistic DFT approaches has recently been discussed by Autschbach.[39] The complete derivation of the *g*-factor using a Douglas–Kroll–Hess approach[40–42] (DKH2) by Neese and Sandhoefer,[43] included the magnetic field in the free-particle transformation, including spin–orbit coupling as a first-order perturbation. A normalised elimination of the small component (NESC) method[44–46] was employed to calculate the hyperfine coupling constants of ^{199}Hg-containing radicals, in comparison to non-relativistic approaches. Autschbach refers to factors of 2–3 increases in calculations ascribable to relativistic effects, which are largely attributable to the sensitivity of the electron-nucleus contact term to increases in the electron and spin densities at heavy nuclei in view of relativistic effects.[39] It is also conceivable that hyperfine coupling constants may be strongly affected by the electron-nuclear distance term, as in heavy atoms such as ^{199}Hg, in which relativistic corrections modify this by up to 15%.[47]

For orbitally non-degenerate states, changes from g_e can be considered to arise from relativistic effects. Indeed, shifts in *g*-tensors tend to be small in magnitude, and are often quoted in parts per thousand or trillion (ppt). In calculations of the *g*-tensors of d^1 metal complexes, higher order spin–orbit effects were found to mediate these quantities to a larger extent than earlier work had expected.[48] It is therefore suggested that calculations in which spin-orbit coupling is estimated as a linear perturbation are not able to reproduce the accurate magnitude of the *g*-tensor. ZORA calculations of shifts in *g* were recently calculated using a gauge-independent atomic orbital approach, with a range of DFT functionals.[49,50] Spin–orbit coupling was also included variationally in the ground state, using: (i) a method which does not allow for spin polarisation in the implementation,[51] and (ii) the inverse system. This method is known as including magnetic anisotropy, and was developed from work concerning theoretical estimation of zero-field splitting.[52] Autschbach has summarised instances in which large shifts in *g* are observed for NpF_6, UF_6^- and UCl_6^-, with electronic configurations of $5f^1$, and which are reproduced with permission from ref. 39 in Table 9.1.

Autschbach describes a shift in the *g*-factor of −2600 parts per thousand away from the recorded experimental value for NpF_6 of −0.6. From Table 9.1, the magnitude and direction of experimental *g*-values have been estimated with CASSCF and ZORA DFT calculations; however, there is some variability in these. Autschbach highlights the intricate issues associated with approximations inherent in certain DFT functional classes with the inclusion of spin polarisation terms. From this assessment, it is recommended to therefore implement the spin polarisation terms within the magnetic

Table 9.1 ESR *g*-factors converted by multiplying by −1. SP indicates spin polarisation considered within the implementation. Experimental results and complete active space self-consistent field (CASSCF) data are extracted from ref. 53. The HF and DFT ZORA data is extracted from ref. 50. Reproduced from ref. 39 with permission from the Royal Society of Chemistry.

Method/System	NpF_6	UF_6^-	UCl_6^-
CASSCF	0.64	0.68	1.33
CASPT2	0.72	0.62	1.21
(SP)HF	0.84	0.65	1.46
(SP)LC-PBE0	0.29	0.43	1.18
HF	1.07	1.06	1.37
LC-PBE0	0.92	1.04	1.32
Experiment	0.60	0.75	Approx. 1.1

anisotropy approach for estimating shifts in *g*-values combined with hybrid DFT functionals, such as the long range-corrected PBE0 implementation.

9.8 Spectral Simulation and Fitting

Based on Matlab, the EasySpin package is arguably one of the most popular implementations for simulation and interpretation of ESR spectroscopic data.[54] The spectral data obtained from ESR experiments requires data to be processed and simulations to be performed, as well as the effective fitting of parameters. Indeed, these simulations are carried out in order to understand the effect of magnetic factors on the spectra, whilst allowing users to identify the benefits of performing a new experiment in terms of providing additional insights, or accurately obtaining experimental factors from spectral data. The EasySpin package includes more than 80 functionalities within Matlab, based on: (i) experimental data treatment; and (ii) spectral simulation of the solid-state, isotropic continuous wave experiments, and ENDOR (see Figure 9.8).

In this example we observe the continuous wave, solution-state simulated spectrum of 6-hydrodipyrido[1,2-c:2′,1′-*e*]-imidazole, which was produced *via* employment of the EasySpin function, *garlic*. This system has an unpaired electron which is coupled to 10 protons and two nitrogen atoms, yielding a total of 9216 resonances in the spectrum. Parameter inputs included a g_{iso} value of 2.00316 and a linewidth of 0.01 mT, with couplings of 12.16 MHz to N; and −6.70, −1.82, 7.88, −0.64 and 67.93 MHz to H, and a microwave frequency of 9.532 GHz. This module in EasySpin will simulate the spectrum to infinite order, and according to the authors, requires less than 0.1 seconds on a 2 GHz Linux facility.

Another option under the designation *resfields* produces transition plots of the resonance field positions, and the effect of the orientation of the paramagnetic part of the system on these positions. This can be observed in Figure 9.9, in which such a plot is presented for a system with $S = 5/2$.

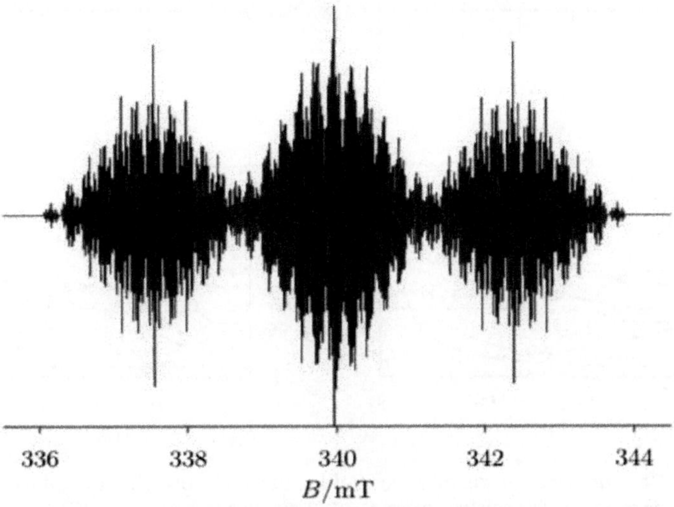

Figure 9.8 An X-band solution-state continuous wave spectrum of the radical cation 6-hydrodipyrido[1,2-c:2′,1′-*e*]-imidazole.
Reproduced from ref. 54 with permission from Elsevier, Copyright 2006.

The axial zero-field splitting for the simulation was 5 GHz for $\phi = 0$. The microwave frequency employed in the simulation was 9.5 GHz. Figure 9.8A shows a plot of B against angle θ, whilst Figure 9.8b is based on the simulation axes in terms of B_z against B_x. Energy levels for the resonance fields are indicated in Figure 9.8b. Finally, we show the simulated ENDOR spectra from Stoll and Schweiger,[54] which is based on a solid-state sample in a varied magnetic field, as shown in Figure 9.10.

This demonstrates the implementation of the *salt* function within EasySpin, which allows a varying B to assist in the estimation of hyperfine coupling and quadrupolar effects in systems with low symmetry. Here, the spin system is coupled to a proton with an orthorhombic hyperfine coupling tensor.

These codes within EasySpin allow for the multifaceted simulation of ESR parameters through the Matlab environment. The authors offer suggestions on improvements, such as the simulation of pulsed ESR spectra; however, this has since been implemented within the code.[32]

Additionally, programmes with bespoke graphical user interfaces, such as XSophe and associated modules[55–58] allow for the simulation of continuous wave ESR spectral parameters from a variety of sample media and orientations, as well as spin systems. XSophe incorporates a simple interface which calculates the parameters of the spin Hamiltonian for continuous wave ESR experiments on single crystal and isotropic samples, and those with random orientations, which are then combined to understand the nature of the paramagnetic centre in more detail. Much as in EasySpin, plots of the resonance field dependency on the paramagnetic centre orientation can be obtained, as well as energy level diagrams. Hanson and co-workers[55]

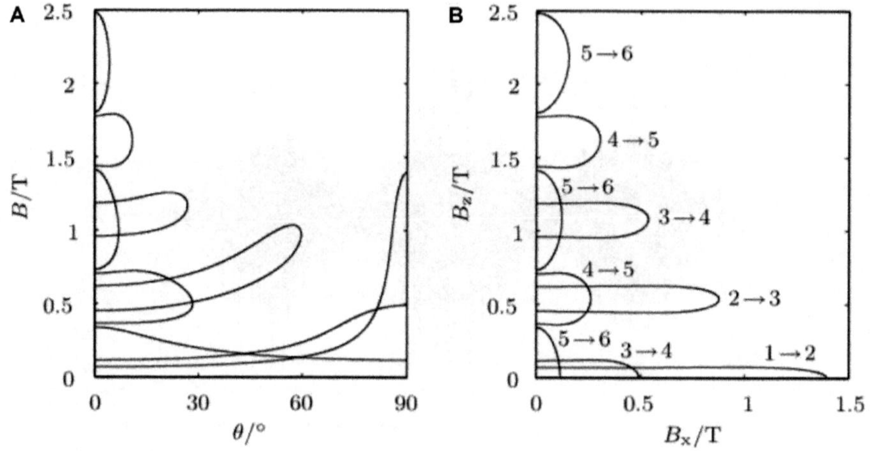

Figure 9.9 Dependency of resonance field positions on the orientation of the paramagnetic centre in an $S=5/2$ system.

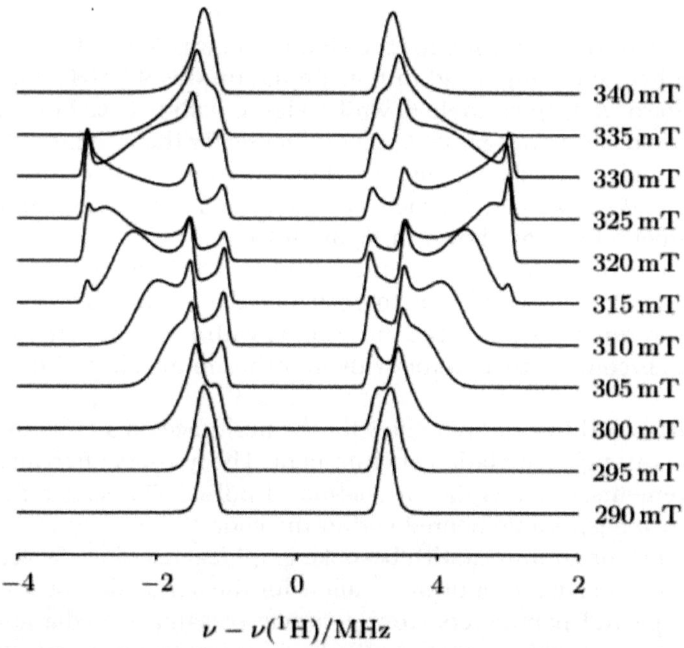

Figure 9.10 Variation of magnetic field strengths applied to an $S=1/2$ spin system.

describe an example simulation of a high spin Cr(III) centre with a zero field splitting of 0.10 cm^{-1}, with a g_e value equivalent to 2.00 and $A_{x,y,z}$ values of 0.012, 0.012, and 0.024, respectively. Following a run of 229.47 seconds, the spin Hamiltonian parameters were obtained with an acceptable SNR index.

9.9 Conclusions

We have surveyed the fundamental aspects of ESR spectroscopy, before describing interpretation and the major requirements for computational tools. QM modelling was described, before identifying DFT approaches to the simulation of hyperfine coupling constants and *g*-factors. Spectral simulation and fitting tools were briefly surveyed with both scripting and graphical user interface (GUI)-based options available. The field of ESR spectroscopy currently relies on computational techniques for spectral simulation, fitting and experimental guidance, with a number of well-established tools available. Many of these have successfully implemented developments with experiments, for example the pulsed-ESR simulation in EasySpin, and will likely continue to do so. Recent identifications of rotational invariance in certain DFT implementations could exert a substantial effect on the related calculation of NMR *J*-couplings. However, their influence on the estimation of ESR parameters remains untested.[59]

Acknowledgements

M.L.M wishes to acknowledge the European Research Council (ERC) for funding this work through the ERC Consolidator Award, TransPhorm, grant number 23432094. P.B.W. wishes to acknowledge The Royal Society for support through Grant RGS\R1\191154.

References

1. M. Che and E. Giamello, *Stud. Surf. Sci. Catal.*, 1990, **57**, B265–B332.
2. S. Schlick and G. Jeschke, *Polymer Science: A Comprehensive Reference*, 2012.
3. Y. Pan and M. J. Nilges, *Rev. Mineral. Geochem.*, 2014, **78**, 655–690.
4. R. V. Parish, *NMR, NQR, EPR, and Mössbauer Spectroscopy in Inorganic Chemistry*, Ellis Horwood Ltd, 1990.
5. E. KÖNIG, *The Organic Chemistry of Iron*, 2012.
6. P. W. Atkins, V. Walters and J. De Paula, *Physical Chemistry*, Macmillan Higher Education, 2006.
7. M. G. Mc Namee, *Food Analysis: Principles and Techniques: Volume 2: Physicochemical Techniques*, 2017.
8. M. J. N. Junk and M. J. N. Junk, *Assessing the Functional Structure of Molecular Transporters by EPR Spectroscopy*, 2012.
9. D. Csontos, U. Zülicke, P. Brusheim and H. Q. Xu, *Phys. Rev. B: Condens. Matter Mater. Phys.*, 2008, **78**, 033307.
10. C. P. Poole and H. A. Farach, *Practical Handbook of Spectroscopy*, 2017.
11. C. Corvaja, *Electron Paramagnetic Resonance: A Practitioners Toolkit*, 2008.
12. E. J. L. McInnes and D. Collison, *eMagRes*, 2016, DOI: 10.1002/9780470034590.emrstm1502.
13. R. Grün, *Quatern. Int.*, 1989, **1**, 65–109.

14. E. J. L. McInnes, *Struct. Bonding*, 2006, DOI: 10.1007/430_034.
15. N. F. Chilton, D. Collison, E. J. L. McInnes, R. E. P. Winpenny and A. Soncini, *Nat. Commun.*, 2013, **4**, 2551.
16. G. R. Eaton, S. S. Eaton, D. P. Barr and R. T. Weber, *Quantitative EPR*, 2010.
17. M. M. Roessler and E. Salvadori, *Chem. Soc. Rev.*, 2018, **47**, 2534–2553.
18. F. Gerson and W. Huber, *Electron Spin Resonance Spectroscopy of Organic Radicals*, 2003.
19. P. Knowles, *Biochem. Educ.*, 1995, **23**, 48.
20. J. A. Weil and J. R. Bolton, *Electron Paramagnetic Resonance: Elementary Theory and Practical Applications*, 2nd edn, 2006.
21. H. J. Hogben, M. Krzystyniak, G. T. P. Charnock, P. J. Hore and I. Kuprov, *J. Magn. Reson.*, 2011, **208**, 179–194.
22. R. A. Butera and D. H. Waldeck, *J. Chem. Educ.*, 2000, **77**, 1489.
23. P. Basu, *J. Chem. Educ.*, 2001, **78**, 666.
24. S. K. Misra, *Multifrequency Electron Paramagnetic Resonance: Theory and Applications*, 2011.
25. J. S. Griffith, *Mol. Phys.*, 1960, **3**, 79–89.
26. F. Neese and M. L. Munzarová, *Calculation of NMR and EPR Parameters*, 2004.
27. A. Bencini and D. Gatteschi, *Electron Paramagnetic Resonance of Exchange Coupled Systems*, 2011.
28. A. D. McLachlan, *Mol. Phys.*, 1959, **2**, 271–284.
29. W. J. Hehre, W. A. Lathan, R. Ditchfield, M. D. Newton and J. A. Pople, *Quantum Chem. Program Exchange*, 1973, **11**, 236.
30. J. A. Pople, D. L. Beveridge and P. A. Dobosh, *J. Chem. Phys.*, 1967, **47**, 2026–2033.
31. F. Neese, *eMagRes*, 2017, DOI: 10.1002/9780470034590.emrstm1505.
32. S. Stoll and R. D. Britt, *Phys. Chem. Chem. Phys.*, 2009, **11**, 6614–6625.
33. P. Hohenberg and W. Kohn, *Phys. Rev.*, 1964, **136**, B864–B871.
34. W. Kohn and L. J. Sham, *Phys. Rev.*, 1965, **140**, A1133.
35. F. Neese, *J. Chem. Phys.*, 2001, **115**, 11080.
36. W. Kohn and L. J. Sham, *Phys. Rev.*, 1965, **140**, 1133–1138.
37. M. L. Munzarová, *Calculation of NMR and EPR Parameters*, 2004.
38. V. Barone, *Chem. Phys. Lett.*, 1996, **262**, 201–206.
39. J. Autschbach, *Philos. Trans. R. Soc., A*, 2014, **372**, 20120489.
40. T. Nakajima and K. Hirao, *Chem. Rev.*, 2012, **12**, 385–402.
41. M. Reiher, *Wiley Interdiscip. Rev.: Comput. Mol. Sci.*, 2012, **2**, 139–149.
42. R. Mastalerz, G. Barone, R. Lindh and M. Reiher, *J. Chem. Phys.*, 2007, **127**, 074105.
43. B. Sandhoefer and F. Neese, *J. Chem. Phys.*, 2012, **137**, 094102.
44. D. Cremer, W. Zou and M. Filatov, *Wiley Interdiscip. Rev.: Comput. Mol. Sci.*, 2014, **4**, 436–467.
45. W. Zou, M. Filatov and D. Cremer, *J. Chem. Phys.*, 2011, **134**, 244117.
46. W. Zou, M. Filatov and D. Cremer, *J. Chem. Theory Comput.*, 2012, **8**, 2617–2629.

47. E. Malkin, M. Repisk, S. Komorovsk, P. MacH, O. L. Malkina and V. G. Malkin, *J. Chem. Phys.*, 2011, **134**, 044111.
48. P. Hrobárik, M. Repiský, S. Komorovský, V. Hrobáriková and M. Kaupp, *Theor. Chem. Acc.*, 2011, **129**, 715–725.
49. J. Autschbach and B. Pritchard, *Theor. Chem. Acc.*, 2011, **129**, 453–466.
50. P. Verma and J. Autschbach, *J. Chem. Theory Comput.*, 2013, **9**, 1052–1067.
51. E. van Lenthe, P. E. S. Wormer and A. van der Avoird, *J. Chem. Phys.*, 1997, **107**, 2488–2498.
52. S. Schmitt, P. Jost and C. van Wüllen, *J. Chem. Phys.*, 2011, **134**, 194113.
53. F.-P. Notter and H. Bolvin, *J. Chem. Phys.*, 2009, **130**, 184310.
54. S. Stoll and A. Schweiger, *J. Magn. Reson.*, 2006, **178**, 42–55.
55. G. R. Hanson, K. E. Gates, C. J. Noble, M. Griffin, A. Mitchell and S. Benson, *J. Inorg. Biochem.*, 2004, **98**, 903–916.
56. A. Mitchell, C. J. Noble, S. Benson, K. E. Gates and G. R. Hanson, *J. Inorg. Biochem.*, 2003, **96**, 191.
57. G. R. Hanson, C. J. Noble and S. Benson, XSophe–Sophe–XeprView and Molecular Sophe: computer simulation software suites for the analysis of continuous wave and pulsed EPR and ENDOR spectra, in *EPR of Free Radicals in Solids I*, Springer, Dordrech, 2013, pp. 223–283.
58. G. R. Hanson, K. E. Gates, C. J. Noble, A. Mitchell, S. Benson, M. Griffin and K. Burrage, *EPR of Free Radicals in Solids: Trends in Methods and Applications*, 2003, pp. 197–237.
59. A. N. Bootsma and S. Wheeler, Popular integration grids can result in large errors in dft-computed free energies, *ChemRxiv*, 2019, 8864204, p. v5.

Subject Index

artificial intelligence
applications, 161–162
chemical analysis, 186–187
decision trees, 171–173
deep learning, 183–186
hierarchical clustering (HC), 179–182
independent component analysis, 182–183
linear regression, 174–176
naïve bayes classifiers, 173–174
neural networks, 166–170
particle swarm optimisation, 187–190
quantum computing, 186–187
self-organising maps (SOMs), 178–179
support vector machines (SVMs), 162–166
κ-means clustering (KMC), 176–178
κ nearest neighbours, 170–171
chemistry and bioanalysis areas, 159–161
historical use of, 155–159

Bethe–Salpeter equation (BSE), 293
binding isotope effects, 143–146

classical molecular dynamics
simulations, 247–248
enhanced sampling simulations schemes, 254–255
constraints, 255–256
enhance the sampling, 256–258
guiding processes, 278
magnesium transport system CorA, 278
interpret experimental measurements
lipids interacting with Na^+/H^+ antiporters, 270–273
secondary active transporter BetP conformational change, 273–277
MD simulation algorithm, 248
barostatting, 251–252
equations of motion, 249–250
periodic boundary conditions, 250
simulation ensembles, 250–252
thermostatting, 250–251
membrane protein function
GPCR-mediated arrestin activation, 265–268
KscA ion channel inactivation, 263–265
M2 muscarinic receptor, 268–270

potential energy function
classical all-atom models, 252–253
coarse-grained models, 253–254
long-range non-bonded interactions, 254
polarizable models, 253
computational spectroscopy
solid state chemistry
absorption spectroscopy, 291–294
inelastic scattering, 294–297
matter *vs.* electromagnetic radiation, 288–289
physical and chemical properties, 289–291
resonance spectroscopy, 297–305
computational vibrational spectroscopy
in complex environments
general vibrational spectroscopy, 97–100
surface-enhanced Raman spectroscopy, 100–102
harmonic approximation, 70–71
applications, 89–97
developments, 89–97
motivation, 76–78
vibrational configuration interaction, 83–85
vibrational coupled clusters, 87–88
vibrational perturbation theory, 85–87
vibrational self-consistent field, 80–83
Watson Hamiltonian, 78–80
intensities calculations, 72–76
current-density functional theory (CDFT), 45

density functional theory (DFT), 43, 44, 301–303, 354–356

electronic structure theory, 79, 80, 83
electron paramagnetic resonance spectroscopy, 58–60
electron spin resonance
density functional theory, 354–356
fine structure in, 342–346
interpretation, 348–353
quantum mechanical consideration of, 346–348
simulation of, 353–354
spectral simulation and fitting, 356–359
theory, 337–342
equilibrium isotope effect (EIE), 126

finite difference time dependent (FDTD), 102

graphene, 148–150

Hamiltonian theory, 93
harmonic approximation, 70–71
applications, 89–97
developments, 89–97
motivation, 76–78
vibrational configuration interaction, 83–85
vibrational coupled clusters, 87–88
vibrational perturbation theory, 85–87
vibrational self-consistent field, 80–83
Watson Hamiltonian, 78–80
Hartree–Fock (HF) theory, 353
β-hexachlorocyclohexane, 142–143
hierarchical clustering (HC), 179–182
hydrogen impurities, 299–301

independent component analysis, 182–183
isotope effects, 125–130
binding isotope effects, 143–146
graphene, 148–150
β-hexachlorocyclohexane, 142–143
kinetic isotope effects, 137–139
QM/MM approaches, 139–142
theory of, 130–132
transition state theory, 132–136
vapor pressure isotope effects (VPIEs), 146–148

kinetic isotope effects, 137–139
κ-means clustering (KMC), 176–178
κ nearest neighbours, 170–171

linear response time-dependent density functional theory (LR-TDDFT), 75
linear spectroscopy
in chemical and biological systems, 209–210
computational perspective, 210–213
DNA, 220
drugs, 220
emission, 213–219
model light absorption, 213–219

mass-independent, 126
MD simulations. *See also* classical molecular dynamics simulations
forward modelling, expectation values, 258–259
reweighting schemes, 259–262
metabolic pathway analysis, 28
metabolomics. *See also* NMR-based metabolomics
brief introduction to, 5–6
experimental design, 6–8
positive signal identification, 13–15
preprocessing of, 8–13
scope of, 33–34
statistical methodologies, 16
modelling circular dichroism spectroscopy, 220–223
Moller–Plesset perturbation theory (MPPT), 86
multivariate analysis, 19–20
partial-least squares discriminatory analysis (PLS-DA), 22–28
principal component analysis, 20–22
multivariate power calculations, 30–33
Muon spin spectroscopy, 301–303

naïve bayes classifiers, 173–174
neural networks, 166–170
NMR-based metabolomics. *See also* Nuclear Magnetic Resonance (NMR)
experimental design, 6–8
statistical methodologies, 16
Nuclear Magnetic Resonance (NMR)
chemical shieldings
environmental effects, 47–54
molecular orbital analysis, 56–58
chemical shifts
calculation of, 43–46
conformational effects, 46–47
relativistic effects, 54–55
experimental ^{1}H NMR spectra, 42

partial-least squares discriminatory analysis (PLS-DA), 22–28
particle swarm optimisation, 187–190
perturbation theory, 46
photochemistry and photobiology
in complex environments, 222–223
adiabatic molecular dynamics simulations, 225–228

computing potential energy surfaces, 223–225
dynamic approach, 225
non-adiabatic molecular dynamics simulations, 228–238
Placzek's theory, 73
principal component analysis, 20–22

QM/MM techniques
computational modelling and, 205–208
embedding, 209
frontier, 208
isotope effects, 139–142
quantum computing, 186–187
quantum-mechanical perturbation theory, 72
quantum mechanics/capacitance-molecular mechanics, 98

Raman scattering calculation, 73
Ramsey equation, 56
real-time time-dependent density functional theory (RT-TDDFT), 76
rotation-translation blocks (RTB), 98

Schrödinger equation, 79, 80, 87, 89
self-organising maps (SOMs), 178–179
solid state chemistry
biocompatible materials
nanoscale materials, 315–318
surface adsorption, 312–315
surface functionalization, 312–315
surface properties, 305–312
computational spectroscopy
absorption spectroscopy, 291–294
inelastic scattering, 294–297
matter *vs.* electromagnetic radiation, 288–289
physical and chemical properties, 289–291
resonance spectroscopy, 297–305
support vector machines (SVMs), 162–166

TCDFT. *See* two-component generalization of density functional theory (TCDFT)
time-dependent density functional theory, 294
time-dependent Schrödinger equation, 87, 227, 230
transition state theory, 132–136
two-component density functional theory, 304–305
two-component generalization of density functional theory (TCDFT), 304

univariate data analysis, 16–18
ANCOVA, 18–19
ANOVA, 18–19
ASCA, 18–19

vapor pressure isotope effects (VPIEs), 146–148
variational transition state theory (VTST), 133
various wave-function, 43
vibrational perturbation theory (VPT), 85–87, 88
VPIEs. *See* vapor pressure isotope effects (VPIEs)
VPT. *See* vibrational perturbation theory (VPT)

Watson Hamiltonian, 78–80
Welch–Satterwaite equation, 17